Purnell's Concise
Dictionary of Science

Offset litho

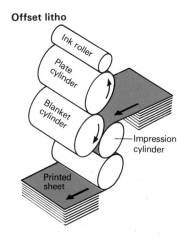

Ink roller

Plate cylinder

Blanket cylinder

Impression cylinder

Printed sheet

Purnell's Concise
Dictionary

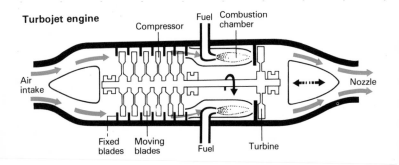

Turbojet engine

Compressor
Fuel Combustion chamber
Air intake
Nozzle
Fixed blades
Moving blades
Fuel
Turbine

# of Science

Robin Kerrod

PURNELL

Preface

We live in an age increasingly dominated by science and technology—the age of the laser and silicon chip, of the computer and robot, of the supersonic airliner and space shuttle. And it is becoming ever more essential that all of us —young and old, at school, at work and in the home—become more familiar with matters scientific.

It is with this very much in mind that *Purnell's Concise Dictionary of Science* has been compiled. It covers broadly the physical sciences—physics, chemistry, geology and astronomy—and the technologies, from civil engineering to astronautics.

With more than 3,000 entries and over 100,000 words, the *Dictionary* describes substances, processes, machines, devices, laws, units and terms likely to be encountered in our scientific age. Chemical formulae and physical data are included where appropriate, and a complete periodic table of the elements and further data appear at the back of the book.

As a quick glance through the pages will reveal, the book is not so much a dictionary in the orthodox sense, but more of a glossary, which goes beyond bare definitions to elaborate and explain. The entries are written in as straightforward a manner as subject, and space, permit. Several hundred illustrations complement and supplement the text.

ISBN 0 361 05346 0
Copyright © 1983 Purnell Publishers Limited
Published 1983 by Purnell Books, Paulton, Bristol BS18 5LQ
a member of the BPCC group of companies
Made and printed in Great Britain by
Hazell Watson & Viney Ltd, Aylesbury, Bucks

Acknowledgements

The author would like to express his gratitude to the many people who have provided assistance in the compilation and production of the *Dictionary*. Particular thanks are due to the chief educational advisers—Barry Fox, Head of Science at Bishop Wordsworth's School, Salisbury, and Margaret Hutchinson, Head of Science at St Leonards-Mayfield School, Mayfield, for their helpful comments; to Ray Burrows and J. & M. Gilkes for preparing the artwork; and to the following organizations for providing the photographs:

Argonne National Laboratory	Central Office of Information
Freeman, Fox and Partners	Cornell University
Kennedy Space Center	Marshall Spaceflight Center
Pest Infestation Control Laboratory	P&O Lines

Abbreviations

For ease of reading, the number of abbreviations used in the text has been kept to a minimum. They include the following:

at no = atomic number	bp = boiling point
mag = magnitude	mp = melting point
rd = relative density	

Weights and measures are given first in metric, or SI units:

cc = cubic centimetres	cm = centimetres
g = grams	kg = kilograms
km = kilometres	km/h = kilometres per hour
m = metres	mm = millimetres

Reflecting their continued use in everyday life, Imperial units follow the metric units in parentheses:

ft = feet	in = inches
lb = pound	mph = miles per hour
oz = ounce	

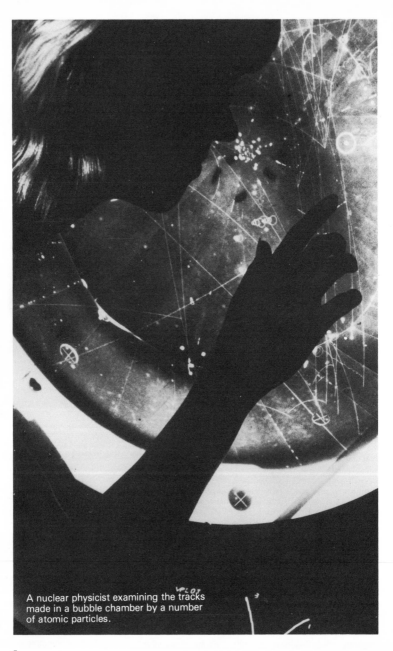

A nuclear physicist examining the tracks made in a bubble chamber by a number of atomic particles.

abacus Ancient calculating device dating back to Babylonian times and still used. The modern abacus consists of beads strung on wires in a wooden frame. Each wire represents the units, tens, hundreds, etc, of a number.

aberration In light, a defect of lenses and curved mirrors that results in a blurred image. Spherical aberration occurs because light rays from different parts of the curved surface are brought to a focus at different points. Chromatic aberration is a defect of lenses, which bring the different colours in light to a focus at different points. A colour-blurred image results. In astronomy, aberration is the apparent change in position of a star due to the Earth's motion in space. It was discovered by James Bradley in 1725.

ablation Melting and boiling away. Spacecraft heat shields dissipate the heat of re-entry by ablation.

abrasive A substance used for grinding and polishing. Common abrasives consist of fine particles of glass, sand or emery glued to paper or cloth.

absolute temperature A temperature scale based on the absolute zero of temperature. The absolute Kelvin scale uses Celsius units; so the freezing point of water is $273 \cdot 16K$.

absolute zero The lowest possible temperature that can be reached ($-273 \cdot 16°C$, or $0K$), when all molecular motion ceases.

absorption The process by which one substance takes up another in its interior; eg water absorbs ammonia gas. Substances can also absorb energy —as heat, light or sound. Green glass absorbs all the colours in light except green.

AC Abbreviation for alternating current.

Ac The chemical symbol for actinium.

acceleration The rate of change of velocity, expressed in units such as cm per second per second. When a body falls to the ground, it experiences an acceleration due to gravity of 981 cm ($32 \cdot 2$ ft) per second per second.

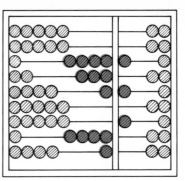

Chinese abacus

accelerator In chemistry, a substance that speeds up a chemical reaction.

accelerator, particle A device used in nuclear research to accelerate charged atomic particles. Linear accelerators accelerate particles in a straight line by applying a progressively increasing electric field. Circular accelerators, such as the cyclotron and synchrotron, accelerate particles in a circle by means of an alternating field.

accumulator A storage battery, or secondary cell. The lead-acid car battery is an accumulator, which can be recharged with electricity.

acetaldehyde (CH_3CHO) Or ethanal: a highly reactive, pungent-smelling liquid obtained by oxidizing ethanol. It is used to synthesize other important chemicals, including acetic acid.

acetate A synthetic fibre made from cellulose acetate, obtained by reacting cellulose with acetic acid.

acetates Salts formed when acetic acid reacts with inorganic compounds such as hydroxides; or esters formed when acetic acid reacts with organic alcohols.

acetic acid (CH_3COOH) Or ethanoic acid; the acid in vinegar. Widely used in the production of organic chemicals—it is made commercially by oxidizing acetylene or ethanol. When pure, it is called glacial acetic acid for it forms crystals near room temperature ($17°C$).

acetone (CH_3COCH_3) Also called dimethylketone and propanone; an important industrial solvent used, for example, in 'spinning' acetate fibres. It is a colourless flammable liquid with a fruity smell.

acetylene (CHCH) Or ethyne; a highly reactive gas that burns with a brilliant white light. Widely used as a fuel in oxyacetylene torches, it is also the starting point for many industrial chemicals and plastics. It can be prepared by adding water to calcium carbide.

acetylsalicylic acid The chemical name for aspirin.

Achernar (α Eridani) The brightest star (mag $0 \cdot 48$) in the constellation Eridanus; ninth brightest in the sky.

achromatic lens A lens made from a combination of crown and flint glass which prevents colour blurring of the image. It corrects chromatic aberration.

acid An important class of chemicals that taste sour and may be corrosive and highly poisonous. Acids react with bases to form a salt and water only. They change the colour of certain indicators, eg turning blue litmus red. All acids contain hydrogen, and when in water solution yield positively charged hydrogen ions (H^+). The strongest acids are the inorganic sulphuric, nitric and hydrochloric acids. Organic acids, such as citric and acetic acids, contain the carboxyl group, $-COOH$.

acoustics The science of sound; the sound properties of, say, a building.

Acrux (α Crucis) The brightest star in the constellation Crux (mag $0 \cdot 9$); 14th brightest in the heavens.

acrylics A group of synthetic fibres and plastics made mainly from acrylonitrile. They include Orlon fibres and Perspex.

actinides A closely related series of radioactive heavy metals, including uranium. Many actinides, including plutonium, are made artificially. See periodic table on page 250.

actinium (Ac) A rare radioactive metallic element (rd $10 \cdot 1$, mp $1,050°C$) found in pitchblende. The prototype of the actinides.

activated charcoal A specially prepared form of charcoal which can absorb vast quantities of gas and remove impurities from liquids.

adhesive A substance that bonds two surfaces together. Fish glues, cellulose paste and rubber solution are widely used adhesives. Stronger adhesives are made from synthetic materials, such as epoxy resins. These resins are mixed with a curing agent to make them set.

'Instant' glues are kinds of acrylic resins which set immediately when they are out of contact with the air.

adiabatic change A change that occurs in a system without any heat being supplied or taken away.

adsorption The process by which a solid attracts molecules of gas and liquid to the surface. Charcoal is highly adsorbent.

Hero's aeolipile

aeolipile A primitive steam turbine designed by the Greek inventor Hero of Alexandria in about AD 100.

aerial Or antenna; a rod or array used to transmit or receive radio, television or radar signals.

aerodynamics The study of the behaviour of bodies moving through the air.

aerofoil A body so shaped that it experiences upthrust, or lift, when it travels through the air. Aeroplane wings have an aerofoil cross-section.

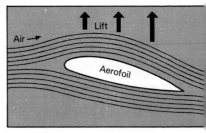

aerolite A stony meteorite.

aeronautics The science that deals with the design, construction and operation of aircraft.

aeroplane A powered heavier-than-air flying machine with fixed wings. The Wright brothers Orville and Wilbur flew the first aeroplane at Kitty Hawk, North Carolina, on 17 December 1903.

aerosol Finely divided particles of solid or liquid suspended in air or another gas. Fog is a kind of aerosol. Many products, from insecticides to hair sprays, are now applied in the form of aerosols from pressurized cans.

aerospace A word coined from 'aeronautics' and 'space'. Aerospace research is concerned with developing vehicles that may operate in space as well as in the air.

affinity Chemical attraction.

Afterburner A device that sprays extra fuel into the rear of a jet engine to provide extra thrust.

Ag The chemical symbol for silver; from the Latin word, argentum.

agate A very hard variety of silica, used for example for the knife edges of balances.

age hardening A property of some alloys, particularly duralumin, to increase their hardness and strength over a period of time.

AGR Abbreviation for advanced gas-cooled reactor.

air The mixture of gases that makes up the Earth's atmosphere. It consists mainly of nitrogen (78%), oxygen (21%) and argon. There are also traces of carbon dioxide and the noble gases, together with variable amounts of water vapour.

air conditioning Treating the air in a room to make it comfortable to live in. An air-conditioning unit controls the temperature and humidity of the air and also rids it of dust and smells.

aircraft Any craft that travels through the air. It may be heavier or lighter than air; powered or unpowered; or have fixed or rotary wings. See **aeroplane, airship, autogiro, balloon, glider** and **helicopter.**

air-cushion vehicle One that glides just above the surface of the ground or water on a 'cushion' of compressed air. It is usually called a hovercraft in Britain.

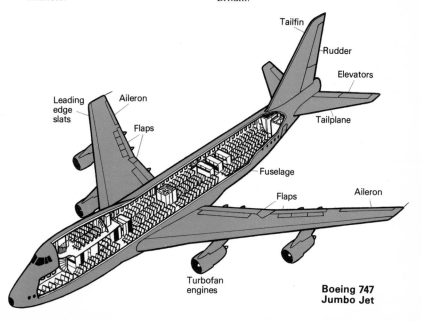

Tailfin

Rudder

Elevators

Leading edge slats

Aileron

Flaps

Tailplane

Fuselage

Flaps

Aileron

Turbofan engines

Boeing 747 Jumbo Jet

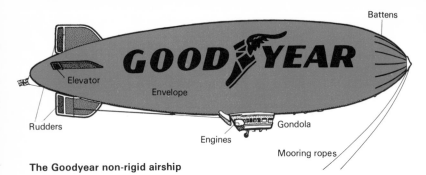

The Goodyear non-rigid airship

airlock A chamber in a caisson, tunnel or spacecraft in which the pressure can be adjusted. Astronauts may use an airlock when they leave their pressurized spacecraft to 'walk' in airless space. It prevents them having to depressurize the whole craft.

airship A powered lighter-than-air craft consisting of a gastight bag, or envelope, containing a light gas such as hydrogen or helium. Henri Giffard built the first successful steam-powered airship in 1862. Count Ferdinand von Zeppelin built the most famous rigid-framed airships, or Zeppelins, from 1900. Airship development virtually ceased after the biggest Zeppelin *Hindenburg* (248 m, 812 ft long) burst into flames on landing in 1937. The few modern airships are non-rigid craft, or blimps.

Al The chemical symbol for aluminium.

alabaster A soft and translucent form of gypsum rock widely used for carving.

albedo A measure of the extent to which a planet or moon reflects sunlight. It is the ratio of the amount of light reflected to the amount of light received. The Moon has a very low albedo (0·07), while that of Venus is high (0·6).

albumin Protein that is soluble in water and coagulates when heated. Egg-white contains albumin.

alchemy An art and craft widely practised for a thousand years or more until the Middle Ages, in which experimenters tried to turn base metals into gold and discover an elixir that would prolong life. The practice was connected with astrology and mysticism.

alcohol The name commonly given to ethanol, or ethyl alcohol, the intoxicating substance in beers, wines and spirits. But ethanol is only one of a class of organic compounds called alcohols, which also includes methanol, glycerol and menthol. They all contain the hydroxyl group (—OH) and combine with organic acids to form esters.

Aldebaran (α Tauri) The brightest star in the constellation Taurus. A red giant with a magnitude of 0·86, it is surrounded by the Hyades star cluster.

aldehydes A class of organic compounds that contain the —CHO group, the simplest being formaldehyde, HCHO.

algebra A branch of mathematics, dating back to Babylonian times, in which letters (a, b, x, y) are used to stand for quantities, and signs and symbols represent words ($\times$, $=$, $>$, $\cap$) and operations. Algebra is often called the language of mathematics.

Algol (β Persei) The second brightest star in the constellation Perseus. The first of a class of variable star called an eclipsing binary to be so identified—by John Goodricke in 1782.

algorithm A sequence of steps required to achieve a particular result. Mathematical operations, such as long division, are algorithms.

alicyclic compounds Organic compounds containing rings of carbon atoms, but not benzene rings.

aliphatic compounds Organic compounds in which the carbon atoms are linked in straight or branched chains.

alizarin A common red dye known to the ancient Egyptians and originally extracted from the madder plant.

alkali A class of strong bases that dissolve in water and turn red litmus blue. In solution alkalis yield hydroxyl ions (OH⁻). Alkalis can be regarded as the chemical opposites of acids, with which they react to form a salt and water—they neutralize acids. The strongest alkalis are the hydroxides of sodium and potassium, termed caustic soda and caustic potash respectively because they can burn.

alkali metals A closely related group of metals which react with water to form hydroxides that are strong alkalis. It includes lithium, sodium, potassium, rubidium, caesium and francium.

alkaline-earth metals A group of metals whose hydroxides are weak alkalis. It includes beryllium, magnesium, calcium, strontium, barium and radium.

alkaloids A class of organic compounds containing nitrogen noted for their powerful effect on the human body. It includes cocaine, morphine, quinine, strychnine and scopalamine.

alkanes See **paraffins.**

alkenes See **olefines.**

alkyl group A radical derived from the paraffin hydrocarbons. The alkyl group methyl (CH_3—) is derived from methane; ethyl (C_2H_5—) from ethane.

allotropy The property of some elements to exist in different physical forms. Carbon exists naturally in two allotropes—soft graphite and hard diamond.

Diamond has a rigid 3D structure

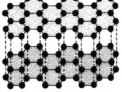

Graphite has layers of molecules which give it a flaky structure

USEFUL ALLOYS	
Alloy	**Composition**
Brass	Copper, zinc
Bronze	Copper, tin
Constantan	Copper, nickel
Cupronickel	Copper, nickel
Duralumin	Aluminium, copper, other metals
German silver	Copper, zinc, nickel
Gun metal	Copper, tin, zinc
Phosphor bronze	Copper, tin, zinc, phosphorus
Solder	Tin, lead
Type metal	Lead, antimony, tin
White metal	Lead, tin, antimony, copper

ALLOY STEELS	
Alloy	**Alloying metals**
Invar	Nickel
Stainless steel	Chromium, nickel
Tool steel	Tungsten, chromium

alloy A mixture of metals or of a metal and a non-metal. Most metals are used in the form of alloys, which are in general stronger and harder than the pure metals. The commonest alloy is mild steel (iron with traces of carbon and other metals).

alluvial deposits Mineral deposits formed from debris transported by flowing water.

Almagest The Arabic name (meaning 'The Greatest') for the astronomical and mathematical encyclopedia compiled by Ptolemy about AD 140. It draws on the knowledge of ancient Greek astronomers such as Hipparchus.

alnico A series of alloys, containing aluminium, nickel and cobalt, used for making powerful permanent magnets.

Alpha Centauri Or Rigil Kent; the third brightest star in the heavens, in the constellation Centaurus. It has a magnitude of −0·2 and, at a distance of 4·3 light-years, is the nearest star to us after its faint companion Proxima Centauri.

alpha-particle A positively charged particle emitted by some radioactive elements. Consisting of two protons and two neutrons, it is identical to the nucleus of the helium atom.

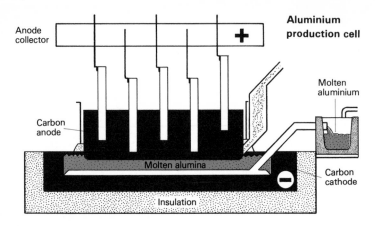

Aluminium production cell

Anode collector

Molten aluminium

Carbon anode

Molten alumina

Carbon cathode

Insulation

Altair (α Aquilae) A 1st magnitude star in the constellation Aquila which forms one of the three stars of the summer triangle.

alternating current (AC) The form of electricity that comes through the mains. It consists of current that is flowing alternately one way and then the other. In Britain the current alternates 50 times a second—its frequency is 50 hertz (cycles per second). In the US mains frequency is 60 hertz.

alternator The usual type of electricity generator in power stations, which produces alternating current. In large alternators the current is generated in the fixed, stator coils.

altimeter An instrument that measures altitude, or height above the ground. One type works by sensing the progressive change in air pressure with altitude. The other works by bouncing radio waves off the ground.

altitude The height of an object above a certain level, usually mean sea level.

alum A double salt consisting of the sulphates of two metals—one monovalent, one trivalent—and water of crystallization. Potash alum is a compound of potassium and aluminium sulphates, $K_2SO_4Al_2(SO_4)_3 . 24H_2O$. It is used as a mordant in dyeing and, with the other alums, for many other purposes in the paper, sugar, paint and water treatment industries. All the alums are isomorphous, or have the same crystalline form.

alumina Aluminium oxide, Al_2O_3, which occurs naturally as the mineral corundum. Rubies and sapphires are a crystalline form of alumina coloured by impurities. Alumina is an excellent refractory material with a high melting point (2024°C). It is obtained in vast quantities by purifying bauxite.

aluminium (Al) The commonest lightweight metallic element (rd 2·7, mp 660°C), first discovered by Hans Christian Oersted in 1825. It is the most abundant metal in the Earth's crust, of which it comprises 8%. It is present in most clays, but can only be extracted profitably from bauxite by the Hall-Héroult electrolytic process. Strong and corrosion resistant, aluminium is most used in the form of alloys. It is an excellent electrical conductor.

AM Stands for amplitude modulation.

Am The chemical symbol for americium.

amalgam An alloy of mercury and another metal. Silver and gold amalgams are used for filling teeth.

amatol A widely used high explosive consisting of ammonium nitrate and TNT.

amber A yellowish-brown fossil resin, prized as a gem.

americum (Am) An artificial radioactive element (rd 11·8, mp 900°C) made by bombarding plutonium with neutrons.

amethyst Violet-tinged quartz crystal, used as a gemstone.

amides Organic compounds of the general formula $RCONH_2$, where R is an alkyl radical. The commonest amide is acetamide.

amines Organic compounds derived from ammonia. They are formed when the hydrogen atoms in ammonia are replaced by organic radicals.

amino acids Organic acids that contain the amino group $(-NH_2)$. They are the prime constituents of proteins and are often termed the building blocks of life. The simplest amino acid is glycine, NH_2CH_2COOH.

ammeter An instrument that measures electric current. The most common type is the moving-coil ammeter.

ammonia (NH_3) A common sharp-smelling gas (bp $-34°C$) made industrially from its constituent elements by the Haber process. It readily liquefies under pressure and is widely used as a refrigerant. It is highly soluble in water, forming the alkali ammonium hydroxide (NH_4OH), which reacts with acids to form ammonium salts. Ammonium nitrate (NH_4NO_3) and sulphate $((NH_4)_2SO_4)$ make excellent fertilizers, while the chloride (NH_4Cl) is widely used in dry batteries.

ammonia-soda process Or Solvay process; the main commercial method of making sodium carbonate, or common soda. In the process, devised in 1865 by Belgian chemist Ernest Solvay, brine is treated in turn with ammonia and then carbon dioxide. This yields sodium bicarbonate, which when heated changes to soda.

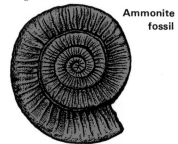

Ammonite fossil

ammonite An extinct mollusc with a coiled shell, which lived between about 350 and 70 million years ago. Fossils of ammonites are common and can measure up to $2 \cdot 7$ m (9 ft) across.

ammonium salts Those derived from ammonia, such as ammonium chloride (NH_4Cl), which yield ammonium ions (NH_4^+) in solution.

amorphous Having no particular shape or form; non-crystalline. Glass is an amorphous substance.

ampere Or amp; the common unit of electric current, named after the French physicist André Marie Ampère.

amphiboles Common silicate minerals of complex composition.

amphoteric A property displayed by some compounds of behaving like an acid or a base under different conditions. Aluminium oxide, for example, is amphoteric. It acts as a base by reacting with hydrochloric acid to form aluminium chloride, but also acts as an acid by reacting with sodium hydroxide to form sodium aluminate.

amplifier In electronics, a device now containing transistors that increases the strength of a signal, which could come from a gramophone pick-up, tape deck or radio circuit.

amplitude The amount by which something that is oscillating or vibrating is displaced from its equilibrium position.

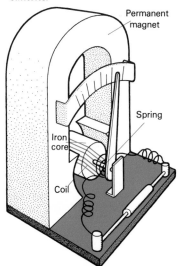

Permanent magnet

Spring

Iron core

Coil

Moving-coil ammeter

amplitude modulation

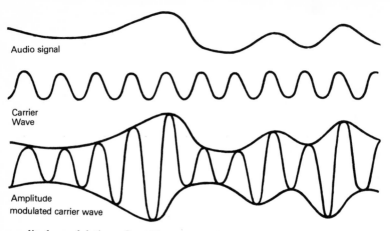

Audio signal

Carrier
Wave

Amplitude
modulated carrier wave

amplitude modulation Or AM; one way in which a radio wave is altered to carry a signal. The amplitude of the radio carrier wave is modulated, or altered, in sympathy with the change in the carried signal.

amygdale An almond-shaped cavity in an igneous rock which has become in-filled with minerals.

anaesthetic A drug used to create loss of feeling in the body. Local anaesthetics such as novocaine and lidocaine are injected to deaden only a small area of the body. General anaesthetics are administered to patients undergoing major surgery to render them completely unconscious. They often take the form of an injection of barbiturates, followed by a gas such as nitrous oxide.

analgesic A drug that helps prevent pain, but does not cause uncon-sciousness. Aspirin is the commonest analgesic.

analog(ue) computer Device for problem solving in which the quantities being considered are represented by changing voltages, currents and so on, which behave in an analogous way. Analog computers are used, for example, to control aircraft and space simulators, and complex industrial processes.

analysis, chemical Determining the make-up of chemical substances. Qualitative analysis seeks to find out which elements and radicals are present, while quantitative analysis seeks to find how much of each there is.

analytical chemistry The branch of chemistry concerned with chemical analysis.

Andromeda nebula (M31, NGC224) Actually a galaxy—one of the few visible to the naked eye—in the constellation Andromeda. The nearest spiral glaxy, it is some two million light-years away and some 200,000 light-years across.

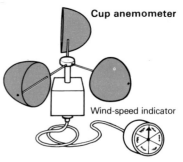

Cup anemometer

Wind-speed indicator

anemometer An instrument for meas-uring wind speed. The usual type features three or more cups which catch the wind and rotate a spindle. The rate of rotation is proportional to wind speed. Another type uses the cooling effect of the wind on a hot wire.

aneroid barometer An instrument for measuring atmospheric pressure. It uses a partly evacuated box to sense pressure changes, the box expanding or contracting as the pressure decreases or increases.

angle The space between two lines, as between the two hands of a clock. angles are usually measured in degrees (°), there being 360° in a complete circle. They may also be measured in radians, there being 2π radians in a circle.

angstrom unit (Å) A unit of length used to measure wavelengths of light. Named after the Swede Anders Jonas Ångstom, it equals 10^{-10} metre.

anhydride A chemical compound obtained by removing water from another compound. Anhydrides, which readily react with water, are useful in chemical synthesis. Sulphur trioxide (SO_3) reacts with water to form sulphuric acid (H_2SO_4).

anhydrite ($CaSO_4$) A form of calcium sulphate, which does not have (like gypsum) any water of crystallization.

anhydrous Without water (of crystallization).

aniline Or phenylamine ($C_6H_5NH_2$); an important organic chemical, derived from benzene, from which many kinds of dyes are made. It is an oily, strong-smelling liquid (bp 184°C).

anion A negatively charged ion, which moves towards the positively charged anode during electrolysis.

anisotropic Having different properties in different directions. Calcite crystals are anisotropic, having a different refractive index along different axes.

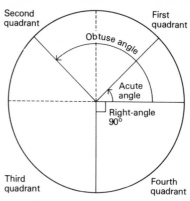

annealing A form of heat treatment in which metal or glass is heated and then allowed to cool at a controlled rate. It helps relieve internal stresses.

anode A positively charged electrode, eg in an electrolytic cell or an electron tube (valve).

anodizing A way of increasing the corrosion resistance of certain metals (eg aluminium and magnesium) by coating them with an oxide film. The process is so called because the metal is made the anode in an electrolytic bath.

Antares (α Scorpii) A 1st magnitude star in the constellation Scorpio. Some 425 light-years distant, it is a red super-giant with over 400 times the diameter of the Sun.

antenna An aerial.

anthracite A hard shiny coal high in carbon which burns with little smoke.

antibiotic A chemical compound produced by microorganisms, such as moulds and bacteria, which is used as a drug. Penicillin was the first antibiotic.

anticline An arch-shaped fold in the strata of sedimentary rocks.

anticyclone Or 'high'; a large region of high pressure in the atmosphere. Winds circulate clockwise around the central, high-pressure region in the northern hemisphere, and anticlockwise in the southern hemisphere.

antidote A substance that counteracts a poison.

antifreeze A chemical compound added, for example, to the water in a car radiator, to prevent it from freezing. Most antifreeze solutions contain ethylene glycol.

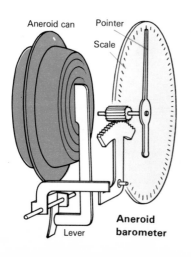

Aneroid barometer

antihistamine A drug administered to sufferers of hay fever and other allergies to combat the effect of excessive amounts of histamine in their bodies.

anti-knock agent A chemical, usually tetraethyl lead, added to petrol to prevent pre-ignition, or 'knocking' in a petrol engine.

antimatter A form of matter made up of elementary particles that are exact opposites of those present in ordinary matter. The atoms of ordinary matter have a nucleus containing positively charged protons, surrounded by a cloud of negatively charged electrons. The atoms of antimatter would contain negatively charged antiprotons surrounded by a cloud of positively charged antielectrons, or positrons. Such particles have in fact been produced in particle accelerators, but they have a very brief life. When a particle meets its antiparticle, both are destroyed, with the release of radiation.

antimony (Sb) A hard, brittle metallic element (rd $6\cdot7$, mp $630\cdot5°C$) frequently alloyed with other metals to increase their hardness. Its chief ore is stibnite.

antiparticle See **antimatter.**

apatite A common mineral containing mainly calcium phosphate. It is the main source of phosphorus and is used in vast quantities to make superphosphate fertilizer. The commonest form is fluorapatite, $3CaP_2O_8.3CaF$.

aperture In optics, an opening that admits light to a lens or optical system. The size of the lens aperture in a camera, controlled by an iris diaphragm, is measured on a scale of 'f-stops', which runs typically from f/1.4 to f/32. The bigger the number, the smaller is the aperture.

aphelion The point in the orbit of a planet or comet farthest away from the Sun. The Earth is at aphelion in early July at a distance of some 152,000,000 km (94,450,000 miles).

apogee The point in the orbit of the Moon or a satellite farthest from the Earth. At apogee the Moon lies at a distance of about 406,000 km (252,000 miles).

Apollo spacecraft The spacecraft that took American astronauts to the Moon between 1968 and 1972. Consisting of three modules, it carried a three-man crew and was lofted into space by the Saturn V rocket. The command module

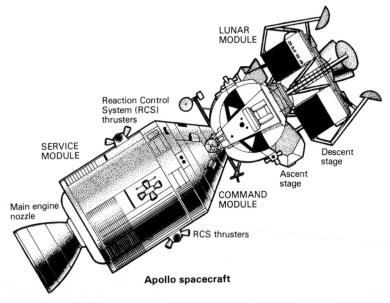

LUNAR MODULE

Reaction Control System (RCS) thrusters

SERVICE MODULE

Descent stage

Ascent stage

Main engine nozzle

COMMAND MODULE

RCS thrusters

Apollo spacecraft

Archimedean screw

was the only part to return to Earth. The lunar module was discarded in lunar orbit, while the service module was jettisoned before re-entry into the Earth's atmosphere.

Appleton layer A region of the ionosphere that reflects radio waves back to the Earth. It was named after its discoverer, the English physicist Edward Victor Appleton.

APT Abbreviation for British Rail's Advanced Passenger Train.

aqua fortis 'Strong water'; a name for concentrated nitric acid.

aqualung The self-contained underwater breathing apparatus (scuba) designed in 1943 by the French diver Jacques Cousteau. Air supply is automatically regulated by a two-stage valve, the demand-regulator.

aquamarine A pale bluish-green form of beryl prized as a gemstone.

aqua regia 'Royal water'; a mixture of concentrated hydrochloric (three parts) and nitric (one part) acids so named because it can dissolve the 'noble' metal gold, and also platinum.

Aquarius The Water-Bearer; a constellation of the zodiac lying between Capricornus and Pisces.

APOLLO LANDINGS

Apollo 11, 16–24 July 1969
N. Armstrong, M. Collins, E. Aldrin
Sea of Tranquillity

Apollo 12, 14–24 November 1969
C. Conrad, A. Bean, R. Gordon
Ocean of Storms

*Apollo 14, 31 January–
9 February 1971*
A. Shepard, E. Mitchell, S. Roosa
Fra Mauro formation

Apollo 15, 26 July–7 August 1971
D. Scott, J. Irwin, A. Worden
Apennines-Hadley rille

Apollo 16, 16–27 April 1972
J. Young, C. Duke, T. Mattingly
Descartes highlands

Apollo 17, 7–19 December 1972
E. Cernan, H. Schmitt, R. Evans
Taurus-Littrow valley

---

aqueduct A man-made channel, tunnel or pipe for carrying water, especially across a valley. The Romans were master builders of aqueducts, as evidenced by the Pont du Gard, a triple-tiered multi-arch aqueduct at Nîmes in Southern France.

aqueous solution One in which water is the solvent.

Ar The chemical symbol for argon.

aragonite A mineral form of calcium carbonate ($CaCO_3$) harder and denser than the more common calcite.

arc, electric A continuous spark, or discharge of electricity between two electrodes, which gives out intense light and heat. It is utilized in arc lamps for ciné projectors and in arc welding.

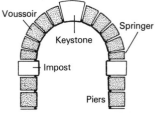

arch A curved structure built across an opening to support a load. The shape of the arch transmits the load from the top down the sides. It is widely used in bridge and dam construction.

archaeology The study of ancient buildings and artefacts.

Archimedean screw A device used in the Middle East and elsewhere to raise water, thought to have been invented in the 200s BC by the Greek scientist Archimedes (287–212 BC).

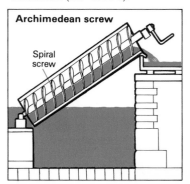

19

Archimedes' principle

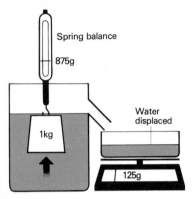

Spring balance

875g

Water displaced

1kg

125g

Archimedes' principle When an object is immersed in a fluid (liquid or gas), the apparent loss in its weight is equal to the weight of fluid it displaces.

Arcturus (α Boötis) The fourth brightest star in the sky, magnitude − 0·1, in the constellation Boötis. It is an orange giant star some 36 light-years away.

are The unit of area in the metric system, equal to 100 sq m or 119·6 sq yd.

argentite The most important ore of silver; silver sulphide, Ag_2S.

argon (Ar) A gaseous element found in air (about 0·9%). One of the noble gases, it is used in filling electric-light bulbs and in inert-welding operations.

Ariane rocket The heavy launch vehicle for the European Space Agency (ESA), which was first flight tested in 1979. Standing some 47·5 m (155 ft) high on the launch pad, its first two stages burn hydrazine and nitrogen tetroxide as propellants, while the third stage burns liquid hydrogen/oxygen.

Aries The Ram; a constellation of the zodiac lying between Pisces and Taurus.

arithmetic The branch of mathematics that deals with counting, measuring and calculating, using such operations as addition, subtraction, multiplication and division. It involves such things as numbers and number systems, fractions, decimals, percentages, ratios, square roots and logarithms.

Arizona meteor crater Or Barringer crater; the world's biggest crater located near Winslow, in Arizona. Made by the impact of a giant meteorite up to 50,000 years ago, it measures some 1,265 m (4,150 ft) across and 175 m (575 ft) deep.

armature The part of an electric generator or motor that rotates between the poles of a permanent magnet or electromagnet. It consists of many coils of wire wound around an iron core.

aromatic compounds Organic chemical compounds related to benzene and containing a benzene-ring structure. Most aromatics have a strong, usually pleasant smell; hence their name.

arsenic (As) A chemical element whose compounds are deadly poisons. It exists in several allotropes, the commonest of which is a brittle grey metal (rd 5·7, sublimes at 610°C). Arsenic occurs naturally as the sulphide in such minerals as realgar and with iron in arsenopyrite, or mispickel. Arsenic oxide (As_2O_3), known as white arsenic, is sometimes used medicinally; other arsenic compounds are used in pesticides.

arsine (AsH_3) An extremely poisonous colourless gas, once used as a military poisonous gas, but now used as a 'doping' agent in the production of semiconductors.

Ariane rocket

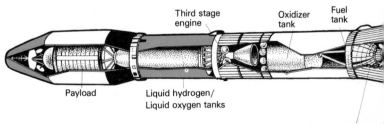

Third stage engine

Oxidizer tank

Fuel tank

Payload

Liquid hydrogen/ Liquid oxygen tanks

Second stage engi

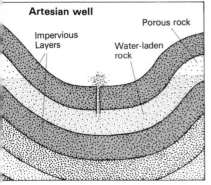

Artesian well

Porous rock

Impervious Layers

Water-laden rock

artesian well A bored well in which the water flows to the surface under natural pressure.

artificial element A chemical element that does not normally exist in nature. Elements with atomic numbers greater than 92 (uranium) do not occur in nature, but have been made artificially by bombarding uranium or other heavy elements with atomic particles, such as neutrons. Elements with atomic numbers up to 106 have been made in this way.

artificial horizon An aircraft instrument that shows the pilot the position of his plane with respect to the horizon. It works by gyroscope.

artificial satellite A man-made body in orbit around the Earth. See **satellite.**

artificial silk A name once applied to the first man-made fibre, rayon.

As The chemical symbol for arsenic.

ASA Short for American Standards Association; a rating for film speed.

asbestos The fibrous form of certain minerals, notably crocidolite and chrysotile. 95% of the world's asbestos is chrysotile, which derives from the mineral serpentine, a complex silicate. Materials made from asbestos are heat-proof and are used for such things as pipe lagging, firemen's suits and safety curtains.

ascorbic acid An alternative name for vitamin C, formula $C_6H_8O_6$.

asphalt A thick black petroleum residue containing more or less mineral matter. It is widely used for road surfacing. The biggest natural deposit of asphalt is Pitch Lake, Trinidad, discovered by Sir Walter Raleigh in 1595.

aspirin The commonest analgesic, or painkiller, whose correct name is acetyl-salicylic acid ($C_9H_8O_4$). It was first isolated from coal tar in 1853 by the French chemist Charles Gerhardt, though its medicinal properties were not realised until the turn of the century.

assay The process of determining the metal content of ores or alloys.

assembly line The basis for the mass-production method of manufacture. Workers stand in line and assemble a product piece by piece as it passes before them. Static assembly lines were in use in the early 19th century—to make ships' pulley blocks for example. In 1918 Henry Ford revolutionized assembly by introducing the moving conveyor belt to make his famous Model T. A more recent innovation has been the intro-duction of robots to the assembly line, particularly in the car industry.

astatine (At) A very rare radioactive element related to iodine that occurs fleetingly when other elements decay.

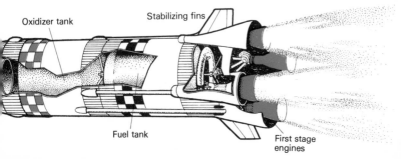

Oxidizer tank

Stabilizing fins

Fuel tank

First stage engines

Ancient Egyptian astronomer's concept of the universe

asteroid Or minor planet; a small rocky body that orbits the Sun. Most asteroids orbit in a 'belt' between the orbits of Mars and Jupiter. The largest asteroid, Ceres, some 1,000 km (620 miles) in diameter, was discovered by Guiseppe Piazzi on the first day of 1801. Over 2,000 asteroids have been identified but thousands more exist, too small to be seen from Earth.

astigmatism A defect of the eye, which prevents it focusing on all parts of the object viewed at the same time. It usually results from irregularities in the shape of the lens or cornea of the eye.

astrolabe An instrument used by early astronomers and navigators to measure the angle of heavenly bodies above the horizon. Developed by the Arabians, it came into widespread use in Europe in the 15th century.

astrology A pseudo science which suggests that human beings are influenced by the relative positions and motions of the heavenly bodies. Astrology, which developed in Chaldea, around the Persian Gulf, some 3,000 years ago, was historically important, however, because in closely studying the heavens astrologers laid the foundations of the true science of astronomy.

astronautics The science of space flight. An astronaut is a person trained to fly in space; the Russians call such a person a cosmonaut.

astronomical unit (AU) The mean distance between the Earth and the Sun: 150 million km (93 million miles).

astronomy The scientific study of the heavenly bodies and of the universe as a whole. Naked-eye astronomy has its origins in ancient Egypt some 5,000 years ago; telescopic observations of the heavens were pioneered by Galileo in 1609. Karl Jansky founded radio astronomy in 1931, picking up radio waves from the heavens. Today astronomers scan the heavens at many wavelengths—gamma-ray, X-ray, ultraviolet and infrared—observing from space (via satellites and probes) as well as from the Earth.

At The chemical symbol for astatine.

atmosphere The layer of gases around a planet, particularly the Earth. Scientists divide the Earth's atmosphere into several layers which, from the ground upwards, are troposphere, stratosphere, ionosphere and exosphere. Among the other planets in the solar system, Venus has a thick atmosphere; Mars has a slight atmosphere; and

Jupiter and the other outer planets have a very thick atmosphere. Whereas Earth's atmosphere consists of air, those of Venus and Mars are mainly carbon dioxide; and those of the outer planets, hydrogen and helium.

atmospheric pressure The pressure exerted by the Earth's atmosphere, which decreases with increasing altitude. At sea level the pressure is 101,325 newtons per square metre (about 14·7 lb/sq in), which is termed a standard atmosphere. It is often expressed as the height of a column of mercury it will support—760 mm (29·9 in)—or, in meteorology, as millibars. 1,000 millibars (= 1 bar) is slightly less than the standard atmosphere.

atmospherics Crackling interference in radio receivers caused by electrical discharges in the atmosphere.

atoll A circular ring of coral that grows up in the sea typically enclosing a shallow lagoon. It eventually develops into a coral island.

atom The smallest part of an element that retains its chemical identity and can take part in a chemical reaction. Atoms are very, very tiny, being typically about one ten-millionth of a millimetre (10^{-7} mm) in radius. Each of the 100 or so chemical elements contains a different kind of atom. Atoms differ in the number of particles they contain.

'Atom' means 'that which cannot be divided', for it was long thought that atoms were indivisible. The Greek Democritos first suggested that substances were made up of atoms in about 400 BC. But the idea was forgotten until John Dalton suggested it again in 1801. Experiments by J. J. Thomson and others near the turn of the century indicated that the atom contains smaller particles. In 1911 Ernest Rutherford built up a new theory of atomic structure which saw the atom as a miniature solar system containing a positively charged nucleus ('Sun'), around which circle negatively charged electrons ('planets'). And, like the solar system, most of the atom consists of space. By the 1930s the modern concept of the atom was established, though it has since been much modified and extended.

In essence the atom consists of a

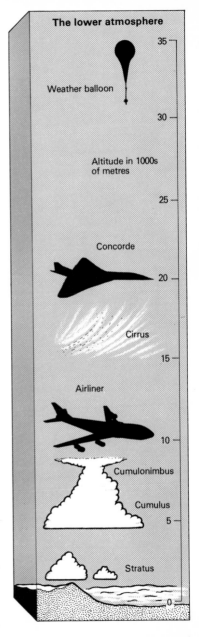

The lower atmosphere

Weather balloon — 35

— 30

Altitude in 1000s
of metres

— 25

Concorde

— 20

Cirrus

— 15

Airliner

— 10

Cumulonimbus

Cumulus

— 5

Stratus

— 0

atomic bomb

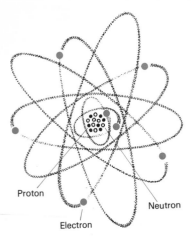

Proton

Neutron

Electron

Nitrogen atom

nucleus containing two main particles—positively charged protons and neutral neutrons—with negatively charged electrons orbiting around it. There are as many electrons as protons, making the atom electrically neutral. Many other atomic particles are known to exist, however; see **baryons, hadrons, leptons, mesons, neutrinos,** and **antiparticles.** And all atomic particles could be made up of different kinds of particles called **quarks**. Nevertheless the structure of matter and the chemistry of the elements can still be best explained in terms of the three main particles—protons, neutrons and electrons. Different atoms have different numbers of these particles.

Most atoms are stable. Others are unstable—they break down and give off particles and radiation. This phenomenon is called radioactivity.

atomic bomb   A bomb with devastating power, which uses the enormous energy given off when certain atoms split (see **nuclear energy**). Only two atomic bombs have ever been dropped in warfare—by the Americans in 1945 on the Japanese cities of Hiroshima and Nagasaki. Each had the explosive force of over 20,000 tons of TNT. They obliterated the two cities and together killed over 100,000 people. See also **hydrogen bomb**.

atomic clock   The most accurate of all clocks, which is regulated by the precise vibration of certain atoms (eg caesium) or molecules (eg ammonia). An atomic clock may gain or lose less than a second in 100,000 years.

atomic energy   See **nuclear energy.**

atomic mass   The mass of an atom expressed on a scale on which the mass of the carbon-12 isotope is exactly $12 \cdot 00$. The relative atomic mass (or atomic weight) of an element is also expressed on this scale, but takes into account the fact that the element may consist of a number of isotopes. Each isotope has a different atomic mass.

atomic number   The number of protons in the nucleus of an atom.

atomic weight   See **atomic mass.**

atomizer   A device that produces a fine spray. In an atomizer like a scent spray, air is blown over a tube dipping into liquid. The liquid is drawn into the air stream and splits up into fine droplets.

atom-smasher   A popular name for a particle accelerator.

atropine   An alkaloid ($C_{17}H_{23}NO_3$) that occurs in deadly nightshade and hen-bane. It attacks the nervous system. It is used in medicine to dilate the pupils of the eyes.

attar   Fragrant oil or perfume distilled from rose petals.

attenuation   In communications, the gradual reduction in strength of a radio wave passing through the air, or of radiation passing through matter.

Au   The chemical symbol for gold; from the Latin, aurum.

audio-frequency   A wave frequency between about 30 and 30,000 hertz (cycles per second). Sound waves of such frequencies can be heard.

aurora   Coloured lights seen in the sky in polar regions. At the north pole the phenomenon is called the aurora borealis, or northern lights; and at the south pole the aurora australis, or southern lights. The light is produced when particles streaming from the Sun collide with molecules of air in the upper atmosphere. Auroras have also been observed on Jupiter.

australite   A form of natural glass found in Australia, thought to be of meteoric origin; a type of tektite.

autoclave   A kind of industrial pressure cooker.

The atomic bomb releases the enormous energy locked inside the atom, equivalent to tens of thousands of tonnes of TNT.

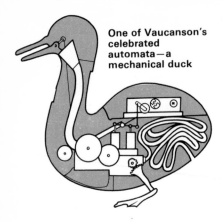

One of Vaucanson's celebrated automata—a mechanical duck

autogiro Or autogyro; an aircraft with a rotary wing which preceded the helicopter. Invented in Spain in 1923 by Juan de la Cierva, the autogiro is propelled by a front propeller. The wing rotor rotates by itself and provides lift when the aircraft travels through the air.

automatic pilot Often nicknamed 'George'; a device which, once set, automatically maintains a plane's course, altitude and speed. It is operated by means of gyroscopes and was first demonstrated by Lawrence Sperry of the US in 1913.

automation The widespread use in industry of self-controlling, automatic machines. It makes use of feedback mechanisms to achieve self-control.

automaton (plural, automata); a device that imitates the actions of a living thing.

autumn Or fall; the season between summer and winter. It begins in the northern hemisphere at the autumn equinox on about September 23, and ends at the winter solstice on about December 22. In the southern hemisphere autumn lasts from about March 21 to about June 21.

average A typical number or member of a group. In arithmetic there are three basic kinds of average—mean, median and mode. The mean of a set of figures is the sum of those figures divided by the number of figures. The median of a set of figures is the middle one, when the figures are placed in order. The mode of a set of figures is the one which occurs most often.

Avogadro constant The number of molecules in one mole of a substance; equal to $6 \cdot 023 \times 10^{23}$.

Avogadro's law Equal volumes of different gases, under the same conditions of temperature and pressure, contain the same number of molecules. The volume occupied by one mole of any gas at STP is about $22 \cdot 4$ litres. The law was first stated in 1811 by the Italian Amadeo Avogadro.

avoirdupois The Imperial system of weights based on the pound (lb), in which 7,000 grains (gr) and 16 ounces (oz) make 1 lb, and 16 drams (dr) make 1 oz.

axis An imaginary line about which a body can be considered to rotate (axis of rotation) or about which a figure is symmetrical (axis of symmetry).

azeotrope A mixture of two liquids which has a constant boiling point. The vapour given off has the same composition as the liquid mixture. The mixture can thus not be separated by distillation.

azide A chemical compound containing a group of three nitrogen atoms $(-N_3)$. Many azides are unstable and can be used as detonating agents to fire high explosives.

azimuth A coordinate for describing the position of a star. It is the angle between the great circle through the star and the meridian.

azo group A group found in organic chemistry, consisting of linked nitrogen atoms, $-N{=}N-$. The most important chemicals containing this group are the azo dyes, which represent more than half the dyes now used.

azurite A distinctive azure-blue mineral often used in jewellery. It is basic copper carbonate, $2CuCO_3 \cdot Cu(OH)_2$.

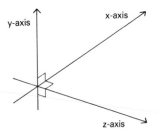

Three-dimensional axes

B

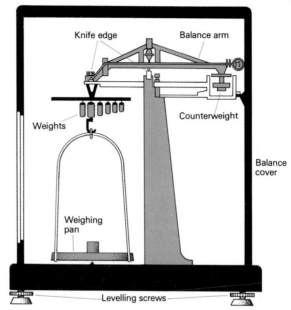

Knife edge

Balance arm

Weights

Counterweight

Balance cover

Weighing pan

Levelling screws

B The chemical symbol for boron.

Ba The chemical symbol for barium.

Babbitt metal An alloy of tin, copper and antimony invented by the American Isaac Babbitt in 1839 for use in bearings. It may also contain lead.

background radiation Radiation from the surroundings that may interfere with measurements, say, of radioactivity. Background radiation may come from cosmic rays or from rocks.

Bailly's beads Bright points of light visible around the limb of the Moon during a total solar eclipse. Named after the English astronomer Francis Bailly, they are caused by shafts of sunlight shining through lunar valleys.

Bakelite The first synthetic plastic, invented in 1909 by the American chemist Leo Hendrik Baekeland. Made from phenol and formaldehyde, it is still widely used to make heatproof materials and electrical insulation. It is a thermosetting material.

baking soda A common name for sodium bicarbonate, $NaHCO_3$. It is so-called because it is included in baking powder to produce carbon dioxide, which makes cakes rise.

balance A sensitive weighing machine used in laboratories. The traditional balance uses a pair of scale pans hanging from an arm that pivots on a knife edge. Most modern balances are single-pan machines and may work electrically. The most sensitive balances can weigh with an accuracy of a millionth of a gram.

ballistics The science that deals with the motion of projectiles through the air.

ball mill A machine that grinds lumps (eg of pigment) into powder, consisting of a rotating drum containing pebbles or steel balls.

balloon A lighter-than-air craft, whose uplift is provided by hot air or a light gas such as hydrogen or helium. The Montgolfier brothers in France flew the first hot-air balloon in 1783, just before their compatriot J. A. C. Charles launched the first hydrogen balloon.

bar A unit of atmospheric pressure used particularly in meteorology. 1 bar = 1,000 millibars.

barbiturate An organic compound widely used in medicine as a sedative and sleeping draught and in large doses as an anaesthetic.

27

barium

barium (Ba) A soft and reactive metallic element (rd 3·6, mp 710°C) related to calcium. One of the alkaline-earth metals, it was isolated by British chemist Humphry Davy in 1808. Its main source is the sulphate mineral barytes ($BaSO_4$). Pure barium sulphate is the main ingredient of the 'barium meal' sometimes given to patients undergoing X-ray treatment. It is highly opaque to X-rays.

Barnard's star A red dwarf star, some six light-years distant, discovered by American astronomer Edward E. Barnard in 1916. It has the largest proper motion of 10·3 seconds of arc per year. There is evidence that one or more planets circle the star.

barometer An instrument that measures atmospheric pressure. Italian physicist Evangelista Torricelli discovered the principle of the barometer in 1643 and made a mercury barometer. This is basically a glass tube full of mercury up-ended in a bowl of mercury. The pressure of the atmosphere balances a column of mercury about 760 mm (30 in) high. Domestic barometers are of the aneroid ('without liquid') type. The pressure sensor in the aneroid barometer is a partly evacuated box, which expands or contracts as the pressure changes. This movement is magnified by levers and moves a pointer over a graduated scale.

barrel A unit of capacity, used particularly for petroleum. A petroleum barrel equals 35 gallons or 159 litres.

Barringer crater See **Arizona meteor crater**.

barycentre The centre of mass, especially of the Earth-Moon system.

baryon The general name of the heaviest nuclear particles, such as protons and neutrons.

barytes A common mineral form of barium sulphate ($BaSO_4$). White in colour and noticeably heavy (rd 4·5), it is also called heavy spar.

basalt A heavy, dark igneous rock consisting of very tiny crystals. It is formed when magma flowing from volcanoes cools quite quickly. When it cools, it often cracks into pentagonal or hexagonal columns. Such columns form the Giant's Causeway in Ireland.

bascule bridge A type of movable bridge formed of one or two leaves which pivot upwards to open. Tower Bridge in London is an example of a twin-bascule bridge.

base In chemistry, a compound that is the chemical opposite of an acid. It combines with an acid to form a salt and water only. It turns red litmus blue. In solution bases yield hydroxyl ions, OH^-. Strong bases are called alkalis. The hydroxides of the alkali metals and of the alkaline-earth metals are strong bases.

base metal The opposite of a noble or precious metal. Base metals, which include tin, zinc and lead, tend to corrode and oxidize readily.

basicity The number of hydrogen atoms in an acid available for replacement by a metal. Phosphoric acid (H_3PO_4) is classed as a tribasic acid; it can form three types of salts, eg the sodium salts Na_3PO_4, Na_2HPO_4, and NaH_2PO_4.

basic oxide An oxide that reacts with an acid to form a salt and water only. Most metal oxides are basic.

basic-oxygen process Now the commonest method of steel-making, it involves blowing oxygen at supersonic speed into molten pig iron. The impurities quickly burn out, leaving steel. It is also called the L-D process after the towns in Austria (Linz and Donawitz) near where the process was developed in 1952.

batch process An industrial process in which a product is made in separate batches. A batch process is usually more costly and time-consuming than a continuous process.

batholith A large body of igneous rock, usually with steep sides, that has intruded into existing (country) rock.

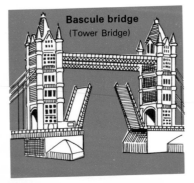

Bascule bridge
(Tower Bridge)

Picard's bathyscaphe 'Trieste'

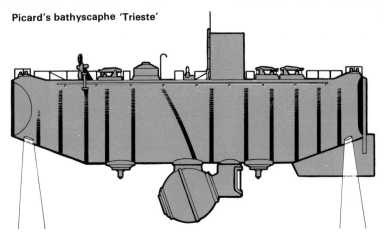

bathyscaphe A deep-diving vessel designed by Swiss scientist Auguste Picard. The bathyscaphe 'Trieste' descended to a record 10,916 m (35,810 ft) in the Mariana Trench of the Pacific Ocean in 1960. It consists of a steel cabin slung beneath a large 'float' containing petrol.

bathysphere A deep-diving vessel designed by William Beebe in 1930.

battery A device that produces electricity as a result of chemical action. Italian physicist Alessandro Volta developed the first battery—the voltaic pile—in 1800. A battery usually consists of one or more units, or cells. Primary cells, such as the dry cell, cannot be recharged; secondary cells, or accumulators, such as the lead-acid car battery, can be recharged.

bauxite A mineral containing mainly alumina, or aluminium oxide (Al_2O_3). Named after the place where it was first found, Les Baux in France, bauxite is the only ore of aluminium. It is purified by the Bayer process.

Bayer process The process by which alumina is extracted from bauxite. The bauxite is treated with caustic soda, and the alumina dissolves and can be separated. It was devised by the German chemist Karl Joseph Bayer in 1888.

Be The chemical symbol for beryllium.

beam In engineering, a horizontal piece of timber, steel or reinforced concrete that supports a load. In a beam bridge it supports the bridge deck.

bearing In mathematics, the direction of something from a given point, stated usually in terms of angle.

Basic oxygen converter

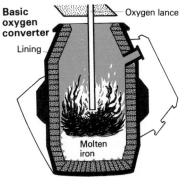

Oxygen lance

Lining

Molten iron

Dry-cell battery

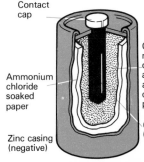

Contact cap

Ammonium chloride soaked paper

Zinc casing (negative)

Carbon, manganese dioxide and ammonium chloride paste

Carbon rod (positive)

bearings

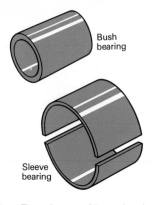

Bush bearing

Sleeve bearing

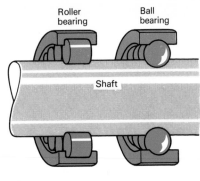

Roller bearing

Ball bearing

Shaft

Common bearings

bearings In a machine, a bearing is a device that typically supports a shaft and permits it to turn. The simplest bearing is the journal or sleeve bearing, which consists simply of a hollow cylinder. In ball bearings the moving surfaces are separated by a ring, or race of steel balls. Often rollers and needles are used in place of the balls. Watches have moving parts that pivot in jewel bearings. Some precision machines have air bearings.

beat In sound, a rhythmic pulsation heard when notes of nearly the same frequency are sounded together.

Beaufort scale A numerical scale for estimating the force of the wind, devised in England by Francis Beaufort in 1805.

Becquerel rays The term once applied to radioactive radiation, first discovered by the French physicist Henri Becquerel in the 1890s.

bedding The layering of sedimentary rocks.

belemnite An extinct mollusc, many fossils of which are found in sedimentary rocks. The fossils are torpedo shaped and represent the internal shell of the organism, which is related to the modern-day squid.

bends, the A painful condition experienced by deep-sea divers who surface too quickly. It is caused by gas bubbles coming out of the blood inside the blood vessels.

THE BEAUFORT SCALE				
Scale no.	**Name of wind**	**Wind speed**		**Effects of the wind**
		km/h	**mph**	
0	Calm	<1	<1	Smoke goes straight up
1	Light air	1–5	1–3	Smoke is slightly bent
2	Light breeze	6–11	4–7	Leaves begin to rustle; wind vane moves
3	Gentle breeze	12–19	8–12	Leaves and twigs are in constant motion
4	Moderate breeze	20–28	13–18	Raises dust and moves small branches
5	Fresh breeze	29–38	19–24	Small trees start to sway
6	Strong breeze	39–49	25–31	Large branches start moving
7	Moderate gale	50–61	32–38	Whole trees sway; walking becomes difficult
8	Fresh gale	62–74	39–46	Twigs are broken off trees
9	Strong gale	75–88	47–54	Chimneys and roofs are damaged
10	Whole gale	89–102	55–63	Trees are uprooted
11	Storm	103–117	64–75	Widespread damage—very rare inland; more common at sea
12	Hurricane	>117	>75	Widespread destruction

benefication Means of concentrating the valuable components in mineral ores.

bentonite A type of clay with many industrial uses, eg as drilling mud for lubricating oil-well drills. Some kinds can absorb large amounts of water and expand enormously.

benzanol See phenol.

benzene (C_6H_6) Also called benzol; a sweet-smelling, colourless liquid (bp $80 \cdot 1°C$) obtained from petroleum. First discovered by Michael Faraday in 1825, it is the starting point of many important organic chemicals. It has a unique ring structure, called the benzene ring, which was first suggested by August Kekulé in 1865. All compounds derived from benzene contain this structure. They form a group of chemicals called the aromatics.

benzene-ring structure See **Kekulé structure**.

Bergius process A process for making oil and petrol from coal, developed by the German chemist Friedrich Bergius in the 1920s. It involves heating a mixture of powdered coal and heavy oil at a temperature of about $450°C$ and under pressure of over 200 atmospheres.

berkelium (Bk) A man-made, radio-active element obtained by bombarding americium with alpha-particles. It was first prepared in 1949 at Berkeley in California.

Bernoulli's theorem states that in a moving ideal fluid the total energy remains constant. This energy is made up of the kinetic energy of motion, the potential energy due to gravity ('head' of fluid), and the pressure energy. The theorem indicates that when the pressure in a fluid decreases, the speed increases and vice versa. This explains, for example, how an aeroplane wing develops lift. The theorem was developed by Daniel Bernoulli in 1738.

beryl A yellowish-green mineral often used as a gemstone and the ore of the metal beryllium. It is often found as huge hexagonal crystals in pegmatites. Emeralds are rare dark green transparent crystals of beryl, which is a complex beryllium aluminium silicate.

beryllium (Be) A rare lightweight element (rd $1 \cdot 8$, mp $1280°C$) related to calcium. It is hard, strong and corrosion resistant and is used most in alloys.

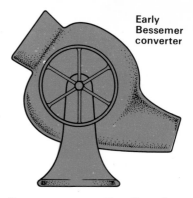

Early Bessemer converter

Bessemer process The first cheap method of making steel, developed by Henry Bessemer in 1856. In the process, now superseded by the open-hearth and basic-oxygen processes, air is blown through pig iron in a conical converter. The impurities in the pig iron are burned out.

beta-particle An electron emitted by certain radioactive atoms. Beta emission is one of the main ways radioactive atoms break down, or decay. Some atoms give out positrons instead.

betatron A circular particle accelerator designed to accelerate beta-particles, or electrons; developed by Donald W. Kerst in 1940.

Betelgeuse (α Orionis) The second brightest star in the constellation Orion. Some 600 light-years distant, it is a red super-giant variable star.

Bi The chemical symbol for bismuth.

bicarbonate A salt containing hydrogen derived from carbonic acid; properly termed 'hydrogen carbonate'. The most familiar example is bicarbonate of soda, or sodium hydrogen carbonate, $NaHCO_3$.

big-bang theory The now-favoured theory of the origin of the universe first put forward in 1927 by Georges Lemaître. It considers that the universe came into existence some 15,000–20,000 million years ago when a highly compressed and intensely hot ball ('primeval atom') of matter exploded. The universe has been expanding ever since and will continue to do so (see **expanding universe**). Compare **steady-state theory**.

billet A piece of rolled steel, thinner and wider than a bloom.

billion Strictly, in Britain 1 billion = 1 million million, but the American usage of 1 billion = 1 thousand million has become common.

bimetallic strip A strip used in thermostats consisting of two different metals joined together. When the temperature changes, the metals expand by differing amounts, causing the strip to bend one way or the other. In a thermostat this makes or breaks a contact.

binary In astronomy, a true double star system consisting of two stars rotating around a common centre of gravity. In an eclipsing binary one star is much darker than the other and eclipses it periodically, causing the brightness of the system to vary. See **Algol**.

binary system In mathematics, a number system that uses only the digits 0 and 1. It has the number base of 2, which means that place values go up in multiples of 2 (as opposed to multiples of 10 in the decimal (base 10) system).

binoculars A compact type of telescope designed for use by both eyes. The standard type are prismatic binoculars, so called because they 'fold' the light path by means of prisms.

biochemistry A branch of science concerned with the chemistry of living things.

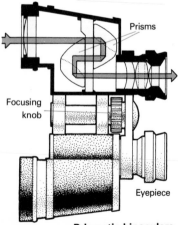

Prismatic binoculars

Prisms

Focusing knob

Eyepiece

biology The science of living things, whose main branches are botany—the study of plants; and zoology—the study of animals.

bioluminescence The emission of light by living things, such as fungi, insects (eg fireflies) and fish.

bionics The study and development of artificial systems designed to imitate those in living things.

biotite A common silicate mineral of the mica family. Black or dark-green in colour, it is also called black mica.

bismuth (Bi) A brittle metallic element (rd 9·8, mp 271·3°C) used most in alloys, such as the low melting point fusible alloys used as plugs in fire-sprinkler systems. Most bismuth is obtained as a by-product from tin, lead and copper refining.

bit In computing, a binary digit, representing a unit of information.

bitumen A black tarry substance consisting of various hydrocarbons, derived from petroleum.

bivalent See **divalent**.

Bk The chemical symbol for berkelium.

black body An ideal substance that absorbs all radiation falling on it. Such a body would also emit radiation of all frequencies.

black hole A region of space whose gravity is so intense that nothing, not even light, can escape from it. It is an object that is thought to result from the catastrophic collapse of a very massive star.

blast furnace A furnace into which hot air is blasted to intensify the heat of combustion. Originally developed to smelt iron, it is now also used to smelt other metals, including lead and zinc.

bleaching The process of whitening paper, cloth or other materials. The traditional bleaching agent is bleaching powder, a substance made by bubbling chlorine through slaked lime. Modern bleaches include sodium hypochlorite and hydrogen peroxide.

blende A common mineral form of zinc sulphide (ZnS). It is also called sphalerite ('deceiver') because it is often deceptively like more valuable ores, such as galena, which often contains silver.

blimp A non-rigid or semi-rigid airship which relies on gas pressure to maintain its shape. Modern airships such as 'Goodyear' are blimps.

Blast furnace

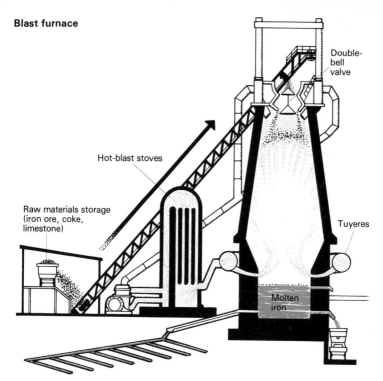

Double-bell valve

Hot-blast stoves

Raw materials storage
(iron ore, coke,
limestone)

Tuyeres

Molten iron

block and tackle See **pulley**.

bloodstone A dark-green variety of the mineral chalcedony which often contains vivid red nodules of jasper.

bloom A rectangular piece of steel formed by rolling an ingot.

blower An air compressor or super-charger.

blowlamp A high-pressure kerosene (paraffin) lamp which produces a hot flame. It has been almost entirely superseded by a gas torch burning bottled gas.

blowpipe A device used to create a fierce flame for glassblowing, soldering and welding. Fuel gas and air or oxygen are fed under pressure to the blowpipe where they mix and burn.

blue john A blue-banded variety of the mineral fluorite.

Blue Lias A sedimentary rock consisting of blue clays and limestones in which fossils are abundant.

blueprint A copy of original plans or drawings. Blueprints are made by a kind of photographic process using light-sensitive paper which, when processed, shows the lines of the plans as white on a blue background.

blue shift In starlight, the shift of the spectral lines towards the blue end of the spectrum. It indicates that the star is travelling towards us.

blue vitriol A popular name for hydrated copper sulphate, $CuSO_4.5H_2O$.

Bode's law In astronomy, an empirical law that gives the approximate distances of the planets from the Sun. Named after the German astronomer Johann E. Bode, the distance (D) in astronomical units is given by the formula:

$$D = 0 \cdot 4 + 0 \cdot 3 \times 2^n$$

where $n = 0, 1, 2$ etc

bog-iron ore A name for the iron oxide ore limonite.

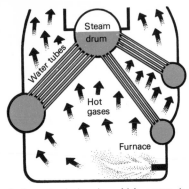

Steam drum

Water tubes

Hot gases

Furnace

boiler A device in which water is converted into steam. The water is heated in the boiler by burning fuel in a furnace beneath. In the water-tube boiler, water is passed through tubes surrounded by the furnace flames and gases. In the fire-tube boiler, the hot gases are drawn through tubes surrounded by the boiler water.

boiling point The temperature at which a liquid bubbles and changes into vapour. Water has a boiling point of 100°C at ordinary atmospheric pressure. The pressure should always be quoted when a boiling point is given, because the boiling point varies with the pressure. It decreases as the external pressure decreases. Boiling points are usually quoted at standard atmospheric presure, 760 mm of mercury.

bolide Also called fireball; a large bright meteor, which may explode.

bolometer A sensitive instrument for measuring radiation, particularly heat. The sensor is usually a blackened metal strip, whose resistance changes when radiation falls upon it.

bonding In physics and chemistry, the means by which one atom links with another to form a chemical compound. The commonest types of atomic bond are the electrovalent bond and the covalent bond. Water contains the hydrogen bond.

bone black A kind of charcoal produced by heating animal bones in a limited supply of air.

bone china An imitation porcelain which contains bone ash. It was introduced by the English potter Josiah Spode in about 1800.

booster In astronautics, the first stage of a multistage rocket, or a rocket that provides additional take-off thrust. The space shuttle has re-usable solid-propellent boosters.

borax The most common compound of boron, known chemically as sodium tetraborate, $Na_2B_4O_7.5H_2O$. Borax is widely used in soaps, water softeners, ointments and eye-washes; and in making glass.

borax-bead test A simple test performed in chemical analysis which can identify certain metals. In the test the end of a platinum wire is dipped in borax and then heated in a Bunsen flame. The borax melts to form a clear bead. This is then dipped into the substance to be tested and heated again. If the bead changes colour, a certain metal is present in the substance tested. A blue colour indicates the presence of cobalt, for example.

borazon A form of boron nitride that is as hard as diamond. It is a widely used abrasive.

Bordeaux mixture A garden fungicide, which contains copper sulphate and lime.

bore A tidal surge that occurs in certain rivers with a gradually widening mouth. The bores of the River Severn in the West of England and the River Ganges in India are particularly well known.

boric acid Also called boracic acid, H_3BO_3; a weak acid that has mild antiseptic properties.

boring Cutting a hole; strictly, the machining operation known as boring means finishing a hole already drilled.

bornite A common reddish-brown copper sulphide mineral, also called peacock ore.

boron (B) One of the metalloid elements (rd 2·5, mp 2030°C) whose most common compound is borax. Boron is used in heat-resistant alloys for jet and rocket engines. It is incorporated in nuclear reactor control rods because it absorbs neutrons readily. Its compounds with hydrogen, boranes, are useful high-energy fuels.

borosilicate glass Glass that contains boric acid to make it heatproof. Pyrex is borosilicate glass.

bort Impure diamond, which is no use as a gem but is an invaluable industrial abrasive.

Bosch process A process for making hydrogen, developed by the German chemist Carl Bosch. In the process a mixture of steam and water gas is passed over a catalyst at high temperature.

boundary layer A thin layer that surrounds a body in a moving fluid. Across the boundary layer the fluid undergoes a marked change in velocity, from zero where it is in contact with the body, to the general fluid flow rate. Study of the boundary layer around aerofoils is an essential part of aerodynamics.

Bourdon gauge See **pressure gauge**.

Boyle's law One of the main gas laws, which describes the relationship between the pressure and volume of a gas. It is named after the English scientist Robert Boyle who first stated it in 1662. Boyle's law states that, when the temperature is kept constant, the pressure (P) of a gas varies inversely with its volume (V). This can be written as: $P \propto 1/V$ or PV = constant. In other words, the law means that if you double the pressure of a gas, you halve its volume.

Br The chemical symbol for bromine.

brake A device used to slow down or stop the movement of a body. Most vehicle brakes work on the principle of friction. They may operate by means of mechanical linkages, or by hydraulic or air pressure. Car brakes operate hydraulically on all four wheels when the brake pedal is pressed. The hand, or parking brake works mechanically just on the rear wheels. Most cars have disc brakes on the front wheels and drum brakes on the rear. In the disc brake, braking pressure forces pads against both sides of a disc that is attached to the wheel. In the drum brake pedal pressure forces a brake shoe against a drum that is attached to the wheel. The pads and shoes have a tough wear- and heat-resistant lining.

Bramah press The original hydraulic press, invented by the English engineer Joseph Bramah in 1795.

brass An alloy of copper and zinc of attractive yellowish colour. It is strong and corrosion resistant and is easy to shape and machine. It is used widely for such things as electrical fittings, plumbing joints and ornaments.

brazing A method of joining metal parts by melting a kind of brass into the joints. It is an operation intermediate between soldering and welding. Bicycle frames have brazed joints.

breccia A rock made up of sharp rock fragments cemented together.

breeder reactor An alternative name for a fast reactor.

breeze A light wind; usually one that blows locally—for example, the land and sea breezes in coastal areas. These breezes occur because the land and the sea heat up and cool down at different rates. The land heats up faster than the sea, but also cools down faster. This means that the land is hotter than the sea during the day, but cooler at night.

Drum brake

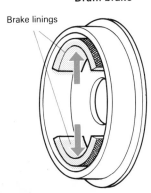

Brake linings

Disc brake

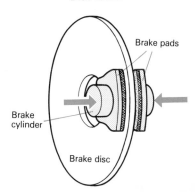

Brake pads

Brake cylinder

Brake disc

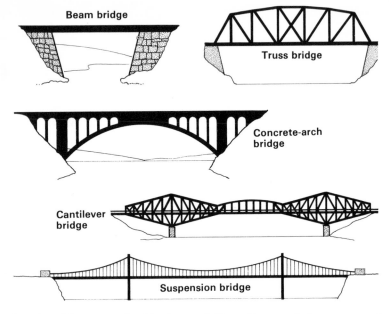

Beam bridge

Truss bridge

Concrete-arch bridge

Cantilever bridge

Suspension bridge

brewing The process of making beer. In brewing, grain (usually barley) is steeped (soaked) in water and then left to germinate. The germinated grain is then dried by heating and the shoots are removed. It is then known as malt. After some weeks the malt is ground and mashed, or mixed with warm water. During this process the starch in the grain is broken down into sugars. The solution resulting, called wort, is filtered and boiled with hops. It is then fermented with yeast. During fermentation the yeast converts the sugars into alcohol, and carbon dioxide bubbles off.

brick A building material, typically of rectangular shape, made of baked clay. Bricks have been used for building for at least 8,000 years. The earliest ones were made of mud baked in the Sun. Such simple bricks, called adobe, are still used in some parts of the world where there is little rainfall.

bridge In electrical engineering, a device for measuring electrical quantites, such as resistance. The first bridge was devised in 1843 by the British scientist Sir Charles Wheatstone.

bridges Structures built to carry a road or railway line over a river, valley or other obstacle. The biggest bridges built today are suspension bridges, in which the roadway hangs from thick cables. Cantilever bridges can also span huge gaps, as can concrete-arch and steel-arch bridges. In arch bridges the shape of the arch transmits the weight of the bridge and its load down to its foundations. The simplest bridge is the beam bridge, which consists essentially of a beam supported at each end. See also **bascule bridge**; **swing bridge**.

brightness Of a star, see **magnitude**.

brine A strong salty solution. Natural brines, which include sea water and underground deposits, are important sources not only of common salt but of other salts as well, and also bromine, iodine and magnesium.

Britannia metal An alloy rather like pewter containing tin, copper and antimony.

British Standards Institution (BSI) The national organization in Britain for establishing standard specifications for materials and methods for use in industry.

British Thermal Unit (Btu) A unit of heat once widely used in English-speaking countries, but now largely superseded by the metric calorie. 1 Btu is the amount of heat required to raise the temperature of 1 pound of pure water through 1°F. 1 Btu = 252 calories; 100,000 Btus = 1 therm.

broaching A machining operation carried out with a cutter which has rows of stepped teeth. Each set of teeth remove only a little metal, but together they remove quite a lot.

bromides The salts derived from hydrogen bromide, HBr. 'Bromide' can also mean a sedative because certain bromides, such as potassium bromide, have a sleep-inducing effect. Silver bromide is light-sensitive and is used in photographic emulsions.

bromine (Br) One of the few elements (rd 3·1, mp – 7·3°C, bp 58·2°C) that is liquid at room temperature. It is dark red in colour and has an unpleasant pungent smell. It is extracted from sea water. One of the halogens, it is closely related to chlorine and is very reactive. Its compound with hydrogen, hydrogen bromide (HBr), is a strong acid which yields salts called bromides, some of which are used as sedatives.

bronze An alloy of copper and tin, sometimes with other elements as well. It has widespread uses because it is strong, hard, corrosion resistant and easy to shape by casting. Its most common use is in coinage, such as the British ½p, 1p and 2p pieces and the American cent. Marine propellers and statues are often cast in bronze. The discovery of bronze in about 3,500 BC was one of the great milestones on the path to civilization.

Brownian movement The constant random motion of tiny particles suspended in a fluid, caused by invisible collisions with the fast-moving molecules of the fluid. The phenomenon was first described by the Scottish biologist Robert Brown in 1827, while examining water containing a suspension of microscopic pollen grains.

brush A conductor, typically made of carbon, which forms the contact between stationary and moving parts. Brushes are found in electric motors and generators to conduct current to and from the rotating armature.

BSI Abbreviation for British Standards Institution.

Btu Abbreviation for British Thermal Unit.

bubble chamber A device used in nuclear research for showing the path of charged atomic particles. The chamber usually contains liquid hydrogen maintained at a temperature just above its boiling point but under pressure so that it does not boil. As particles pass through the chamber, the pressure is released, allowing the hydrogen to boil. The particles act as centres for boiling to occur, and leave a line of bubbles in their wake, which can be photographed. See the photograph on page 8.

buffer solution A solution in which the acidity or pH tends to remain constant, even when diluted or when small amounts of acid or alkali are added. A common buffer solution is a solution of acetic acid and sodium acetate.

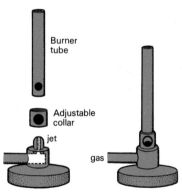

Burner tube

Adjustable collar

jet

gas

Bunsen burner The common gas burner in the laboratory, named after the German chemist Robert Bunsen who devised it in 1855. Gas is introduced to the burner through a jet in the base. A rotating collar can cover or uncover a hole at the base to allow more or less air to enter and mix with the incoming gas. The hottest (blue) flame is produced by opening up the collar fully; the coolest (yellow) by closing the collar fully.

buoyancy The upthrust, or upward force, experienced by a body when it is immersed in a fluid. By Archimedes' principle, it is equal to the weight of fluid displaced.

burette

burette A piece of apparatus used in chemistry to deliver a measured volume of liquid. It consists of a graduated glass tube, open at the top, but closed at the bottom with a tap. The volume of liquid delivered is measured by the difference between readings on the scale before and after delivery.

burin Or graver; an engraving tool with a narrow blade. The modern tool, used for example in wood working, has a metal shaft with a diamond-shaped point at the tip. The burin was one of the first specialized hand tools, developed some 25,000 years ago. It consisted then of a narrow blade of flint.

burning See **combustion.**

bush A cylindrical sleeve, placed around a shaft; a simple bearing.

butadiene A gaseous hydrocarbon ($CH_2CH=CHCH_2$) obtained from refinery processes; used to make synthetic rubbers.

butane A hydrocarbon gas found in natural gas. Its boiling point is $-0.5°C$, which means that it can readily be liquefied under slight pressure. In liquid form it is sold as bottled gas, for use with portable stoves and lamps. Butane is one of the paraffin hydrocarbons (C_4H_{10}).

butte A small, flat-topped hill rising abruptly from a plain. Buttes are features of the dry plateaux of the American West.

butt welding A form of resistance welding. Current is passed through the pieces to be welded, which are in contact. As the metal at the contact point fuses, the two pieces are forced together, forming the weld.

butyl group The alkyl radical (C_4H_9-) derived from butane.

buzz bomb See **V-1, V-2.**

buzzer See **electric bell.**

BWR An abbreviation for Boiling Water Reactor, one type of nuclear reactor.

by-product In a chemical process, a useful substance that is produced in addition to the main product. The main object in cracking petroleum, for example, is to produce petrol. But large quantities of ethylene gas, an extremely valuable raw material, are produced as a by-product.

byte In electronics, a unit of information consisting of several binary digits.

C

c Usual symbol for the velocity of light, as in Einstein's equation, $E = mc^2$.

C The chemical symbol for carbon.

Ca The chemical symbol for calcium.

cable An insulated electric conductor, usually consisting of a number of copper wires stranded together. Telephone cables have many cores, or conductors, each insulated from one another. Overhead transmission lines often have no insulation, but are isolated from the pylons that carry them by glass or porcelain insulators. They are usually made from aluminium, and are strengthened with a steel wire. See also **coaxial cable.**

cable car A kind of tram hauled by a continuous moving cable beneath the rails. The best known cable-car system is in San Francisco, introduced by Andrew S. Hallidie in 1873.

cadmium (Cd) A soft metallic element (rd 8.6, mp 321.1°C) chemically related to zinc. It is corrosion resistant and is electroplated on other metals to protect them. Another use is in nickel-cadmium storage batteries. Many of its compounds are poisonous and are a major source of industrial pollution.

caesium (Cs) A light reactive alkali metal (rd 1.9, mp 28.6°C). It has the lowest melting point of all the solid elements. Its vapour is used in the atomic clock.

caffeine ($C_8H_{10}N_4O_2$) A mild drug that stimulates the central nervous system, found in coffee and tea. It is an alkaloid.

cairngorm A yellow or smoky-brown crystal of quartz that is used as a gemstone.

caisson A box-like or cylindrical structure that is used during excavations, particularly for the foundations of bridge piers. It is usually made of steel or reinforced concrete. Box caissons, closed at the bottom but open at the top, are sunk in position on the river bed and concrete is poured into them. Open caissons are open both ends. They have a cutting edge at the bottom, allowing them to sink into the river bed as material is excavated from

38

Pneumatic caisson

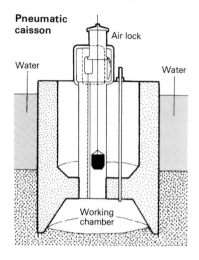

Air lock

Water

Water

Working chamber

inside. The pneumatic caisson has an enclosed lower chamber, which is supplied with compressed air so that workmen can excavate inside. The air pressure prevents the water entering.

caisson disease See **bends, the.**

calamine A mixture of zinc carbonate and iron oxide which is used in a lotion to treat skin ailments, including sunburn. It is an alternative name for zinc carbonate minerals such as smithsonite.

calcination Heating substances strongly to oxidize them or to drive off moisture or carbon dioxide. Ores are often calcined before being processed.

calcite The common mineral form of calcium carbonate ($CaCO_3$) in chalk, limestone and marble. It forms beautiful hexagonal crystals, such as dog-tooth calcite. A very pure calcite, known as Iceland spar, is transparent and exhibits double refraction—when you look through it, you literally see double.

calcium (Ca) An alkaline-earth metal (rd $1 \cdot 5$, mp 850°C), calcium is one of the commonest metals in the Earth's crust. It is too reactive to be found free and occurs mainly as the carbonate (in chalk and limestone), as sulphate (in gypsum) and as the phosphate. Calcium has no metallurgical importance, but its compounds find widespread application.

calcium carbide (CaC_2) A grey powder obtained by processing quicklime and coke in an electric furnance. It is useful because when water is added to it, acetylene gas is produced. It can also be made, by treatment with nitrogen, into the fertilizer cyanamide, or nitrolime.

calcium carbonate ($CaCO_3$) The commonest compound of calcium. It occurs naturally in chalk, limestone and marble in two mineral forms—calcite and aragonite.

calcium chloride ($CaCl_2$) A deliquescent powder—one that readily absorbs moisture. It is widely used as a drying agent in the laboratory. It is obtained as a by-product of the ammonia-soda process for making sodium carbonate.

calcium fluoride See **fluorite.**

calcium hydroxide See **slaked lime.**

calcium oxide See **quicklime.**

calcium phosphate The main inorganic material in animal bones. It also occurs widely in mineral form, such as apatite. Phosphate rock is treated with sulphuric acid to form the fertilizer superphosphate whose main ingredient is calcium hydrogen orthophosphate, $Ca(H_2PO_4)_2 \cdot H_2O$.

calcium sulphate See **gypsum.**

calculator, digital A device for performing arithmetic operations. The abacus is the oldest and simplest type of calculator. The modern electronic digital calculator is a mini-computer, made pocket-size thanks to the ingenious silicon chip.

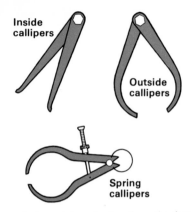

Inside callipers

Outside callipers

Spring callipers

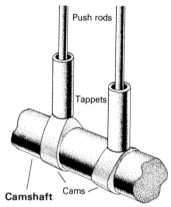

Push rods

Tappets

Camshaft Cams

calculus A branch of mathematics that allows the solution of complex problems. There are two main forms—differential and integral calculus. Calculus was pioneered in the 17th century by Sir Isaac Newton in England and Gottfried Liebniz in Germany.

calendar A means of marking the passage of time, dividing up the year into months and days. Early calendars were based on the Moon. They did not work well because the lunar month—the time it takes to go through its phases (29½ days)—does not go exactly into one year. The present calendar is based on the solar year of 365¼ days. The odd ¼ day each year is allowed for by adding one extra day to the year every fourth year—a leap year. The present calendar is known as the Gregorian calendar, after Pope Gregory XIII, who introduced it in the 1580s.

calendering An industrial process in which a material such as paper or fabric is passed between heavy rollers. It produces a smooth surface.

calibration Preparing an instrument so that measurements can be made with it. A mercury thermometer, for example, would be calibrated by noting the mercury levels when the thermometer is immersed in ice water and steam. Water freezes and turns to steam at definite temperatures known as fixed points (0° and 100° respectively on a centigrade scale). The thermometer would be calibrated by attaching a scale which divides the distance between the two mercury levels into 100 equal parts.

caliche A sodium nitrate ($NaNO_3$) mineral, found in Chile.

californium (Cf) A man-made radioactive element (at no 98); one of the actinides.

calliper brake The common brake used on bicycles. It works by scissor action.

callipers An instrument used in engineering and other professions to measure dimensions. They look much like a pair of compasses, and have two legs, which may be curved. They are particularly useful for measuring the inside and outside diameters of pipes.

Callisto The second largest of Jupiter's many moons, with a diameter of 4,820 km (2,995 miles). It has an ancient, heavily cratered surface.

calomel A chloride of mercury (HgCl) that is sometimes used in medicines for its antiseptic or purgative actions. It is, however, poisonous in large doses. It is found in nature in association with mercury ores such as cinnabar.

caloric A hypothetical fluid introduced in the 18th century to explain heat and combustion.

calorie A unit of heat in the metric system. It is the heat required to raise the temperature of 1 g of water through 1°C (strictly between 15°C and 16°C). It is equal to 4·18 joules. The large calorie, or kilocalorie, is 1,000 calories.

calorific value Of a fuel, the quantity of heat given out when a certain mass of fuel is burned. It is usually expressed in terms of calories per gram or Btus per pound.

calorimeter A piece of apparatus designed to measure the heat given out in a chemical reaction. It commonly consists of a closed container surrounded by a water jacket. The reaction takes place in the container, and the heat produced is measured from the rise in water temperature.

cam A component of machines that often takes the form of an eccentric projection on a shaft, such as a car engine camshaft. The cam is used to impart a regular motion to another machine part through a so-called follower.

camber The gradual curve in the cross-section of a road, to allow the water to drain away, introduced by the Romans.

Cambrian Period A geological time period, which lasted from about 570 million to 500 million years ago. Rocks laid down during this time contain the first abundant traces of fossils.

camera In photography, a device for forming and recording a photographic image. It was developed from a camera obscura. It is essentially a light-tight box, which has a lens in one side to focus a sharp image on a piece of light-sensitive film opposite. Cameras vary in their complexity. The simplest have few controls, simply a button to open a shutter briefly to expose the film, and a knob or lever to wind on the film for a fresh exposure. Better cameras have a shutter whose speed can be varied. They have a diaphragm whose aperture (opening) can be varied. And they have a means of moving the lens in and out to focus a sharp image on the film. The best modern cameras are reflex cameras, which use a mirror to form an image on a ground glass screen for focusing. The mirror flips up when the shutter is released. See also **reflex camera; television.**

camera lucida A simple optical device incorporating a prism, once used by artists to make accurate sketches. It was invented by W. H. Wollaston in 1807.

Single lens reflex camera

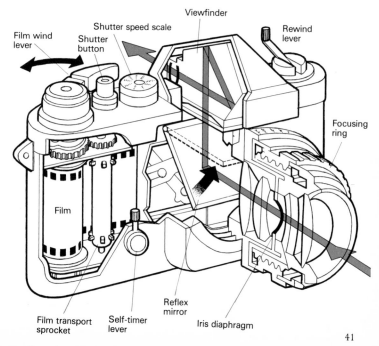

camera obscura

camera obscura Literally 'dark room', a small dark room into which light was admitted through a small hole. It threw an upside-down image of the scene outside on the opposite wall. From the 16th century it was used as an aid to drawing.

camphor ($C_{10}H_{16}O$) A compound obtained from the camphor tree that has a strong odour. It is useful as an insect repellant and is incorporated in some medicines. One of its main uses is as a plasticizer for celluloid.

canal An artificial waterway. Canals have been built since the beginnings of civilization, the earliest in the Middle East over 5,000 years ago. The present Suez Canal in Egypt between the Red Sea and the Mediterranean, probably follows the same course as one built by the ancient Egyptians. The Suez Canal, which is about 160 km (100 miles) long, was built by French engineer Ferdinand de Lesseps and was completed in 1869. De Lesseps also planned, started (1881) but never completed, the more demanding 80-km (50-mile) long Panama Canal linking the Pacific and Atlantic Oceans. It eventually opened in 1914. See also **lock.**

Cancer The Crab; a faint zodiacal constellation that lies between Gemini and Leo. Cancer contains a notable star cluster, called Praesepe, or the Beehive, which contains several hundred stars and is visible to the naked eye.

candela Or new candle; a unit of luminous intensity (intensity of light).

candle power The intensity of a light source, expressed in terms of candelas.

Canopus (α Carinae) The brightest star in the southern constellation Carina and the second brightest (magnitude -0.7) in the whole heavens. It is a yellow supergiant that lies about 100 light-years away.

cantilever A beam that is fixed at one end only and may be supported some way along its length. A diving board is a simple cantilever.

cantilever bridge A bridge constructed of cantilever beams. Typically it consists of twin cantilevers fixed on either bank with their free ends meeting in the middle. Scotland's Forth Bridge, completed in 1890, is one of the most famous cantilever designs, with twin main spans of 521 m (1,710 ft).

canyon Also spelt cañon; a deep valley with steep sides, such as the Grand Canyon in the western United States.

capacitance The capacity of an electrical device such as a capacitor for storing electric charge. Capacitance is measured in farads.

capacitor Also called condenser; a device designed to store electric charge. A simple capacitor consists of two electrical conductors separated by a layer of insulation, such as waxed paper, mica or just air. A variable capacitor, used to tune radio circuits, has fixed and movable metal vanes. The capacitance of the device varies according to the extent the vanes overlap.

Capella (α Aurigae) The brightest star in the constellation Auriga and the sixth brightest in the sky (magnitude 0.1). Nearly 50 light-years away, it is a giant star with a dim red companion, forming with it a spectroscopic binary.

capillarity A property of liquids that enables water, for example, to rise in narrow tubes. It occurs as a result of surface tension, a force that exists at the surface of liquids. Water rises in a narrow capillary tube because its molecules are attracted more by those of the tube than by one another, and this attraction overcomes gravity. By contrast mercury falls in a capillary tube because its molecules are attracted more by one another than by those of the tube. Capillarity is vitally important in nature, being responsible for the transport of moisture through the soil and in plants.

Forth Rail Bridge

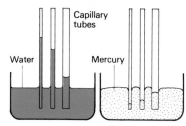

Capillary tubes

Water   Mercury

Capricornus The Sea Goat; a faint zodiacal constellation in the Southern Hemisphere. It lies between Aquarius and Sagittarius.

capstan A winding device used especially on board ship—for hauling up the anchor, for example. It consists of a drum around which the hauling rope or cable is wound. The drum may be driven by electricity or steam, or turned manually by means of handles (bars) that fit into sockets in the drumhead. A ratchet mechanism at the base prevents the capstan moving back.

capsule In astronautics, the name given to the cramped crew compartment of early spacecraft, such as Mercury and Vostok.

carat A measure used when weighing gems, or to express the proportions of gold in an alloy. For weighing gems, 1 carat equals 200 milligrams. In expressing the purity of gold, 1 carat equals a 24th part. So pure gold is 24 carats. 18-carat gold contains 75% gold.

carbamide See **urea**.

carbides Compounds of elements with carbon, such as calcium carbide.

carbohydrates An important class of organic compounds which act as a source of energy for plants and animals. They include cellulose, starch and sugar. They are so called because they usually contain hydrogen and oxygen in the ratio 2:1, as in water.

carbolic acid See **phenol**.

carbon (C) Arguably the most important of all elements (rd 2·3, mp 3,500°C) because it is vital to life. Its atoms have the unique property of combining with one another to form long chains and rings. Complex carbon compounds with chain and ring structures form the basis of animal and plant life. The study of carbon compounds, all except the simplest (such as the carbon oxides and carbides), forms the basis of organic chemistry. Carbon occurs native in the Earth's crust in two very different forms, as very soft graphite and very hard diamond. In the rocks it occurs widely in carbonates ($-CO_3$). Combined with hydrogen in hydrocarbons, it occurs in coal, petroleum and natural gas. In the combustion of these fuels carbon combines with oxygen to form carbon dioxide (CO_2), or monoxide (CO) if the supply of oxygen is limited.

carbonado Black or discoloured diamond, used as an industrial abrasive.

carbonates The salts of the weak carbonic acid, H_2CO_3.

carbon black A fine black pigment used in inks and as a filler for rubber, obtained by the incomplete combustion of oil or natural gas.

carbon cycle The continual exchange of carbon (as carbon dioxide) between living things and the atmosphere. Plants take in carbon dioxide from the atmosphere during photosynthesis, and give off oxygen. Animals take in oxygen and give out carbon dioxide as they breathe.

carbon dating A method of archaeological dating that relies on the proportion of the carbon-14 isotope in a specimen. Carbon-14 is a radioactive isotope of carbon that is present in the carbon dioxide of the atmosphere. It is also present in living tissue. But after living things die, the level of the carbon-14 falls at a known rate. So by finding what proportion of carbon-14 remains, the age can be determined.

carbon dioxide

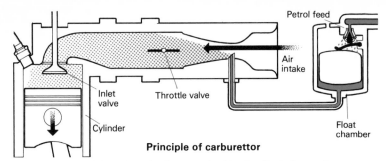

Petrol feed
Air intake
Inlet valve
Throttle valve
Cylinder
Float chamber

Principle of carburettor

carbon dioxide (CO_2) A colourless, odourless gas formed when carbon burns. It is heavier than air and, though non-toxic, can cause suffocation. It is present in small quantities (about $0 \cdot 03\%$) in the atmosphere. Plants absorb carbon dioxide during photosynthesis, while animals give it out during respiration. There is a worry that the level of carbon dioxide in the atmosphere, currently rising, may result in a greenhouse effect that will result in appreciable warming of the climate. Carbon dioxide is readily liquefied under pressure and in this form is used in fire extinguishers. It is slightly soluble in water, forming the weak carbonic acid, H_2CO_3, from which the carbonates are derived. See also **dry ice**.

carbon disulphide (CS_2) A poisonous, flammable and unpleasant smelling liquid of great industrial importance as a solvent. It is used, for example, in the manufacture of viscose rayon.

carbon fibre A form of extruded carbon now widely used for reinforcing plastic, ceramic and metallic materials. It is exceptionally strong and stiff for its size and weight and has found widespread uses in the aerospace field.

carbonic acid (H_2CO_3) A weak acid existing only in solution; the source of the carbonates and hydrogen carbonates, or bicarbonates.

Carboniferous Period A geological time span covering the period from about 345–280 million years ago. The first great forests thrived during this time, which gave rise to our coal seams.

carbon monoxide (CO) A highly toxic, colourless and odourless gas present in car exhaust fumes. It is formed by the incomplete combustion of carbon. It causes death by replacing the oxygen in the blood, effectively causing oxygen starvation. It is flammable and is a useful fuel, occurring in water gas.

carbon tetrachloride (CCl_4) Or tetrachloromethane; a useful industrial solvent that was once widely used as a domestic dry-cleaning agent. Its vapour is, however, toxic in large quantities. It is non-flammable and often used in fire extinguishers.

carbonyl group The organic group $>C=O$, found in such compounds as aldehydes, ketones and carboxylic acids.

carborundum A very hard abrasive material; chemically silicon carbide (SiC).

carboxylic acids The most important class of organic acids, which contain the carboxyl group, – COOH. Formic ($HCOOH$) and acetic (CH_3COOH) are two of the most common carboxylic acids.

carburettor A device used in a petrol engine to mix petrol with air to form an explosive vapour. It incorporates a float chamber, which supplies a constant head of petrol to a jet that sprays it into the air stream.

Carnot cycle An idealized cycle for operating a heat engine. Named after the French engineer Nicolas Sadi Carnot, it illustrates Carnot's principle that the maximum efficiency of a heat engine depends on the temperatures between which it operates. The four stages in the cycle are isothermal expansion, adiabatic expansion, isothermal compression and adiabatic compression.

carrier wave A radio wave used in radio and television broadcasting to carry an audio or video signal. The signals are superimposed on the carrier wave by the process of modulation.

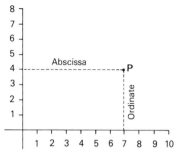

Cartesian coordinates In mathematics, the common system of co-ordinates used to fix the location of a point in a plane. It uses two axes, x and y, at right-angles to each other. The point is located by reference to its perpendicular distance from the two axes. In the diagram the location of point P in Cartesian coordinates would be $(7,4)$. The system is named after the French mathematician René Descartes.

case-hardening A means of hardening the outer surface of steel components. It can be done by heating the metal in contact with charcoal. Carbon diffuses into the surface and hardens it.

casein The main protein found in milk, from which it can be precipitated with rennet. It can be converted into a plastic and is also used in some paints and adhesives.

Cassegrain telescope A reflecting telescope containing an auxiliary mirror to reflect the image back through a hole in the main mirror. This design makes for a compact instrument. It is named after the 17th century French physicist N. Cassegrain.

cassette A plastic magazine for holding a reel of magnetic recording tape. Video cassettes hold tapes that will play back pictures through a television receiver.

Cassini division The most prominent gap in Saturn's rings. First discovered by Jean Dominique Cassini in 1675, the division is about 4,000 km (2,500 miles) wide.

cassiterite Or tinstone; the main ore of tin, consisting of tin oxide (SnO_2). It is noticeably heavy (rd about 7).

casting In metallurgy, the process of shaping molten metal in moulds. The metal takes the shape of the mould when it cools and hardens. Casting is commonly done in sand moulds, the moulds being broken up to release the casting. Permanent metal moulds, or dies, may also be used.

cast iron An iron alloy widely used in casting operations. It is partly refined pig iron containing up to about $4 \cdot 5\%$ carbon. It is very fluid when molten and casts easily. It is hard when cold, but very brittle. It is widely used for rigid machinery parts, such as car engine blocks and lathe beds.

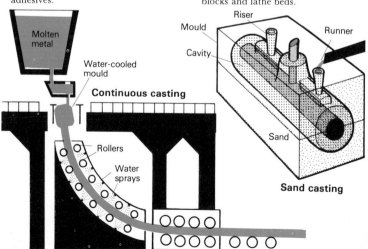

Castor

Castor (α Geminorum) A first magnitude (1·6) multiple star in the constellation Gemini. It has at least six components.

catalyst A substance that promotes or inhibits a chemical reaction, while remaining chemically unchanged itself. Catalysts are widely used in industry to speed up chemical processes. Platinum is a particularly useful catalyst for many reactions, as are nickel, iron and other transition metals. Catalysts that are used to slow down or prevent reactions taking place are called inhibitors.

catamaran A vessel with twin hulls, which originated in India.

cataract In geology, a waterfall with a large flow of water. Niagara Falls is a cataract.

catenary In mathematics, the curve formed by a rope or chain hanging between two points.

caterpillar tracks See **crawler tracks.**

cathode In electrochemistry, the negative electrode. In electronic devices, such as valves, it is a source of electrons.

cathode-rays Streams of electrons emitted by a cathode when heated. See **thermionic emission.**

cathode-ray tube (CRT) An evacuated tube that has at one end a source of electrons (cathode-rays) and at the other a fluorescent screen on which electron traces can be displayed. The path of the electrons can be deflected by means of magnetic coils. The CRT is the basis of the oscilloscope and the television receiver.

cation In electrolysis, a positively charged ion (usually a metal) which moves towards the cathode.

cat's-eyes Self-cleaning reflecting studs used to mark the centre of the road or traffic lanes. They were invented in 1934 by the Yorkshireman Percy Shaw.

cat's-whisker A fine wire used in early radio sets to make contact with the crystal detector.

caustic curve A curve of light formed by reflection from a spherical mirror or shiny surface. It can be seen on the surface of a liquid in a cup after reflection from the curved walls.

caustic potash The common name for the strong alkali potassium hydroxide, KOH.

caustic soda The common name for the strong alkali sodium hydroxide, NaOH.

cavitation The formation of cavities of vapour (bubbles) in a liquid that is rapidly accelerated. Cavitation occurs behind ships' propellers, for example. When the bubbles impinge on the propellers, they cause erosion.

cc The abbreviation for cubic centimetres.

Cd The chemical symbol for cadmium.

Ce The chemical symbol for cerium.

celestial mechanics The branch of astronomy concerned with the movement of the heavenly bodies.

celestial sphere An imaginary sphere enveloping the Earth, on the inside of which the stars appear to be fixed. Though it is an ancient and incorrect concept, it is nevertheless a useful means of visualizing the heavens. On the celestial sphere are the celestial poles, which are points immediately above the Earth's geographical poles. The celestial equator is the projection on the celestial sphere of the Earth's equator. Stars can

Cathode-ray tube

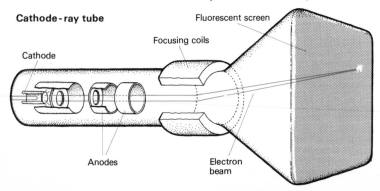

Fluorescent screen
Focusing coils
Cathode
Anodes
Electron beam

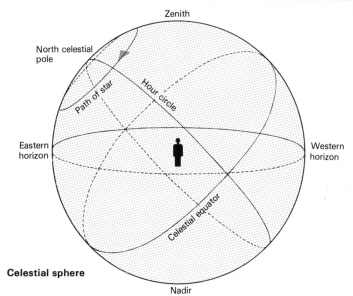

Zenith

North celestial pole

Path of star

Hour circle

Eastern horizon

Western horizon

Celestial equator

Celestial sphere

Nadir

be pinpointed on the celestial sphere by reference to the angular distance from the celestial equator (declination) and the distance along the celestial equator from the First Point of Aries (right ascension). This is equivalent to terrestrial latitude and longitude.

cell In electricity, a device in which electricity is produced, usually by chemical action. A primary cell (as in a dry battery) cannot be recharged with electricity but a secondary cell, or accumulator, can. Other types of cells include the fuel cell and the solar cell.

cellophane A transparent packaging film made from the cellulose in woodpulp in much the same way as rayon, but as a sheet rather than a fibre.

celluloid The first man-made plastic, perfected by John W. Hyatt as a substitute for ivory in 1869. It is cellulose nitrate containing traces of camphor as a plasticizer. Its main drawback is that it is highly flammable.

cellulose The main material in the cell walls of plants. It is a carbohydrate containing long-chain molecules. It is a natural polymer made up of many glucose units and is the raw material for celluloid, acetate, cellophane and rayon. Celluloid is modified cellulose nitrate;

acetate is cellulose acetate; and cellophane and rayon are pure regenerated cellulose.

Celsius scale The common scale of temperature, named after the Swedish astronomer Anders Celsius. On this scale the freezing point of water is 0° and the boiling point 100°. It is a centigrade scale, the interval between 0° and 100° being split up into 100 divisions (centi-). Temperatures are expressed in degrees Celsius (°C).

cement Or Portland cement; a grey powder that is mixed with sand, gravel and water to make concrete. It is made by firing a mixture of limestone, clay, iron ore and gypsum in a rotary kiln.

cementation An early method of producing steel by heating iron with charcoal.

cementite A form of iron carbide (Fe_3C) found in cast iron and some steels. It is very hard and brittle.

Cenozoic Era The most recent geological era, which began about 65 million years ago. It has been marked by the evolution of the mammals.

centigrade scale See **Celsius scale**.

centimetre (cm) One hundredth of a metre. 1 cm = 0·39 inch; 2·54 cm = 1 inch.

centre of gravity (mass) An imaginary point within a body at which all the weight (mass) of the body appears to be concentrated.

centrifugal force See **centripetal force.**

centrifuge A rapidly rotating device that uses centrifugal force to separate substances from mixtures. Suspensions of solids in liquids, or liquids of different densities, can be separated by centrifuging. Ordinary centrifuges operate at about 5,000 rpm. The ultra-centrifuges used in research laboratories to separate viruses, for example, may operate at 100,000 rpm.

centripetal force A force that compels a body to move in a circle, which acts towards the centre of the circle. If the body has mass m and travels in a circle radius r at a velocity of v, then the centripetal force $F = mv^2/r$. The reaction to the centripetal force, apparently acting away from the centre, is known as the centrifugal force.

Cepheids Variable stars whose brightness varies predictably, and whose period of variation is related to their luminosity, or absolute brightness. This relationship, which is useful in determining astronomical distances, was discovered by H. Leavitt in 1912.

ceramics Materials made by firing clay or other substances at high temperatures. Pottery, bricks, porcelain, cement, cermets and glass are different kinds of ceramics. Refractories are ceramic materials chosen for their exceptional temperature resistance.

Cerenkov radiation Radiation produced when high-energy atomic particles pass through a medium with a velocity greater than the velocity of light in that medium. The ethereal blue glow of Cerenkov radiation can be seen in water tanks containing waste radioactive materials.

Ceres The largest asteroid, or minor planet, discovered on the first day of 1801 by Italian astronomer Giuseppe Piazzi. It measures about 1,000 km (600 miles) across.

cerium (Ce) One of the most useful of the rare-earth elements (rd 6·8, mp 804°C). It is the main constituent of the alloy misch metal, from which lighter flints are made.

cermet A composite material formed from a ceramic and a metal, such as aluminium oxide and nickel. It is a refractory used, for example, in jet and rocket engines. It is usually made by powder metallurgy.

Cf The chemical symbol for californium.

cgs units Units based upon the centimetre as a standard of length, the gram as a standard of weight and the second as a standard of time. They have now been largely superseded by the SI units.

chain A series of links, usually of metal, forming a flexible band, typically for transmitting power. Chains are used to drive bicycles and motorcycles, for example, via toothed sprockets.

chain, surveyor's A chain used for measuring distances, 66 ft long.

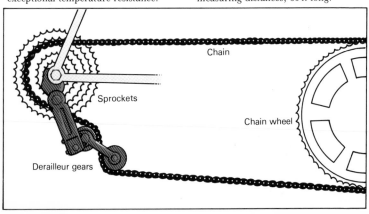

Chain

Sprockets

Chain wheel

Derailleur gears

Chain reaction

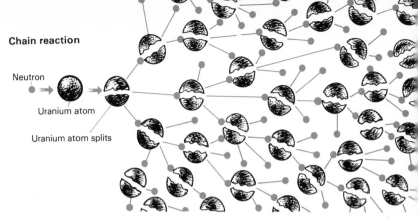

Neutron

Uranium atom

Uranium atom splits

chain reaction In nuclear physics, a process in which a neutron splits a uranium atom (for example) and produces at least two more neutrons, which in turn split other uranium atoms, producing new neutrons, which split still more uranium atoms. A chain reaction cannot be sustained until a certain critical mass of uranium is present. Otherwise too many neutrons escape. In the atomic bomb a chain reaction occurs in a fraction of a second, with the release of a fantastic amount of energy. In a nuclear reactor the chain reaction is controlled.

chalcedony An attractive form of impure quartz, showing coloured bands or fibres. Agate, onyx and bloodstone are varieties of chalcedony.

chalcocite A common copper sulphide ore (Cu_2S), dark grey or black in colour.

chalcopyrite Or copper pyrites; the most common ore of copper, being a mixed copper and iron sulphide ($CuFeS_2$). It is yellowish and resembles pyrite, with which it is often found.

chalk One of the commonest sedimentary rocks, being relatively pure calcium carbonate ($CaCO_3$). White and fine grained, it is composed of the remains of minute marine organisms. It has abundant uses in the manufacture of cement, lime and fertilizers and as a filler in paints and cosmetics.

chalybite An alternative name for the iron carbonate mineral siderite.

change of state The change that occurs when a solid changes to a liquid, a liquid changes to a gas, or vice versa. A solid may change directly from a solid into a gas and back again, a phenomenon called sublimation. When a substance is changing state, its temperature remains constant. Heat is given out or taken in when substances change state. See **latent heat.**

charcoal An impure form of carbon made by heating wood or bone in a limited supply of air. Charcoal is very porous and adsorbent. Specially prepared activated charcoal is used to remove odours from gases and impurities from liquids, as in life-support systems and sugar refining.

charcoal-block test A quick method used in chemical analysis for identifying certain metals in unknown mixtures. The technique involves heating fiercely a small amount of mixture on a block of charcoal with a blowpipe. If compounds of metals such as iron or lead are present, then the charcoal reduces them to metal.

charge See **electric charge.**

Charles' law One of the gas laws, which describes the effect of a change of temperature on the volume of a gas when the pressure is kept the same. The law states that at constant pressure the volume of a gas expands by $1/273$ of its volume at $0°C$ for each $1°C$ rise in temperature. Or at constant pressure, the volume (V) of a gas is proportional to its absolute temperature (T); ie $V \propto T$, or V/T = constant.

charm In nuclear physics, a property exhibited by one of the subatomic particles called quarks.

Charon The recently discovered (in 1978) moon of Pluto, the most distant planet in the solar system. Its diameter is about 800 km (500 miles).

chemical analysis

chemical analysis The branch of chemistry concerned with determining the composition of substances. It includes qualitative analysis, finding out what elements and radicals are present; and quantitative analysis, finding out how much of each is present.

chemical compound A substance formed when two or more elements combine together chemically. In a given chemical compound the elements always combine together in the same proportions. In common salt, for example, sodium (Na) and chlorine (Cl) always combine together in the ratio 1:1, giving a chemical formula of NaCl.

chemical elements The fundamental substances—building blocks—of which all matter is made up. Elements cannot be broken down into simpler substances. In the early 1980s 106 elements were known to exist. Ninety of them exist in detectable quantities in nature. The remainder have been made by man—they are unstable, radioactive elements. The elements can be classified by their atomic structure and properties into families in a so-called periodic table. See the full periodic table on page 250.

chemical engineering The branch of engineering concerned with the design and construction of plant and machinery to carry out chemical reactions on a large scale. Chemical engineers design equipment to carry out chemical processes, such as oxidation and polymerization; and physical operations, such as distillation and filtration. See **unit operations; unit processes.**

chemical equation A shorthand method of representing a chemical reaction. It uses symbols to represent the elements taking part and show the relative numbers of atoms and molecules involved. The reaction between dilute hydrochloric acid (a combination of hydrogen and chlorine atoms in the ratio 1:1, symbolized by HCl) and iron sulphide (a combination of iron and sulphur atoms in the ratio 1:1, symbolized by FeS) can be written as follows:

$$2HCl + FeS \rightarrow H_2S + FeCl_2$$

This shows that 2 molecules of hydrochloric acid react with 1 molecule of iron sulphide to produce 1 molecule of hydrogen sulphide and 1 molecule of

THE CHEMICAL ELEMENTS AND THEIR SYMBOLS

(Elements shown in *italics* are radioactive)

Actinium, Ac	Mercury, Hg
Aluminium, Al	Molybdenum, Mo
Americium, Am	Neodymium, Nd
Antimony, Sb	Neon, Ne
Argon, A	*Neptunium, Np*
Arsenic, As	Nickel, Ni
Astatine, At	Niobium, Nb
Barium, Ba	Nitrogen, N
Berkelium, Bk	*Nobelium, No*
Beryllium, Be	Osmium, Os
Bismuth, Bi	Oxygen, O
Boron, B	Palladium, Pd
Bromine, Br	Phosphorus, P
Cadmium, Cd	Platinum, Pt
Caesium, Cs	*Plutonium, Pu*
Calcium, Ca	*Polonium, Po*
Californium, Cf	Potassium, K
Carbon, C	Praseodymium, Pr
Cerium, Ce	Promethium, Pm
Chlorine, Cl	*Protactinium, Pa*
Chromium, Cr	*Radium, Ra*
Cobalt, Co	*Radon, Rn*
Copper, Cu	Rhenium, Re
Curium, Cm	Rhodium, Rh
Dysprosium, Dy	Rubidium, Rb
Einsteinium, Es	Ruthenium, Ru
Erbium, Er	*Rutherfordium, Rf*
Europium, Eu	Samarium, Sm
Fermium, Fm	Scandium, Sc
Fluorine, F	Selenium, Se
Francium, Fr	Silicon, Si
Gadolinium, Gd	Silver, Ag
Gallium, Ga	Sodium, Na
Germanium, Ge	Strontium, Sr
Gold, Au	Sulphur, S
Hafnium, Hf	Tantalum, Ta
Hahnium, Ha	*Technetium, Tc*
Helium, He	Tellurium, Te
Holmium, Ho	Terbium, Tb
Hydrogen, H	Thallium, Tl
Indium, In	*Thorium, Th*
Iodine, I	Thulium, Tm
Iridium, Ir	Tin, Sn
Iron, Fe	Titanium, Ti
Krypton, Kr	Tungsten, W
Lanthanum, La	*Uranium, U*
Lawrencium, Lr	Vanadium, V
Lead, Pb	Xenon, Xe
Lithium, Li	Ytterbium, Yb
Lutetium, Lu	Yttrium, Y
Magnesium, Mg	Zinc, Zn
Manganese, Mn	Zirconium, Zr
Mendelevium, Md	

iron chloride. Two molecules of hydrochloric acid are required to balance the equation. Balancing the number of atoms on each side is necessary since atoms can neither be created nor destroyed in a chemical reaction. To finish the equation for the above reaction the state of the reactants and products should be stated as either solid (s), aqueous (aq) or gaseous (g). The final equation thus becomes:

$$2HCl(aq) + FeS(s) \rightarrow H_2S(g) + FeCl_2(aq)$$

chemical kinetics A branch of physical chemistry concerned with studying the rates and mechanisms of chemical reactions.

chemical reaction A process in which substances interact and undergo a chemical change to form different substances. The atoms of the reacting substances (reactants) combine together differently to form the new substances (products).

chemical symbols Symbols used to represent the chemical elements and compounds derived from them. They are derived from the name of the element, for example, S for sulphur, O for oxygen, Cl for chlorine. Some are derived from the Latin name of the element, for example, Pb (lead) from the Latin plumbum; Fe (iron) from the Latin ferrum.

chemiluminescence The emission of light during a chemical reaction, usually oxidation. Phosphorus glows noticeably when it is exposed to air. Some animals and insects (such as fireflies) emit light as a result of chemical processes occurring inside them; this is bioluminescence.

chemistry The branch of science concerned with study of properties of substances and of the way they interact. The main branches of chemistry are inorganic chemistry—concerned with all the chemical elements except carbon; organic chemistry, concerned with carbon compounds; analytical chemistry, concerned with analysing substances; physical chemistry, concerned with the physical changes that occur in chemical reactions; biochemistry, concerned with the chemistry of life processes.

chemotherapy The treatment of infections and diseases by means of chemicals.

chemurgy The application of chemistry to the production and development of new and existing agricultural products.

chert An impure form of quartz; light-coloured flint.

Chile saltpetre Impure sodium nitrate ($NaNO_3$), mined in large quantities in Chile and widely used as a fertilizer.

china clay An alternative name for kaolin.

chinook A warm, dry wind that blows eastwards from the Rocky Mountains in North America mainly in winter. It is an example of a foehn wind.

chloral (CCl_3CHO) A chlorine derivative of acetaldehyde, used to make the insecticide DDT. It reacts with water to form a hydrate, which is widely used as a sedative.

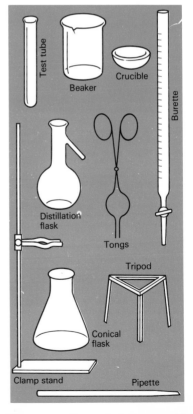

51

chlorates

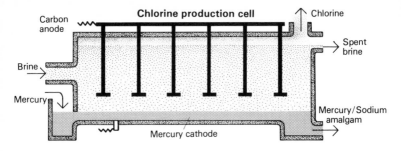

Chlorine production cell

Carbon anode
Chlorine
Spent brine
Brine
Mercury
Mercury/Sodium amalgam
Mercury cathode

chlorates The salts of the hypothetical chloric acid. Potassium and sodium chlorates ($KClO_3$ and $NaClO_3$) are powerful oxidizing agents.

chlorides The salts of hydrochloric acid (HCl).

chlorine (Cl) A poisonous, yellowish gas (bp $-34 \cdot 1°C$) with a choking taste. It is the best known of the halogen family of elements. Its most common compound is sodium chloride (NaCl, common salt), which occurs dissolved in sea water and as the mineral rock salt, or halite. Chlorine is prepared industrially by the electrolysis of sea water. It is a strong bleaching agent and disinfectant, used for example to purify drinking water. It combines with hydrogen to form the gas hydrogen chloride, which in solution is the powerful acid, hydrochloric acid (HCl).

chloroform ($CHCl_3$) Or trichloromethane; a colourless heavy liquid with a sweet smell that has anaesthetic properties.

chlorophyll The green substance in plants that absorbs the energy in sunlight during photosynthesis.

choke In electricity, a coil that has a high impedance (resistance) to alternating current. Also called an inductor.

choke valve A butterfly valve in the carburettor of a petrol engine that controls the amount of air entering the engine. It is closed when starting from cold to allow a rich mixture into the cylinders.

cholesterol ($C_{27}H_{46}O$) One of the steroid family of organic compounds, found in the human body, where it performs vital functions. But in excess it may cause artereosclerosis, or hardening of the arteries.

chondrite The common kind of stony meteorite, which contains rounded masses of minerals. Some chondrites contain carbon, and are called carbonaceous chondrites.

chord In mathematics, a straight line joining two points on a circle.

chromates Salts of the hypothetical chromic acid, typified by potassium chromate, K_2CrO_4.

chromatic aberration See **aberration.**

chromatography Literally 'colour writing'; a method of separating substances in a mixture, pioneered by the Russian chemist Mikhail Tsvett in 1906. It uses the different speeds of travel of the components in the mixture when they are flushed with a fluid (mobile phase) through a suitable medium (stationary phase). Often the components are coloured, which explains the name of this technique. See **gas chromatography; paper chromatography.**

chrome alum One of the common alums, being chromium potassium sulphate, $K_2SO_4Cr_2(SO_4)_3 \cdot 24H_2O$.

chromite The main ore of chromium, a mixed iron chromium oxide mineral ($FeCr_2O_4$), black in colour and metallic in lustre.

chromium (Cr) A hard, corrosion-resistant metallic element (rd $7 \cdot 2$, mp $1900°C$), widely used as a protective coating on steel (chromium plating). One of its main uses is in alloys, such as stainless steel. It is a transition metal and is so called because many of its compounds are coloured. Lead chromate ($PbCrO_4$), for example, is used as a yellow pigment. Potassium dichromate ($K_2Cr_2O_7$) is a powerful and widely used oxidizing agent. Other chromium salts are used in tanning.

chromium plating Coating steel and other metals with a thin layer of chromium by electroplating. Usually the chromium is plated on top of a thicker layer of nickel, which is often itself plated on a layer of copper on the base metal.

chromophore A group of atoms in a coloured organic compound that is responsible for its colour.

chromosphere The bottom layer of the Sun's atmosphere, visible as a red-tinged band only when the Sun is totally eclipsed by the Moon.

chronology The science of dating; or a sequence of events in the order in which they occurred.

chronometer An accurate timepiece. The original chronometers were developed to aid navigation at sea. The first accurate instrument was invented by Yorkshireman John Harrison in 1759.

chrysotile A fibrous variety of the mineral serpentine which is the most important type of asbestos. The fibres may be up to 15 cm (6 in) long.

ciné camera A camera that takes a series of still photographs in rapid succession. When the photographs are projected in equally rapid succession, they convey the impression of movement. This happens because of the persistence of vision.

cinnabar The main ore of mercury, mercury sulphide (HgS). Red in colour, it often occurs as hexagonal crystals.

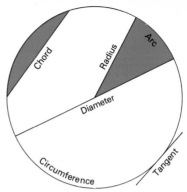

circle A closed curve drawn so that all points on it are at the same distance (radius) from a single point, the centre. A line pasing through the centre, dividing the circle in two, is called the diameter. It divides the circle into two semicircles. The diameter (d) is twice the radius (r). The distance around the edge of the circle is called the circumference (c). The ratio of the circumference to the diameter equals the number known as pi, π. The area of a circle equals πr^2; π is approximately 22/7 or $3 \cdot 142$.

circuit In electricity, a path along which the electricity flows. A typical circuit consists of a source of electricity (eg a battery), electrical conductors (wires) to carry the current, and devices that use the current (bulbs, valves, motors).

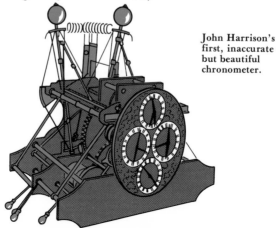

John Harrison's first, inaccurate but beautiful chronometer.

circuit-breaker A kind of switch employed in some electrical circuits which is designed to open if the current in that circuit exceeds a certain value. It acts rather like a fuse. Many work by means of solenoids.

circumference The distance around the curve of a circle. It is equal to π times the diameter.

circumpolar stars Those stars which are always above the horizon throughout the year from a particular latitude. From the latitude of Britain, for example, the stars in the constellations Ursa Major (Great Bear) and Cassiopeia are circumpolar.

cirrus One of the principal cloud types, observed at high altitudes as fine streamers, popularly called mares' tails. They are made up of ice crystals.

cis-trans isomerism A form of isomerism displayed by some organic compounds having double bonds. The classic examples are maleic and fumaric acids, which have the same molecular formula HCCOOH.HCCOOH. These compounds differ by having the functional groups on the same (*cis-*) or on opposite (*trans-*) sides of the double bond:

HCCOOH   HOOCCH
‖   ‖
HCCOOH   HCCOOH
maleic acid (*cis-*)   fumaric acid (*trans-*)

citric acid ($C_6H_8O_7$) A weak organic acid found particularly in citrus fruits such as lemons. It is one of the carboxylic acids. It is an important intermediate in the breakdown of sugar in animals. It is widely used in flavouring foods and is also used for cleaning metals.

citrine A transparent yellowish-brown variety of quartz used as a gemstone.

civil engineering The branch of engineering concerned with the construction of massive structures such as dams, bridges, skyscrapers, tunnels and roads. In every project a great deal of earthmoving, or 'muck shifting', has to be done, nowadays by huge machines such as scrapers and bulldozers. Sites and routes have to be accurately measured by surveying; and firm foundations have to be laid. Knowledge of the properties of the soil is also needed. See **soil mechanics.**

Cl The chemical symbol for chlorine.

clastic rocks Rocks made up of fragments of earlier rocks. Conglomerates and breccias are examples.

Claude process A process by which air can be liquefied. It involves successive compressions and expansions of the air. It was devised by the French engineer Georges Claude in 1902.

Clavius The Moon's largest crater, over 230 km (140 miles) across. Its wall and floor are heavily cratered.

clay Soil made up of very fine particles; or rock made up of massive amounts of such soil. Clay becomes plastic when wet, which accounts for its use as a material for making pottery and other ceramics. One of the purest clays is kaolin. The composition of clays varies, but they are usually made up of aluminium and magnesium silicates.

World climates

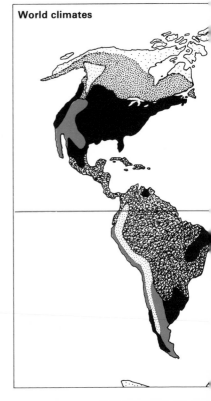

cleavage The tendency of a crystal to split along certain planes, leaving a smooth surface. It splits in a characteristic way that is related to its crystal form. Mica cleaves readily.

clepsydra Or water clock; a device dating back at least to ancient Egyptian times which used the flow of water from a vessel to measure the passage of time.

climate The average weather conditions of a region over a long period. Broadly speaking the climate of a region is related to its latitude—its distance from the equator. It is also affected by its proximity to water—coastal regions tend to experience a different climate from inland regions at the same latitude.

clinical thermometer A mercury thermometer designed for medical use, which has a constriction near the base of the thermometer tube.

clock A device for indicating the passage of time. The mechanical clock has a power source, such as a falling weight or spring, whose energy is released slowly and at a measured rate by a regulator, such as a balance wheel or pendulum. The gradual movement of the spring or fall of the weight is conveyed by a train of gears to hands that indicate the time. The first mechanical clocks appeared in Europe in the 13th century. But clocks did not become reasonably accurate until the mid-1600s when Christiaan Huygens first used a pendulum as a regulator. In the 1670s the introduction of a balance wheel for regulation led to the development of compact portable clocks, or watches. See also **atomic clock; chronometer; clepsydra; quartz watch; sundial.**

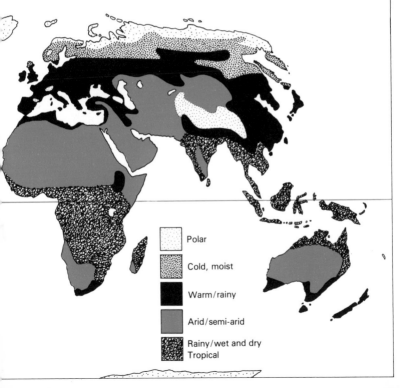

Polar

Cold, moist

Warm/rainy

Arid/semi-arid

Rainy/wet and dry Tropical

55

closed-circuit television (CCTV) A localized television system in which camera, receiver and controls are linked by cable.

cloud chamber A device for detecting the presence of charged atomic particles, developed in 1911 by the British physicist C.T.R. Wilson. It consists of a chamber that is saturated with vapour. Just as the particles are introduced into the chamber, the vapour is allowed to expand. It condenses into tiny droplets around the particles as they are moving. So the particle tracks can be seen.

clouds Visible masses of water droplets or ice crystals that form in the atmosphere when water vapour condenses there. The main types of clouds are cirrus, cumulus, nimbus and stratus.

cloud seeding Introducing particles into clouds to bring about precipitation, particularly rain. Experiments in seeding using dry ice and silver iodide have achieved some success.

cluster, star See **globular cluster; open cluster.**

clutch A device for disconnecting rotating shafts in power transmission systems, such as a car transmission. A friction clutch uses a pair of mating discs or cones, which are forced together and apart by springs and levers.

Cm The chemical symbol for curium.

Co The chemical symbol for cobalt.

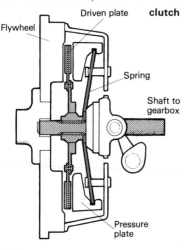

Driven plate **clutch**

Flywheel

Spring

Shaft to gearbox

Pressure plate

coagulation In chemistry, a process in which a dispersed or dissolved substance changes into a jelly or is precipitated out. The protein in egg-white, albumin, is coagulated by heat, changing into an insoluble white mass.

coal A carbon-rich fossil fuel, resulting from the decay and consolidation of plant materials over aeons. The coal seams mined today, often at considerable depths, were laid down up to 300 million years ago during the Carboniferous Period. Coals may contain more than 90% carbon, the highest grade being anthracite, which is hard, shiny and jet black. Bituminous coals contain less carbon and more moisture. They are softer, duller and dirtier to the touch. The lowest grade of coal is lignite, or brown coal.

coal gas Gas produced from coal by destructive distillation—heating in the absence of air. It consists mainly of methane, hydrogen and carbon monoxide. It has been largely superseded as a domestic fuel by natural gas.

Coal Sack A well-known dark nebula in the Southern Hemisphere, located in the constellation Crux, the Southern Cross.

coal tar A tarry substance obtained by condensing the vapours given off when coal is destructively distilled to produce coal gas. It is a valuable source of organic chemicals, including benzene, aniline, phenol, toluene, xylene, anthracene and other aromatics.

coaxial cable An electric cable that has a central core separated by insulation from an outer sheath, which is usually connected to earth; or a cluster of such cables. The cable connecting a television aerial to the receiving set is of the coaxial type. Coaxial cables are designed to carry high-frequency signals.

cobalt (Co) A hard metallic element (rd 8·9, mp 1,492°C) closely related to iron. Like iron it is magnetic. It is combined with aluminium and nickel in the alloy alnico, which is used to make powerful permanent magnets. Cobalt is also incorporated in high-temperature alloys for use in jet and rocket engines. Cobalt blue is a vivid blue-green pigment containing cobalt oxide and alumina. The radioactive isotope cobalt-60 is widely used in radiotherapy as a source of penetrating radiation.

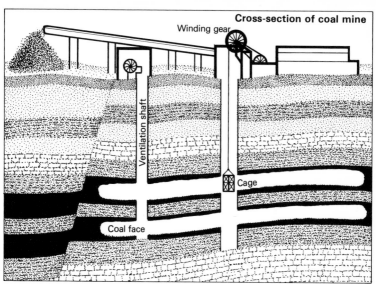

Cross-section of coal mine

Winding gear

Ventilation shaft

Cage

Coal face

COBOL An abbreviation for a simplified computer language, the Common Business Orientated Language, designed specifically for use by businessmen who will not necessarily have a technical background.

cocaine ($C_{17}H_{21}NO_4$) A powerful habit-forming drug, one of the alkaloids, which has been widely used as an anaesthetic. It is obtained from the leaves of the tropical coca shrub.

cochineal A red dye obtained from the dried bodies of female scale insects, which live on cacti. It is still widely used to colour cosmetics and drinks.

cochlea See **ear**.

codeine ($C_{18}H_{21}NO_3$) A pain-killing drug related to, but weaker than, morphine.

coefficient In algebra, the constants, or scalars in an equation. In the algebraic expression $ax + by + cz$, a, b and c are the coefficients of the variables x, y and z.

cofferdam In engineering construction, a temporary dam placed in a river or lake to divert the water while construction of, say, a bridge pier takes place.

coherent light Very pure light of single wavelength, whose waves are exactly in phase (in step). Lasers produced coherent light.

cohesion In physics, attractive forces that exist between two parts of the same substance, for example, between adjacent molecules in a liquid. It contrasts with adhesion, the attractive force between the molecules of different substances.

coil See **ignition coil**.

coke The porous residue remaining after coal has been gasified, consisting of impure carbon. It is a major fuel used, for example, in blast furnaces, where it also functions as a reducing agent, reducing iron oxide to metallic iron.

collagen The tough, fibrous protein that forms the main material in skin, ligaments and hides. Tanning converts it to leather. When it is boiled, it yields glue and gelatin.

collector One of the electrodes in a transistor.

collimation Aligning the optical axis of a telescope.

collimator An optical device attached to a microscope or spectroscope for producing a parallel beam of light. It consists essentially of a slit placed at the focus of a convex lens.

collodion A solution of nitrocellulose in ethanol and ether; used to make photographic emulsions in the early days of photography.

colloid

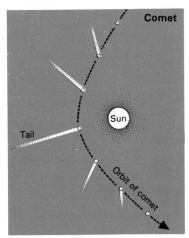

colloid A kind of mixture in which one substance (dispersed phase) in a finely divided state is dispersed in another (dispersion medium). A colloid is midway between a suspension and a solution. Smoke is a colloid, consisting of fine carbon particles suspended in air. Fog is a colloid of fine water droplets in air, a type we call an aerosol. Milk is essentially a colloid of fine fat globules in water, a type we call an emulsion. A liquid-in-solid colloid is called a gel; a colloidal solution is called a sol.

collotype A printing process using light-sensitized gelatin, which can reproduce photographs without the use of a half-tone screen.

colorimeter An instrument designed to measure or compare the intensity of colour. Colorimeters are used in chemical analysis to identify substances and measure their concentration.

colour The way the eye, or rather the brain, interprets different wavelengths of light. Light is the visible part of the electromagnetic spectrum. Ordinary white light is made up of many different wavelengths. We see each wavelength as a different colour. All colours can be produced by combining light of three primary colours—blue, red and green—in different proportions. Combined in equal proportions they produce white light. Two of the primary colours combined give a complementary colour.

columbite A hard black mineral that is the main ore of niobium. It is a mixed iron, manganese and niobium oxide.

columbium An early name for niobium.

coma In astronomy, a nebulous cloud around the nucleus of a comet. In optics, coma is an optical defect, particularly of mirrors, that produces a comet-like image.

combinations See **permutations and combinations.**

combustion Or burning; a chemical reaction in which a substance combines with oxygen, with the production of light and heat. It is a form of rapid oxidation, contrasting with the slow oxidation of rusting.

comet A small member of the solar system that becomes visible only when it reaches the vicinity of the Sun. It is composed of ice and dust, rather like a dirty snowball. When it nears the Sun, the pressure of the solar wind evaporates the ice and drives off the dust. This usually produces a visible tail to the comet, pointing away from the Sun. Comets orbit the Sun and may have periods of a few years (like Encke's comet), of a few decades (like Halley's comet) or very long periods of tens of thousands of years.

communications satellite An artificial satellite that relays telephone, telegraph, radio and television signals. Early communication satellites (eg Telstar) had low orbits and had to be tracked across the sky. Most modern satellites are located in stationary orbits above the equator, where they move in step with the Earth and thus appear stationary.

commutator A device used in dynamos, for example, to reverse the direction of electric current. In its usual form it is a ring consisting of pairs of conductors insulated from one another.

compass A device for determining direction. The simple magnetic compass, invented by the Chinese about 1,000 years ago, uses the directional properties of the Earth's magnetic field. It consists of a magnetized needle pivoted so that it can swing freely. It aligns itself with the Earth's field, and points approximately North and South. Aircraft and ships now generally use the gyrocompass, which is unaffected by local magnetic fields.

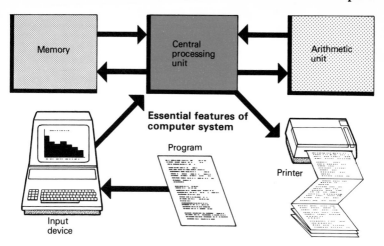

Essential features of computer system

Memory

Central processing unit

Arithmetic unit

Program

Input device

Printer

compasses An instrument used for drawing arcs and circles and measuring off distances. Usually called a pair of compasses, it has two hinged legs, one tipped with a point and the other with a pencil or pen. When it has two points, it is called a pair of dividers.

complementary colour. A colour produced by combining a pair of primary colours. The complementary colours in light are yellow (formed by combining green and red), cyan (blue and green) and magenta (red and blue).

complex In chemistry, an ion in which a central metal cation is associated with a number of anions or neutral molecules. Potassium ferrocyanide, $K_4[Fe(CN)_6]$, is a complex salt containing the complex ion $[Fe(CN)_6]^{4-}$.

compound See **chemical compound.**

compound engine A steam engine in which the exhaust steam from one cylinder is used in another cylinder.

compressibility The extent to which a substance can be compressed. Solids can scarcely be compressed at all. Liquids are slightly compressible; gases can be compressed with ease.

compression-ignition engine See **diesel engine.**

compression ratio Of an internal combustion engine, the extent to which the air or fuel mixture is compressed in the cylinders. A petrol engine has a compression ratio of about 8:1; a diesel engine, one of about 15:1.

compressive strength The strength of a material under compression. Concrete has high compressive strength, but low tensile strength.

compressor A machine designed to compress a gas, particularly air. A reciprocating compressor has pistons and works like a pump. A rotary compressor has a vaned impeller that rotates in a chamber. The compressor in a gas-turbine engine consists of a many bladed rotor.

computer A calculating device. The abacus and slide rule are simple computers, but the term is generally applied these days to electronic devices, such as the digital computer and analog computer. Pocket calculators are tiny computers made possible by the tiny dimensions of the silicon chip.

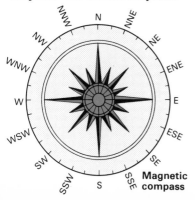

Magnetic compass

computer language

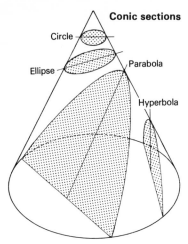

Conic sections

Circle

Ellipse

Parabola

Hyperbola

computer language An abbreviated language by which it is possible to communicate directly with a computer without first having to translate information and instructions into binary digits. Common computer languages include BASIC, COBOL and FORTRAN.

computer program See **program.**

concave Having a surface that curves inwards, like the inside of a saucer. Concave lenses and mirrors have many uses.

concrete One of the commonest building materials, made by mixing together cement, sand, gravel or stone chips and water. Reinforced concrete contains steel rods to increase its tensile strength. The setting of concrete is a chemical reaction in which heat is given out. This must be allowed for when using concrete on a large scale.

condensation The change of vapour into liquid, usually when it comes into contact with a cool surface. The opposite of evaporation. Water vapour in the atmosphere condenses into droplets of water, forming clouds.

condensation reaction A reaction between two substances in which a molecule of water or another simple molecule is eliminated.

condenser In chemistry, a device for condensing a vapour. The simple laboratory condenser is the Liebig condenser.

condenser, electrical See **capacitor.**

conductance A measure of the extent to which a body conducts electricity. It is the reciprocal of the electrical resistance, and it is measured in siemens (mhos). The conductance of a metre cube of a substance is called the electrical conductivity.

conduction One of the three main methods by which heat can be transmitted. It occurs by means of the vibrations of adjacent molecules, not by large-scale molecular movement as in convection. Conduction is the slowest form of heat transfer. The rate of heat transfer by conduction is called the thermal conductivity.

conductor A body that will allow heat (thermal conductor) or electricity (electrical conductor) to flow readily through it. Most metals are both good electrical and good thermal conductors. Such materials as wood, asbestos and plastic are bad conductors.

cone A solid figure formed by rotating a straight line which passes through a fixed point.

conglomerate Or puddingstone; a sedimentary rock containing rounded rock fragments, such as pebbles.

congruent figures In geometry, figures that are identical in all respects.

conic sections Curves obtained by slicing through a cone.

conjunction In astronomy, two or more bodies are in conjunction when they are lined up in space. A planet is in superior conjunction when it is lined up with the Earth and the Sun and is on the other side of the Sun. It is in inferior conjunction when it lies directly between the Sun and the Earth.

conservation laws Fundamental laws that govern the behaviour of the physical universe. Two basic laws are the conservation of mass (matter can neither be created nor destroyed) and the conservation of energy (energy can neither be created nor destroyed). These two laws hold for the normal physical and chemical world, but not for the peculiar world of nuclear physics, where mass is 'destroyed' and energy 'created' during fission. But mass and energy are equivalent (Einstein's equation). So the two laws can be combined into one—that the sum of mass and energy in a system remains constant.

constant An invariable quantity, such as π and c, the velocity of light.

constant composition, law of A chemical compound always contains the same elements in the same fixed proportions by weight.

constellations Patterns made by groups of stars in the night sky. But the stars are not physically associated. Though they are moving rapidly, the stars are so far away that they appear from Earth to remain in the same relative positions century after century. Ancient astronomers saw the skies much as we do today and named the constellations after creatures, objects and mythical characters that they thought they could see in the star patterns. The English and Latin names of major constellations are given in the table.

contact-breaker A kind of switch in the distributor of a petrol engine that interrupts the current flowing to the ignition coil. This induces a high enough voltage (15,000 volts) to produce a spark at the sparking plugs.

contact lens A thin plastic lens fitted directly on the eyeball to correct vision, first made by A. E. Fick in 1887.

contact process The main method of manufacturing sulphuric acid. In the process sulphur dioxide gas (SO_2) is passed with oxygen over a heated vanadium or platinum catalyst. It is oxidized into sulphur trioxide (SO_3), which is then absorbed in water (actually dilute acid) to form sulphuric acid:

$$SO_3 + H_2O \rightarrow H_2SO_4$$

continental drift The theory, first advanced by Alfred Wegener in the 1930s, that the land masses on Earth were originally one vast continent (Pangaea), which gradually split and drifted apart. And the continents are still drifting apart (see **plate tectonics**).

continental shelf The portion of land on the margins of the continents that is submerged under shallow seas, typically about 100–200 m (330–660 ft) deep.

control rods Rods incorporated as a control and safety feature in nuclear reactors. They are made of materials such as cadmium and boron, which absorb neutrons readily. They are moved in or out of the reactor core to slow down (remove neutrons) or speed up the nuclear chain reaction.

MAJOR CONSTELLATIONS

Latin name	English name
Andromeda	Andromeda
Aquarius	The Water-Bearer
Aquila	The Eagle
Ara	The Altar
Aries	The Ram
Auriga	The Charioteer
Boötes	The Herdsman
Camelopardus	The Giraffe
Cancer	The Crab
Canes Venatici	The Hunting Dogs
Canis Major	The Great Dog
Canis Minor	The Little Dog
Capricornus	The Sea Goat
Carina	The Keel
Cassiopeia	Cassiopeia
Centaurus	The Centaur
Cepheus	Cepheus
Cetus	The Whale
Coma Berenices	Berenice's Hair
Corona Borealis	The Northern Crown
Corvus	The Crow
Crater	The Cup
Crux	The Southern Cross
Cygnus	The Swan
Delphinus	The Dolphin
Dorado	The Swordfish
Draco	The Dragon
Eridanus	Eridanus
Gemini	The Twins
Hercules	Hercules
Hydra	The Water Serpent
Leo	The Lion
Lepus	The Hare
Libra	The Scales
Lyra	The Lyre
Ophiuchus	The Serpent-Bearer
Orion	Orion
Pegasus	The Flying Horse
Perseus	Perseus
Pisces	The Fishes
Piscis Austrinus	The Southern Fish
Puppis	The Poop
Sagittarius	The Archer
Scorpio	The Scorpion
Serpens	The Serpent
Sextans	The Sextant
Taurus	The Bull
Triangulum	The Triangle
Triangulum Australe	The Southern Triangle
Ursa Major	The Great Bear
Ursa Minor	The Little Bear
Vela	The Sails
Virgo	The Virgin

convection One of the three main methods of heat transfer. Convection occurs in a fluid and involves the bodily transfer of fluid molecules.

converging lens A convex lens that brings light rays to a point.

converter In electricity, a device which converts alternating current into direct current. In metallurgy, a vessel in which steel is refined, such as the Bessemer converter.

convex Curving outwards, like an upturned saucer. A convex lens converges light rays.

conveyor A device designed to move materials. It may take the form of an endless belt, a spiral screw, a row of buckets, an overhead chain, and so on.

cooling tower A huge tower seen at steam power stations, used to cool the condenser cooling water. Water is trickled down inside the tower, and is cooled as some of it evaporates.

coordinates A set of numbers that pinpoints the position of a fixed point in space. In ordinary geometry Cartesian coordinates are used. Stars are pinpointed on the celestial sphere by the stellar coordinates of declination and right ascension.

coordinate bond A form of covalent bond in which a pair of electrons is donated by only one of the combining atoms. Also called a dative bond.

Copernican system The Sun-centred concept of the universe, advanced in the face of opposition from the church, by Nicolaus Copernicus in 1543.

copolymer A polymer formed from two different monomers. Nylon is a copolymer. Compare **homopolymer.**

copper (Cu) One of the first metals known and the first to be smelted, at least 6,000 years ago. It is a native element (rd 8·9, 1,083°C), which also occurs abundantly in minerals as oxides (such as cuprite), sulphides (chalcopyrite) and carbonates (malachite). It is a soft reddish metal, and many of its compounds are coloured. Copper is an excellent conductor of both heat and electricity, and most copper is used by the electrical industry. It resists corrosion and is incorporated into corrosion-resistant alloys such as brass and bronze.

copper glance An alternative name for the copper ore chalcocite.

copper pyrites An alternative name for the copper ore chalcopyrite.

copper sulphate ($CuSO_4$) Or cupric sulphate; one of the best-known compounds of copper, which forms vivid blue hydrated crystals, $CuSO_4.5H_2O$. When these crystals are heated, the water of crystallization is given off, and the blue colour disappears.

coral A tiny marine creature that forms an external skeleton of limestone. The skeletons of huge colonies of these animals form coral reefs and islands.

cordite A smokeless propellant powder and explosive made from nitrocellulose and nitroglycerine.

Coriolis effect The apparent deflection of the path of a moving object when observed from a moving point of reference. An object moving above the Earth appears to an observer (who is rotating with the Earth) to travel in a curved path. It appears to be experiencing a force that deflects it from a straight line. The concept of this 'Coriolis force' is useful in calculations. The Coriolis effect, first described by Gaspard de Coriolis in 1835, explains why the prevailing winds and ocean currents move as they do.

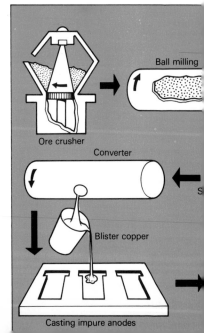

Ball milling

Ore crusher

Converter

Blister copper

Casting impure anodes

corona The solar corona is the outer atmosphere of the Sun, usually visible only at the time of a total eclipse. The 'halo' often seen round the Moon, caused by the effect of thin clouds in the atmosphere, is also called the corona. In electricity, corona refers to a particular type of electrical discharge, which is seen as a visible glow. See **St Elmo's fire.**

corpuscular theory A theory advanced by Isaac Newton to explain the nature of light. He thought that light consisted of streams of minute balls, or corpuscles, travelling rapidly. The theory was later abandoned in favour of a wave theory. But there is evidence to show that under certain conditions light does behave as if it were made up of streams of individual particles (photons).

corrosion The eating away of a metal or other material by chemical action. The rusting of iron and steel is the commonest form of corrosion, caused by oxygen and moisture in the air attacking the iron. Corrosion is often an electro-chemical process, resulting from the formation of tiny electric cells on the metal surface.

cortisone ($C_{21}H_{28}O_5$) A hormone that is secreted by the adrenal gland. Belonging to the steroid family, it is now produced synthetically and is used as a drug for treating rheumatoid arthritis, with derivatives such as hydrocortisone.

corundum A mineral form of aluminium oxide (Al_2O_3), which is the next hardest mineral after diamond. Coloured transparent varieties of corundum include the gems ruby and sapphire.

cosine (cos) See **trigonometry.**

cosmic radiation Radiation coming from outer space, particularly from the Sun. It consists mainly of high-energy protons and electrons. When cosmic rays strike the air molecules in the upper atmosphere, other atomic particles are produced, including mesons.

cosmology The study of the nature and evolution of the universe as a whole. The term cosmogany is sometimes used in connection with the study of the origins of the universe, but this study is generally included under cosmology. The most important theories of the origin of the universe have been the widely accepted big-bang theory and the now less favoured steady-state theory.

cosmonaut The Russian term for an astronaut. Yuri Gagarin became the first cosmonaut on 12 April, 1961, when he pioneered manned spaceflight in Vostok 1.

cosmos Another name for the universe.

cotton A textile fibre, obtained from the seed boll of the cotton plant, genus *Gossypium*. Cotton fibres, which are up to about 5 cm (2 in) long, are separated from the boll by the cotton gin.

cotton gin A machine invented by Eli Whitney in 1793 for separating the fibres from cotton seeds.

cotton linters Short fuzzy fibres found in the cotton boll, which are too short for making into cloth. Being nearly pure cellulose, however, they are useful as a raw material for making rayon and cellulose plastics.

coulomb A unit of electric charge, being the quantity of electricity transported by a current of one ampere in one second. It is named after the French physicist Charles Coulomb.

Coulomb's law Of electrostatics; the force between two electric charges is proportional to their magnitudes and inversely proportional to the square of their distance apart.

Copper refining process

Flotation

Smelting

Copper matte

Cathode

Pure copper cathodes

de

ctrolytic refining

Hydraulic truck crane

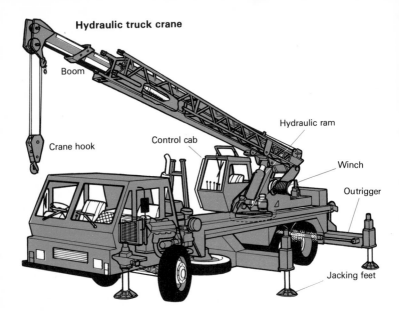

Boom

Crane hook

Control cab

Hydraulic ram

Winch

Outrigger

Jacking feet

countdown In astronautics, the counting backwards of time before the launch of a spacecraft. For space shuttle launches, for example, the countdown generally begins at T – 72, or take-off time minus 72 hours. It is actually longer than three days because of built-in holding periods, or holds.

counterglow See **gegenschein.**

counter, particle See **Geiger counter; scintillation counter.**

couple In mechanics, two equal forces that act in parallel, but opposite directions. The action of a couple is to produce (or prevent) a turning motion. The turning effect (moment) of a couple is the product of either of the forces and the perpendicular distance between the lines along which they act.

covalency One of the main ways in which atoms can bond together. In a covalent bond the atoms share electrons with one another. In the carbon dioxide molecule the carbon atom (which has 4 electrons) shares its electrons with two atoms of oxygen (each of which has 6 electrons). By electron sharing both atoms attain the required 8 electrons to complete their outer electron shell.

Cr The chemical symbol for chromium.

Crab nebula A bright nebula (M1, NGC1952) in the constellation Taurus. It is the remnants of a supernova recorded by Chinese astronomers in AD 1054. Some 5–10 light-years across, it lies about 5,000 light-years away.

cracking One of the main oil refinery processes in which heavy hydrocarbon oils are broken down into more useful lighter fractions, such as petrol. Thermal cracking is carried out in the presence of steam at high temperatures and pressures. Catalytic cracking uses a special clay catalyst to bring about cracking at lower temperatures.

crane A machine for hoisting heavy loads. The main elements of a crane are a power winch, which winds a rope or cable, at the end of which is a pulley block with a hook for attaching the load. The rope may hang from a movable arm or jib or from a trolley that travels along a fixed arm.

crank A simple, but essential part of many machines. It is an arm attached at right-angles to a shaft with which it can turn. It is a means of translating reciprocating (back-and-forth) motion into rotary motion, and is seen most commonly in the bicycle and the car.

crater A bowl-shaped depression found at the top of a volcano. Or the hole in the ground made by the impact of a meteorite. The most famous meteorite crater on Earth is the Arizona meteorite crater in the United States. Most of the craters visible on other planets and their moons have been caused by meteorite bombardment.

crawler tracks Also called caterpillar tracks; endless 'belts' of metal plates found on tanks, cranes and tractors instead of wheels. Crawler tracks distribute the weight of a vehicle over a large area and enable it to travel over soft terrain which wheeled vehicles would find impassible.

creep In engineering, the tendency of a solid (particularly a metal) to deform slowly under a steady load.

creosote Most commonly, a liquid used to preserve timber, obtained by distilling coal tar. It is a mixture of hydrocarbons. The same name is given to a substance obtained from wood tar, which is used as an antiseptic in medicine.

cresol ($CH_3C_6H_4OH$) An aromatic liquid derived from toluene that is a useful raw material in making plastics, explosives and dyes. It has antiseptic properties.

Cretaceous Period The geological period between about 136 million and 65 million years ago, during which most chalk rocks were laid down. The end of the Cretaceous Period is marked by the rapid and mysterious extinction of the once-dominant dinosaurs.

crinoids Or sea lilies; primitive sea animals that anchor themselves to the seabed. They have a distinctive limy skeleton. Fossil crinoids are found widely in ancient rocks.

critical mass In nuclear physics, the minimum mass of uranium or other fissile material that can support a nuclear chain reaction. If the mass is less than critical, then too many neutrons escape.

critical path analysis A method of analysing a complex project (eg a large-scale construction operation) so as to define which activity is critical at any time to the overall production schedule.

critical point In physics, the conditions (temperature, pressure and density) under which a liquid and its vapour become identical.

critical temperature The temperature above which a gas cannot be liquefied just by applying pressure.

Crookes tube A discharge tube devised in 1876 by the English physicist Sir William Crookes with which he discovered cathode rays.

crucible A refractory container, often made from pure silica or graphite, in which high-temperature reactions can be carried out.

crude oil Petroleum as it comes from the ground. It needs to be refined before it becomes useful.

cryogenics A branch of science concerned with the production, study and applications of very low temperatures.

cryolite Also called ice-stone; a crystal-clear mineral that has a refractive index nearly identical to that of water. When you place it in water, it virtually disappears. Chemically it is sodium aluminium fluoride, Na_3AlF_6.

cryotron An electronic device that can be used as an amplifier or a switch. It works on the principle of superconductivity at very low temperatures.

Crank

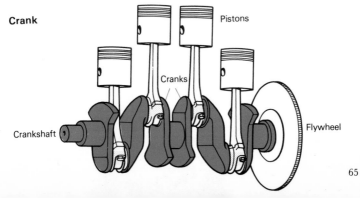

Pistons

Cranks

Crankshaft

Flywheel

crystal

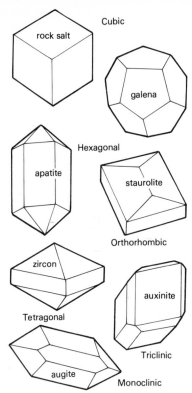

Cubic — rock salt, galena

Hexagonal — apatite, staurolite

Orthorhombic — zircon

Tetragonal — auxinite

Triclinic — augite

Monoclinic

crystal The regular shape assumed by a substance when it solidifies. Most pure substances form crystals, which have an unvarying and distinctive geometric shape. This shape reflects the symmetry of their internal structure—the way the atoms are arranged in the so-called space lattice. There are six basic crystal lattices, which give rise to about 32 distinctive crystal types.

Cs The chemical symbol for caesium.

Cu The chemical symbol for copper, from the Latin cuprum.

cube A solid figure with six square sides; a regular hexahedron. In mathematics, the cube of a number is that number multiplied by itself three times: $x^3 = x \times x \times x$.

cumulonimbus The typical storm cloud, appearing like a huge mountain with a broad black base. The top may spread into a characteristic anvil shape.

cumulus The distinctive fluffy white cloud of summer, looking rather like a ball of cotton wool.

cupellation a method of purifying precious metals. Hot air is blasted over the molten metal, which is held in a flat dish, or cupel. Impurities are oxidized and absorbed or swept away.

cupric The divalent state of copper; now often written as copper(II).

cuprite A major oxide ore of copper (Cu_2O).

cupronickel A common alloy of copper and nickel, from which most modern 'silver' coins are made. It usually contains about 75% copper and 25% nickel.

cuprous The monovalent state of copper; now often written as copper(I).

curie A unit of radioactivity, named after the pioneer of radioactive research Marie Curie, who discovered and isolated radium.

curium (Cm) An artificial radioactive element (at no 96).

current, electric See **electric current.**

cyanide process A process that uses a cyanide solution to extract silver and gold from their ores.

cyanides Compounds containing the – CN group that are extremely poisonous. Inorganic cyanides, such as sodium cyanide (NaCN), are salts derived from hydrocyanic acid (HCN). Organic cyanides are called nitriles.

cyanogen (C_2N_2) A poisonous, colourless and flammable gas.

cybernetics A branch of science that is concerned with communication and control systems in both machines and animals. The word was coined in 1948 by the American mathematician Norbert Wiener. Cybernetic systems are involved in automation.

cyclamates Sweetening agents more than 30 times sweeter than ordinary cane sugar. They are the sodium and calcium salts of an organic acid (cyclohexylsulfamic acid). They were once widely used as a sweetener in foods and drinks, but were found to be potentially harmful and are now seldom used.

cycle A series of changes that is repeated over and over again, as in the carbon cycle or nitrogen cycle.

cycles per second See **Hertz.**

cyclic compound A compound such as cyclohexane that contains a ring structure.

cyclohexane (C_6H_{12}) A hydrocarbon compound whose molecules contain six atoms of carbon joined in a ring. It is a flammable liquid used mainly as a solvent.

cycloid The arch-shaped curve traced by a point on the circumference of a circle when it rolls in a straight line. Some parts of gear teeth are cut with a cycloid curve.

cyclone A low-pressure region in the atmosphere in which winds spiral inwards towards the centre. The winds spiral anticlockwise in the northern hemisphere and clockwise in the southern hemisphere. Violent tropical cyclones are called hurricanes or typhoons, depending on their location.

cyclotron One kind of atomic particle accelerator in which the particles are made to spiral. It was invented by the American physicist Ernest O. Lawrence in 1930. The cyclotron has two D-shaped electrodes (dees) located between the poles of a powerful magnet. Charged particles are accelerated between the dees by an alternating electric field. They spiral outwards because of the surrounding magnetic field.

cylinder In geometry, a solid with the shape of a can. The area of a cylinder, height h and base radius r is $2\pi r(h + r)$; the volume is $\pi r^2 h$.

cylinder block and head See **petrol engine.**

Principle of cyclotron

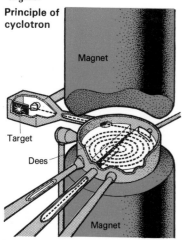

Magnet

Target

Dees

Magnet

D The symbol for deuterium, a heavy isotope of hydrogen.

Dacron A trade name for polyester synthetic fibre.

daguerreotype The first practical photographic process, introduced in France by Louis-Jacques Mandé Daguerre in 1839. In the daguerreotype process the light-sensitive surface was copperplate coated with silver iodide. It was developed in mercury vapour and fixed in salt solution.

Dalton's law See **partial pressures, law of.**

dam A structure built to check the flow of water, used to control flooding, assist irrigation and provide hydroelectric power. A gravity dam remains in place by virtue of its weight. It may be made of concrete, earth or masonry. Arch dams derive their strength from their arch shape—they are curved towards the water and transmit pressure along the arch to the sides.

damper See **shock absorber.**

damping Reducing the amplitude of a vibration, oscillation or wave. A car shock absorber is a damping device.

Daniell cell A primary electric cell, which was a standard source of electricity from its invention by the English chemist J.F. Daniell in 1836 until the development of the Leclanché cell in the 1860s. It has a zinc cathode in a solution of sulphuric acid separated by a porous partition from a copper anode in a solution of copper sulphate. Its output is $1 \cdot 1$ volts.

data processing A term used to describe the operations a computer performs on the information, or data, it is fed.

date line See **international date line.**

dative bond See **coordinate bond.**

Davy lamp A miner's safety lamp invented by the English chemist Humphry Davy in 1815. It is an oil lamp whose flame is surrounded by a gauze. The gauze conducts away the heat of the flame and prevents the ignition of fire-damp, or methane, which might otherwise cause an explosion.

day

day The solar day of 24 hours is the time it takes the Earth to rotate once on its axis with respect to the Sun. The sidereal day—the time it takes the Earth to rotate on its axis relative to the stars is fractionally shorter: 23 h 56 min 4 sec. Astronomers work with sidereal time.

dc Abbreviation for direct current.

DDT The powerful insecticide dichloro-diphenyl-trichloroethane $(C_6H_4Cl)_2CHCCl_3$. It is effective against a wide range of insect pests, but its use is now limited because it is potentially lethal to higher life forms, being retained in body tissue.

dead reckoning A simple method of navigation that uses speed and direction of travel to determine the distance covered.

deadweight tonnage The maximum amount of cargo a ship can safely carry.

decay In nuclear physics, the breakdown of a radioactive element into other (daughter) elements. The rate of radioactive decay is measured in terms of the half-life.

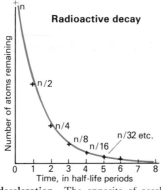

Radioactive decay

Number of atoms remaining (vertical axis). Time, in half-life periods (horizontal axis). n, $n/2$, $n/4$, $n/8$, $n/16$, $n/32$ etc.

deceleration The opposite of acceleration; properly termed retardation.

decibel A unit used to measure noise levels, or relative loudness. A whisper is rated at about 20 dB (decibels), the noise 33 m (100 ft) from a jet engine may exceed 140 dB and cause physical pain.

decimal system The normal number system which uses powers of 10 and the digits 0, 1, 2, 3, 4, 5, 6, 7, 8, and 9.

declination In astronomy, one of the two main celestial coordinates. It is a measure of the angular distance of a heavenly body from the celestial equator. See **celestial sphere.**

decomposition In chemistry, a reaction in which a substance is split up into simpler substances. Heat, electricity and even light may bring about decomposition. Water is decomposed by electrolysis into its elements hydrogen (H) and oxygen (O):
$$2H_2O \rightarrow 2H_2 + O_2$$
A double decomposition, or ionic precipitation, occurs when two substances split up into ions, which then recombine in a different way, as in the reaction between lead nitrate, $Pb(NO_3)_2$, and sodium carbonate, Na_2CO_3:
$$Pb(NO_3)_2 + Na_2CO_3 \rightarrow PbCO_3 + 2NaNO_3$$
or essentially:
$$Pb^{2+}(aq) + CO_3^{2-}(aq) \rightarrow PbCO_3(s)$$

decrepitation The bursting of certain crystals when they are heated.

deflagration Sudden burning.

deformation The change in dimensions of a body when it is acted upon by mechanical forces. The study of deformation phenomena is called rheology.

degaussing Removing the magnetism from a magnetized object.

degree A unit of measurement on, for example, a temperature scale, such as degrees Celsius (°C). In mathematics, degree is a measure of angle, there being 360° in a complete circle. Degree is also a measure of arc, there being 360° of arc around the circumference of a circle. Each degree is subdivided into 60 minutes, each divided into 60 seconds.

degrees of freedom The minimum number of independent variables needed to describe a system completely. A point in space has three degrees of freedom, because three coordinates (x, y and z) are needed to establish its location.

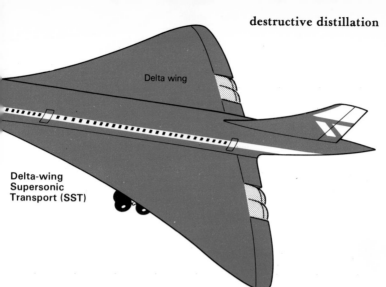

Delta wing

**Delta-wing
Supersonic
Transport (SST)**

dehydration The removal of water from a substance by chemical means. Sulphuric acid is a powerful dehydrating agent.

Deimos The smallest of the two moons of Mars, an irregularly shaped body measuring a maximum of 15 km (9 miles) long.

deliquescence A property displayed by some substances, which absorb moisture from the air and gradually form a solution. Sodium hydroxide and calcium chloride are deliquescent.

delta A flat plain formed near the mouth of a river by mud carried downstream and deposited by the river. The Mississippi delta, for example, covers some 40,000 square km (15,000 square miles).

delta-wing An aircraft wing shaped like the Greek letter capital delta (Δ). The supersonic airliner Concorde and the space shuttle both have delta wings.

dendrite The tree-like form in which metal crystals grow.

Deneb (α Cygni) The brightest star (mag 1·3) in the constellation Cygnus, a supergiant some 1,400 light-years distant.

density The mass of a unit volume of a substance, expressed usually in grams per cc or kilograms per cubic metre. Densities are often given relative to the density of water (= 1), this being termed relative density.

deoxyribonucleic acid See **DNA**.

depolarizer See **polarization**.

depression In meteorology, a low-pressure region in the atmosphere; also called a cyclone.

depth of field In photography, the distance between which objects are clearly in focus. Depth of field increases with decreasing aperture.

derrick A simple lifting apparatus consisting essentially of a movable pole, or jib, from which hangs the lifting rope and tackle.

desalination Removing the salt from seawater, usually to provide drinking water. Large-scale desalination plants obtain pure water by distillation.

desert A land region where there is little life, generally because of the shortage of water. Any regions with less than about 25 cm (10 in) of annual rainfall are considered deserts. In hot deserts water is not available because of evaporation. In cold deserts, as near the poles, water is present only as ice.

desiccator A device used to dry substances.

destructive distillation Heating a substance in the absence of air so that it breaks down without undergoing combustion. The destructive distillation of coal yields combustible gases, coke and vapours. The vapours can be condensed into coal tar, which is a valuable source of organic chemicals.

detector

detector Also called demodulator; a device in a radio receiver that separates the audio signal from its radio carrier wave.

detergent A cleansing substance, including soap and synthetic materials derived from petroleum. A detergent works because its molecules have water-loving (hydrophilic) and water-hating (hydrophobic), or dirt-loving ends. During washing the hydrophobic ends of the molecules are attracted to dirt particles, leaving the hydrophilic ends exposed, enabling the particles effectively to dissolve.

determinant In mathematics, a square array of numbers (elements) used sometimes in solving equations. An example is:

$$\begin{vmatrix} 9 & 5 \\ 4 & 7 \end{vmatrix}$$

which means $(9 \times 7)-(4 \times 5)$.

detonator A small explosive charge used to set off a main charge of high-explosive. Mercury fulminate and lead azide are often used in detonators, being ignited by an electric current or the heat from a fuse.

deuterium (D) A heavy isotope of hydrogen, which occurs in water as deuterium oxide (D_2O), heavy water. In nature hydrogen contains about 0.015% deuterium.

developing In photography, the process by which a visible image is formed on film. When the film is exposed to light, invisible changes occur in the silver salts of the light-sensitive emulsion. An invisible, latent image is formed. Treatment with a chemical developer changes the latent image into metallic silver, which forms a black deposit. Unchanged silver salts then have to be removed by fixing.

Devonian Period The geological time period between about 395–345 million years ago, marked by the evolution of the fishes.

dew Droplets of water that form on grass and plants on cold nights, caused by condensation of water vapour in the air. Condensation occurs when the temperature reaches the dew point, the temperature at which the air becomes saturated with water vapour.

Dewar flask See **vacuum flask.**

dextrorotatory See **optical activity.**

dextrose Another name for glucose, which is ordinarily dextrorotatory.

dialysis A means of separating dissolved ions or molecules from colloidal particles or macromolecules using a semipermeable membrane. The dissolved ions and molecules diffuse through the membrane more quickly. In kidney machines dialysis is used to remove waste products from the bloodstream.

diamagnetism A very weak kind of magnetism displayed by all materials due to the distortion of electron orbits.

diamond Naturally occurring carbon in a crystalline form, the hardest of all natural substances (Mohs hardness, 10). It is one of two allotropes of carbon (the other is graphite). When expertly cut, it is prized as a gem for its brilliance and 'fire'. Poor quality diamonds are used as abrasives.

diastase An enzyme found in malt, which changes starch into sugar (maltose) during the brewing of beer.

diatomaceous earth Also called kieselguhr and fuller's earth; fine, porous siliceous material formed from the shells of microscopic plants known as diatoms. It is used as a filter medium, as a fine abrasive in polishes and toothpaste and for insulation.

dibasic acid One which has two replaceable hydrogen atoms and thus gives rise to two kinds of salts—normal salts, where both hydrogen atoms have been replaced; and acid salts, where only one hydrogen atom has been replaced. Sulphuric acid is dibasic, giving rise to normal sulphates (eg Na_2SO_4) and bisulphates or hydrogen sulphates (eg $NaHSO_4$).

dichlorodifluoromethane See **Freons.**

dichroism A property displayed by some crystals (eg tourmaline) which makes them appear different colours when viewed from different directions.

die A mould used to shape metals and other materials. Hot metal is forced through a die in extrusion; cold metal is drawn through a die to make wire. Molten metal is cast in a die in diecasting. Many toys, for example, are made by injecting molten alloys into water-cooled dies. Dies are also used in stamping and forging operations.

diecasting See **die.**

dielectric An electrical insulator.

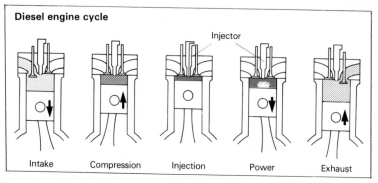

Diesel engine cycle

Injector

Intake   Compression   Injection   Power   Exhaust

diesel engine An internal combustion piston engine that works by the compression-ignition of oil; named after its German inventor Rudolf Diesel who patented it in 1892. Diesel engines are widely used to power lorries, buses, locomotives and ships. They are in general cheaper to run than petrol engines but require more robust construction because of the higher pressures developed inside their cylinders. In the four-stroke diesel-engine cycle, air is drawn into the engine cylinder on the first downstroke of the piston. It is compressed and is raised to a high temperature as the piston moves up. Oil is then injected into the cylinder and ignites immediately, forming gases that force the piston down on its power stroke. On the next (upward) piston stroke the piston expels the burnt gases. Two-stroke diesel engines are also widely used.

differential gears The gears in the transmission system of a vehicle that, during cornering, permit the inside and outside wheels to travel at different speeds.

diffraction The spreading out of waves when they pass through an aperture or round an obstacle. Sound, light and water waves display this phenomenon. The diffraction of light prevents an object casting a sharp shadow—the light is slightly bent around the object.

diffraction grating A device that uses the phenomenon of diffraction to produce a spectrum from light. A transmission grating consists of a glass plate on which parallel lines are drawn very close together, diffraction taking place as the light passes through. In a reflection grating the lines are ruled on a sheet of metal and are diffracted into a spectrum after reflection. A grating may have over 1,500 lines per mm.

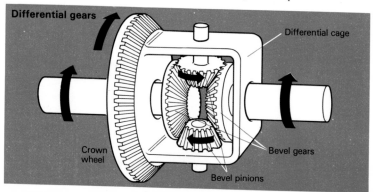

Differential gears

Differential cage

Crown wheel

Bevel gears

Bevel pinions

diffusion The mixing together of different liquids or gases due to random molecular motion. It occurs most rapidly in gases and is independent of gravity.

digital computer A computer that handles information in the form of numbers, or digits. It deals with numbers in the form of binary digits (bits), numbers expressed in the binary number system.

digital watch See **quartz watch.**

dihedral angle The slight angle at which the wings of an aeroplane are inclined upwards to reduce the tendency of the plane to roll from side to side.

DIN The abbreviation for Deutsche Industrie Normen, the German industrial standards. Film speeds are commonly expressed on a DIN scale.

diode A two-electrode vacuum tube or semiconductor, which is used as a detector in radio circuits and for rectifying alternating current.

dioptre A measure of the power of a lens. The power in dioptres is the reciprocal of the focal length in metres. The power of a converging lens is positive; that of a diverging lens, negative.

dip circle Also called inclinometer; an instrument that measures the angle of dip of the Earth's magnetic field. It consists essentially of a magnetized needle pivoted so that it can swing vertically.

dipole In electricity, a pair of equal and opposite electric charges separated by a short distance. The water molecule forms an electric dipole.

direct current (dc) One-way electric current, such as that produced by a battery. It contrasts with two-way, alternating current (ac) produced by an electric generator.

discharge tube A tube containing gas at very low pressure in which an electric discharge takes place. Tubes containing neon emit an intense orange-red glow.

disaccharide A sugar of formula $C_{12}H_{22}O_{11}$, which can be split into monosaccharides. See **lactose; maltose; sucrose.**

dispersion In chemistry, the even distribution of fine particles in a medium, as in a colloid. The particles are known as the disperse phase, and the medium the disperse medium. In light, dispersion refers to the splitting of white light into its component wavelengths, or colours—as a spectrum.

dissociation Reversible decomposition of a substance. Heat decomposes ammonium chloride (NH_4Cl) into ammonia (NH_3) and hydrogen chloride (HCl). But when these gases are cooled, they recombine to form ammonium chloride: $NH_4Cl \rightleftharpoons NH_3 + HCl$

distillation The process of heating a liquid to boiling point and condensing the vapour back into liquid. It is a common method of purifying liquids (such as sea water) and separating the components of a liquid mixture. If the boiling points of the components are far apart, simple distillation can separate them, the more volatile (lowest boiling point) distilling off first. If the boiling points of the components are close together, then fractional distillation must be used.

distributor A device in the ignition system of a petrol engine that distributes high-voltage pulses to the sparking plugs. It feeds the pulses to the plugs in the right order at exactly the right time to produce the sparks that ignite the fuel mixture. It also contains the contact-breaker, which interrupts the battery current to the coil and causes the coil to produce the high-voltage pulses. It is driven from the camshaft.

diurnal motion The apparent movement of the heavenly bodies from east to west across the sky, caused by the Earth's rotation from west to east.

divalent Or bivalent; having a valency of two.

Dihedral angle

Simple distillation

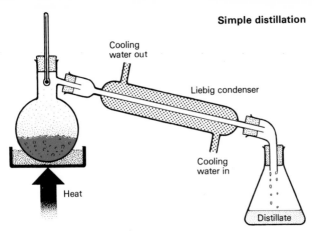

divergent Spreading out from a common point. A divergent, or concave lens, for example, causes a parallel beam of light passing through it to spread out, as if from the focus of the lens.

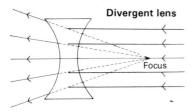

Divergent lens

dividers A drawing instrument used to divide lines or to transfer dimensions from, say, a ruler to a drawing. The standard pair of dividers looks rather like a pair of compasses, except that it has points on each leg.

diving bell A bottomless chamber, sometimes shaped like a bell, in which men can work underwater. The bell is supplied with compressed air, which prevents the water entering.

DNA The compound deoxyribonucleic acid, a complex organic chemical with a spiral, double-helix structure. It is the 'brains' behind the workings of the living cell, and is made up of chains of nucleotides. See also **RNA**.

Dog star An alternative name for Sirius, brightest star in the sky, which lies in the constellation Canis Major, the Great Dog.

doldrums Areas of light winds and calms experienced in equatorial regions. They are also marked by strong upward air currents.

dolomite A common mineral, being a mixed calcium and magnesium carbonate, $CaMg(CO_3)_2$. It is often found with limestone and marble.

dopant In electronics, a substance deliberately introduced as an impurity into a semiconductor to give it characteristic properties. Phosphorus and arsenic are common dopants for silicon.

Doppler effect The apparent change in frequency of sound or light waves that occurs when the source of the waves and the observer are moving relative to one another. It is named after Christian Johann Doppler, who first described it in 1842. The effect is most commonly experienced with sound, as for example when an ambulance or police car passes by with siren blaring. The frequency, or pitch of the sound when the vehicle is approaching is noticeably higher than when it is receding. The sound waves approaching are effectively bunched closer together, or shortened (higher frequency), while those receding are effectively drawn out, or lengthened (lower frequency). The same effect occurs with starlight. The light from an approaching star has an apparently shorter wavelength, or looks bluer than it should. The light from a receding star has an apparently longer wavelength, or looks redder (see **red shift**).

73

double bond A double covalent bond between two atoms. Carbon frequently forms double bonds in organic compounds, such as ethylene, $H_2C = CH_2$.

double decomposition Or metathesis; see **decomposition.**

double refraction The property of certain crystals, notably Iceland spar, to split one incoming light ray into two. Looking through such a crystal, you can literally see double.

double star Or binary; two stars that revolve around each other, or rather around a common centre of gravity. Visual binaries appear to the naked eye to be a single star, but can be resolved into two through a telescope. Spectroscopic binaries are so close together that they can be detected only via the shift of lines in their spectrum.

drag The resistance to the motion of an object through a fluid. It is minimized by streamlining the object.

drawing In metallurgy, pulling metal rods through a die so as to make wire.

dredging Excavating material from navigation channels and harbours. The bucket dredger, for example, excavates by means of an endless chain of buckets that angles down into the water.

drier A substance incorporated in paint to assist the drying process.

drill A tool for making holes, or the machine for holding the tool. For drilling metal, the twist drill is invariably used, which cuts at the tip and has spiral grooves (flutes) to allow the cut material to escape. Oil wells are dug by rotary drills, which have tips (bits) consisting of rotating toothed gears. Rock drills work by percussive (hammer) action and are powered by compressed air (see **pneumatic tools**).

drop forge A forging machine in which a heavy ram, lifted by steam or air pressure, is allowed to fall on a hot metal workpiece.

drug A chemical that affects biological processes, generally taken to treat or prevent disease. Many drugs, including cannabis, cocaine and amphetamines, are habit-forming and can cause addiction.

drumlin A rounded, elongated hill that formed beneath a moving glacier. Drumlins are generally found in groups.

dry cell See **battery.**

dry ice Solid carbon dioxide, at a temperature of about $-78°C$. It is so called because it sublimes—turns directly into gas without first melting into a liquid.

ductility The property of a material, particularly metal, that allows it to be drawn out into wire. Gold, silver and copper are highly ductile.

Duralumin A lightweight aluminium alloy containing copper, magnesium and manganese which, after heat treatment, gradually hardens and becomes as strong as steel. This is called age hardening. Duralumin is most widely used in aircraft construction.

Dy The chemical symbol for dysprosium.

dyes Substances used for colouring textiles. Plants such as madder and insects such as cochineal were once the only source of dyes. But since William Perkin discovered the first coal-tar dye, mauve, in 1856, synthetic dyes have dominated the market. Among the most important classes of dyes are the aniline dyes, the anthraquinones and the azo dyes. Synthetic dyes are more vivid and more light-fast (resistant to fading) than the natural products they replaced. Some fibres must be treated with a mordant before a dye will take.

dyke A sheet-like body of igneous rock that cuts through other rock formations.

dynamics The branch of mechanics that deals with the action of forces on bodies and the change in motion they produce.

dynamite A widely used high-explosive, consisting of nitroglycerin absorbed in kieselguhr or other inert material. It was invented by the Swede Alfred Nobel in 1867. It is much safer than nitroglycerin itself and requires a detonator to set it off.

dynamo An electric generator that incorporates a commutator and produces direct electric current.

dynamometer An instrument for measuring the power of a machine.

dyne A unit of force in cgs units, being the force that, when acting upon 1 g of matter, will produce an acceleration of 1 cm per second per second. 1 newton = 100,000 dynes.

dysprosium (Dy) One of the rare-earth elements (rd 8·5, mp 1,500°C), or lanthanides.

E

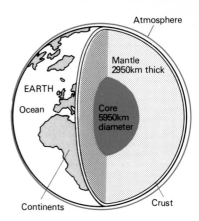

E = mc² See **Einstein's equation.**

ear The sense organ in higher animals by which they hear and keep their balance. The visible outer ear channels sounds to the middle ear, where they vibrate a membrane, the eardrum. Three small bones, the hammer, anvil and stirrup, transmit the vibrations to the inner ear. The vibrations make fluid in the cochlea move, which stimulates nerves in the hearing organ, the organ of Corti, to send appropriate messages to the brain. Sense of balance is provided by means of the liquid-filled semi-circular channels of the inner ear.

Earth The planet on which we live—the only one in the solar system that experiences just the right conditions to sustain life. The Earth is the third planet from the Sun and lies on average 150 million km (93 million miles) from the Sun, around which it circles in 365 ¼ days. It spins on its axis every 24 hours with respect to the Sun (solar day), or some 4 minutes shorter with respect to the stars (sidereal day). Its axis is tilted 23½ ° with respect to the plane of its orbit around the Sun, which gives rise to the seasons. The Earth is slightly oblate, or flattened at the poles. Its diameter at the equator is 12,756 km (7,926 miles),

while its polar diameter is some 43 km (26 miles) less. The Earth has a solid outer layer, or crust, up to about 30 km (20 miles) thick. The crust is split up into a number of segments, or plates, that are moving relative to one another—this concept being known as continental drift. Beneath the crust is a semi-fluid mantle and beneath that an outer and inner core made up of iron and nickel. The moving iron core is thought to be the source of the Earth's magnetic field. The overall relative density of the Earth is about 5·5. The Earth has one satellite, the Moon.

earth Or ground; an electrical connection to the ground.

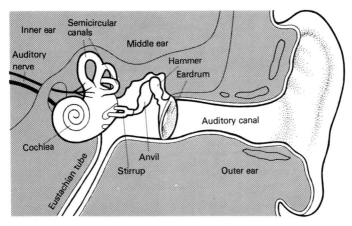

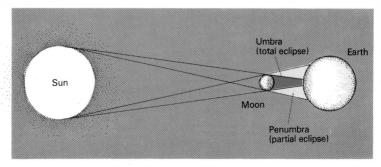

earthenware Pottery fired at a relatively low temperature so that it is not vitrified but still remains porous. For most uses earthenware pottery is glazed to make it watertight.

earthquake Violent shaking of the Earth's crust caused by rupturing of the rocks. Shock waves travel from the site of the rupture and shake the ground far from the epicentre, the point on the surface immediately above the site. Earthquake shock waves are called seismic waves and are detected by instruments called seismometers.

earthshine Faint light reflected from the Earth on to the Moon when the Moon is new, making the whole Moon just visible.

ebonite A hard black insulating material made from vulcanized rubber.

Echo An early type of American communications satellite, consisting of a huge balloon which reflected radio waves from one ground station to another.

echo Sound reflected from an object back to its source. Echoes occur not only in the air but also in water, where they are utilized in sonar devices such as the echo sounder. The reflected microwave signals used in radar are also called echoes.

echo location Using the direction and time of return of echoes from objects to pinpoint their location. Bats use echo location to navigate through the air. Radar is an echo-location technique.

echo sounder A sonar device used in boats to measure water depth. It emits a sound pulse down to the seabed, which reflects the pulse back to a receiver. From the time of travel of the pulse, the water depth can be calculated.

eclipse The partial or complete concealment of one heavenly body by another. An eclipse of the Sun occurs at times of the new Moon when the dark disc of the Moon passes directly in front of the Sun. The size of the Moon's disc is almost exactly the same as the Sun's disc. This can lead to a total eclipse, when the Sun is completely obscured. A total eclipse can never last more than 7½ minutes. During a total eclipse the Sun's pearly white corona can be seen. Sometimes the eclipse is annular, with a narrow ring of Sun surrounding the Moon's disc. Sometimes the eclipse is partial, the Moon's disc covering only part of the Sun. A lunar eclipse occurs when the Moon passes into the Earth's shadow.

ecliptic The apparent path of the Sun through the heavens each year. It is marked as a great circle on the celestial sphere and passes through 12 constellations—the constellations of the zodiac.

eddy currents Electrical currents set up, or induced, in electrical equipment by a changing electric or magnetic field. They oppose the changes that are taking place and constitute a considerable waste of energy in transformers and electromagnets, for example.

effervescence The escape of small gas bubbles from a liquid as a result, for example, of chemical action. Fruit salts effervesce when they are added to water.

efficiency Of a machine, the ratio of its useful energy output to the amount of energy put into it, expressed as a percentage. Mechanical efficiency can be very high. Thermal efficiency is generally very low. The thermal efficiency of a petrol engine is about 25%.

efflorescence The loss of water from hydrated crystals when they are exposed to air. Ordinary washing soda (sodium carbonate, $Na_2CO_3.10H_2O$), effloresces when exposed to the air, changing into a white powder, $Na_2CO_3.H_2O$.

einsteinium (Es) A radioactive artificial element (at no 99).

Einstein's equation An equation demonstrating the equivalence of mass (m) and energy (E): $E = mc^2$, where c is the velocity of light. It was formulated by Albert Einstein as part of his special theory of relativity.

elasticity The ability of a body under stress to return to its original state when the stress is removed. Once the stress rises above a certain limit—the elastic limit—the body suffers permanent deformation, or strain.

elastomer A substance, usually a polymer such as rubber, that tends to return to its original size after being stretched.

electric arc See arc, electric.

electric bell A common household device that uses an electromagnet to vibrate a clapper against a bell. When the bell push is pressed, current flows to the electromagnet which attracts the clapper. Movement of the clapper breaks the circuit which cuts off the electromagnet and thus releases the clapper. And the cycle begins again. An electric buzzer works in much the same way, but there is no bell.

electric charge A quantity of electricity, which may be positive or negative. Electrons are regarded as having a unit negative electric charge; protons, a unit positive electric charge. A flow of electric charges constitutes an electric current. Study of isolated charged bodies is known as electrostatics.

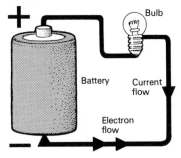

electric current The flow of electricity in a conductor, which occurs when electrons flow through it. The current may flow either one way (direct current) or alternately one way, then the other (alternating current). By convention the electric current in a circuit is said to flow from positive to negative. The force that drives electric current through a circuit is called the electromotive force (emf) and is measured in volts. The current is measured in amperes. Conductors have an inherent property of resisting the flow of electric current called their electrical resistance, measured in ohms. In a circuit, voltage (V), current (i) and resistance (R) are related by Ohm's law: $V = iR$.

Electric bell

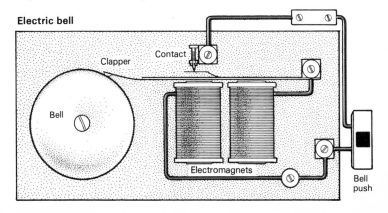

electric discharge

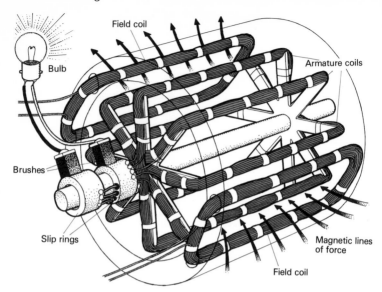

Field coil

Bulb

Armature coils

Brushes

Slip rings

Magnetic lines of force

Field coil

electric discharge The passage of electricity through gases, which are usually insulators. It takes place between two points when a sufficient potential difference exists between them. Lightning is a massive electric discharge between clouds and the ground. Electric discharge can also occur in gases at low pressure, as in discharge lamps.

electric eye See **photoelectric cell.**

electric field The region around an electric charge in which other charged particles are affected. The force the particles experience is proportional to the size of the electric charge and is inversely proportional to the distance from the charge.

electric furnace One that uses electricity to produce heat. In steel refining, for example, the electric-arc furnace is widely used. It uses the heat produced when an arc is struck between carbon electrodes and the metal to be melted. It produces high-quality steel, uncontaminated by furnace fumes. Another type of electric furnace is the induction furnace. A high-frequency electric current is sent through coils surrounding the furnace and sets up eddy currents in the metal inside, which heats it up.

electric generator A machine that produces electricity. An alternator produces alternating current; a dynamo, direct current. Electric generators work on the principle that a current is set up in a conductor when it is moved in a magnetic field. They are made up of coils of wire wound on an armature, which is made to rotate between the poles of a magnet. The current is taken from the armature windings via slip rings (in the alternator) or a commutator (in the dynamo).

electricity The branch of science concerned with the phenomenon associated with positively and negatively charged particles at rest and in motion. The study relating to charged particles at rest is called static electricity, or electrostatics as opposed to current electricity when the particles are in motion.

electric lamp See **discharge lamp; electric-light bulb; fluorescent lamp.**

electric-light bulb The modern version of the incandescent filament lamp, developed in Britain by Joseph Swan (1878) and in the US by Thomas A. Edison (1879). It consists of a thin glass bulb filled with nitrogen and argon gas. The filament, made of fine tungsten wire, gives out light when electricity is passed through it.

electric motor A device that changes electrical energy into mechanical energy to drive all kinds of machines. It is constructed much like an electric generator but operates in the opposite way. It uses the principle that when electric current is passed through a conductor in a magnetic field, the conductor moves.

electrocardiograph An instrument that monitors the electrical activity of the heart. It records the activity as a trace on a moving paper roll—an electrocardiogram (ECG).

electrochemical series Or electromotive series; a series in which elements are listed in order of their electrode potentials. It is indicative of the ease with which the elements can gain or lose their electrons. In general the elements are more reactive the higher up the series they are.

electrochemistry The branch of physical chemistry concerned with the chemical changes that electricity can bring about, as in electrolysis; and the production of electricity by chemical reactions, as in batteries and fuel cells.

electrode The terminal in an electric cell or in an electron tube that conducts electricity in or out. The positive electrode is called the anode; the negative electrode, the cathode.

electrode potential Of an element, the electric potential (voltage) developed by an electrode of the element when immersed in a molar solution of its ions.

electroencephalograph An instrument designed to monitor the electrical activity of the brain, via electrodes on the scalp. The activity is recorded as traces on a moving paper roll, this being called an electroencephalogram (EEG).

electrolysis The splitting up of a chemical compound by passing electricity through it. Compounds may be split up while molten or in solution. Only compounds that conduct electricity, ie are electrolytes, can be split up in this way. Sodium chloride, for example, is made up of sodium ions (Na^+) and chlorine ions (Cl^-). In the electrolysis of molten sodium chloride, sodium ions migrate to the cathode, pick up electrons and form sodium metal. The chlorine ions migrate to the anode, where they lose electrons and form chlorine gas. Electrolysis is a useful way of preparing several elements, particularly metals. Aluminium is prepared by the electrolysis of molten alumina—aluminium oxide. Sodium is prepared by the electrolysis of molten sodium chloride. Electrolysis is the basis of electroplating. See **aluminium; Faraday's laws.**

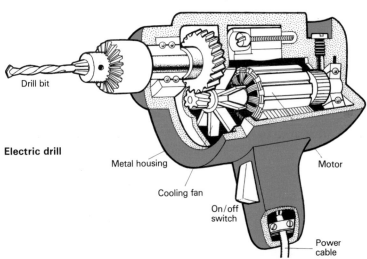

Drill bit

Electric drill

Metal housing

Cooling fan

On/off switch

Motor

Power cable

electrolyte

electrolyte A substance that splits up into ions and conducts electricity when molten or in solution. Acids, bases and salts are good electrolytes.

electrolytic refining Using electrolysis to extract and purify metals. Copper is purified by electrolysis.

electromagnet A temporary magnet formed by passing electric current through a coil wound around a soft iron core. Its magnetism disappears when the current is switched off.

electromagnetic radiation Radiation such as X-rays, light, heat, microwaves and radio waves, which consists of varying electric and magnetic fields.

electromagnetic spectrum See **spectrum.**

electromagnetism The branch of science concerned with the close relationship between electricity and magnetism. When an electric current passes through a wire, a magnetic field is produced. When a wire moves in a magnetic field, an electric current is set up in it.

electromotive force A measure of the tendency to cause the flow of electricity in a circuit; the voltage produced by a source of electricity, such as a battery or generator. Commonly abbreviated to emf.

electron One of the fundamental particles of matter, present in all atoms, circling around the nucleus. It has a unit electric charge, which is deemed negative. Its mass is 1/1836 that of the proton. Electrons can conveniently be considered to orbit the nucleus of atoms in electron 'shells' at different distances from the nucleus. Cathode-rays and beta-rays are streams of electrons. See also **positron.**

electronegativity The tendency of an atom to attract electrons. Fluorine is the most electronegative element. Caesium is the least electronegative, or most electropositive element.

electron gun The part of electronic devices such as the television camera and the cathode-ray tube which produces a beam of electrons.

electronic flash A compact discharge tube used as a photographic flash gun. When a high-voltage pulse is applied to electrodes in the tube, which contains an inert gas such as krypton, a brilliant flash results.

electronics The branch of physics concerned with the manipulation and control of electrons in devices such as semiconductors, vacuum tubes and valves.

electron microscope A microscope that manipulates beams of electrons rather than light rays to achieve magnification. It uses magnetic coils rather than glass lenses to bend the electron beam and form enlarged images. Whereas an optical microscope can magnify objects up to a few thousand times, an electron microscope can magnify hundreds of thousands of times and even view large molecules.

electron shell In an atom the electrons circling around the nucleus can conveniently be considered as occupying different shells at different distances from the nucleus. Each shell can hold a maximum number of electrons. It is the electrons in the outermost, or valency, shell that usually take part in chemical bonding. In general, elements whose atoms have a similar outermost shell tend to have similar chemical properties.

Electron shells of carbon

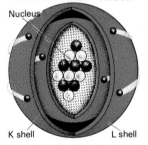

Nucleus

K shell   L shell

electron tube See **thermionic valve.**

electron-volt The unit of energy generally used in nuclear physics. It is equal to the energy gained by an electron passing through a potential difference of 1 volt. 1 electron-volt (eV) $= 1 \cdot 6 \times 10^{-12}$ erg. 1 MeV $= 1$ million eV; 1 GeV $= 1,000$ million eV.

electrophoresis The movement of charged particles through a fluid under the influence of an electric field. It can be used to separate and analyse proteins and other large molecules, which have different rates of movement. If a semi-permeable membrane is used as well, the technique is known as electroosmosis.

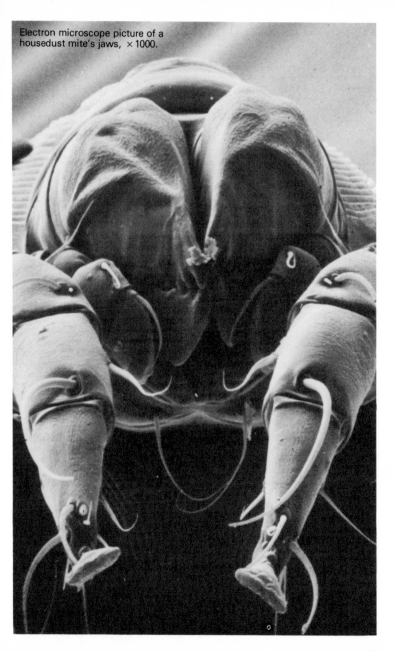

Electron microscope picture of a
housedust mite's jaws, × 1000.

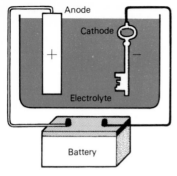

Anode
Cathode
Electrolyte
Battery

SOME ELEMENTARY PARTICLE DATA

Particle	Charge	Spin	Mass(amu)
Electron	−1	½	0·0005
Neutrino	0	½	0
Muon	−1	½	0·1134
Proton	+1	½	1·0073
Neutron	0	½	1·0087
Xi particles	0	½	1·4116
	−1	½	1·4185
Omega particle	−1	³⁄₂	1·7955
Pions	+1	0	0·1499
	0	0	0·1449
	−1	0	0·1499
Psi particle	0	1	3·3227

(1 atomic mass unit (amu) = $1·66 \times 10^{-27}$ kg)

electroplating The coating of a metal on another material (usually another metal) by means of electrolysis. In plating copper on iron, for example, the iron is made the cathode, pure copper the anode, and the electrolyte is a copper salt such as copper sulphate. During electrolysis, copper dissolves from the anode and is simultaneously deposited on the iron cathode.

electropositive A term applied to metals with the tendency to form positive ions. It is the opposite of electronegative.

electroscope An instrument for detecting electric charge. In the gold-leaf electroscope, two strips of gold leaf repel each other when it is charged.

electrostatic generator A device for building up high electric charges. See **van de Graaff generator.**

electrostatic precipitator A device used in industry to clear dust and other particles from the air—in furnace flues for example. It uses electrostatic attraction to attract the particles to an electrode.

electrostatics The study of static electricity.

electrotype A duplicate printing plate made by electroplating a mould of the made-up type.

electrovalent bond Or ionic bond; one formed by the transfer of electrons between atoms, as between sodium and chlorine atoms when they combine to form sodium chloride.

electrum A naturally occurring alloy of gold and silver, which was used to make the first coins in the 600s BC.

element, chemical See **chemical element.**

elementary particles Also called fundamental particles and sub-atomic particles; the basic particles from which all matter is made up, found in the atom. With the exception of hydrogen, which lacks neutrons, all atoms contain protons, neutrons and electrons. Other particles, such as mesons, play a part in holding the atom together. Every particle has an equivalent antiparticle. Every particle might be made up of different combinations of quarks.

elements, the The forces of nature; the weather.

elevator See **lift.**

elevators On an aeroplane, the control surfaces at the rear of the tailplane. On a delta-winged craft they also act as the ailerons and are then termed elevons.

ellipse An oval-shaped geometric figure, which is one type of conic section. It has two foci. An ellipse can be drawn by pinning the ends of a piece of string at two points (the foci) and, keeping the string taut, moving a pencil as shown in the diagram.

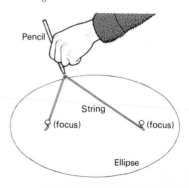

Pencil
String
(focus)
(focus)
Ellipse

elongation In astronomy, the angular separation between the Sun and a heavenly body. The maximum elongation for Mercury is about 28° and for Venus, 48°.

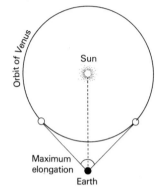

Orbit of Venus

Sun

Maximum elongation

Earth

emanation See **radon**.

emerald A valuable bright green gem, a variety of the mineral beryl.

emery A blue-grey mineral widely used, glued on cloth, as an abrasive. It is an impure variety of corundum containing magnetite and other minerals.

emf Abbreviation for electromotive force.

emission spectrum Or bright-line spectrum; the spectrum of light emitted by an element or compound when its atoms become 'excited' (eg by heating or electric discharge). Light of characteristic wavelength is emitted when the atoms return to a lower energy state. An emission spectrum consists of bright lines on a black background.

empirical formula The simplest chemical formula of a compound, which shows the relative numbers of atoms present, but not necessarily their quantities. Benzene, whose molecules contain six atoms each of carbon (C) and hydrogen (H), has the empirical formula CH. Its correct molecular formula is C_6H_6.

emulsion A type of colloid consisting of minute liquid droplets dispersed in another liquid. Milk is an emulsion—of fat droplets in a water solution.

emulsion, photographic The light-sensitive coating on film. It consists of gelatin in which are dispersed tiny grains of silver bromide and silver iodide.

enamel A kind of glaze used on ceramics, coloured by metal oxides; or a very hard paint dried by heating.

Encke's comet A faint comet with the shortest orbital period known—only 3·3 years. It is named after the German astronomer Johann Encke.

Encke's division A narrow gap in the outermost (A-) ring of Saturn's rings, first observed by Johann Encke.

endoscope An optical instrument that consists of a flexible lighted tube or shaft which can be inserted in a body opening and used to view the body internally. Similar instruments are used by engineers to check machines internally.

endothermic reaction One in which heat is absorbed.

energy The capacity for doing work. Heat, electricity and radiation are forms of energy. Fuels release chemical energy when they burn. Kinetic energy is the energy possessed by moving objects. Potential energy is energy that is stored in a system—an object that can fall has gravitational potential energy. In a closed system, the overall energy content remains constant no matter what chemical or physical changes take place therein: this being termed the conservation of energy. Only in nuclear reactions does the conservation law not apply. Then energy is created by the destruction of matter, for the two are equivalent, related by Einstein's equation $E = mc^2$. The standard unit of energy is the joule.

energy level Or energy state; one of a number of states in which electrons or other atomic particles can exist. In each state a particle or atom possesses a certain energy. When it acquires sufficient extra energy, it moves into another higher-energy state. In this state it is said to be excited.

engine A machine designed to convert energy into useful work. Most engines used today are heat engines, converting heat energy into work. See **diesel engine; gas turbine; jet engine; petrol engine; rocket; steam engine; steam turbine.**

engineering Broadly, the application of the sciences for the benefit of man. The main branches are aeronautical, chemical, civil, electrical and mechanical.

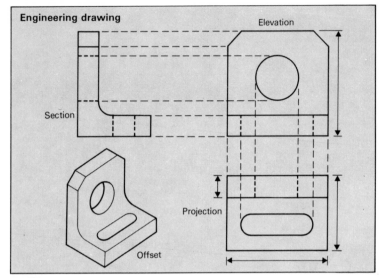

Engineering drawing A technical drawing that shows to scale details of structures and machines and their component parts, and explains how they are put together. Draughtsmen prepare engineering drawings showing different views of the same object in order to define its exact shape and size.

engraving Producing a printing plate or block by cutting a design in a metal or wood surface. Designs may also be etched in plates by acid. Photoengraving uses a photographic technique to transfer a design to the plate.

enthalpy A measure of the heat content of a system.

entropy A concept used in thermodynamics. The entropy of a closed system is a measure of the degree of disorder in that system. Entropy either remains the same or increases.

enzymes Proteins that act as catalysts for the chemical reactions involved in life processes. For example, pepsin and rennin, found in the stomach, are enzymes that aid digestion. Lipase, made in the pancreas, helps break down fats.

Eocene Epoch The period of geological time that lasted from about 54 million until 38 million years ago. Its name refers to the 'dawn' of recent life, such as the modern mammals.

ephemeris A table that lists the positions of the heavenly bodies for certain dates, based upon their predictable orbital motions. Annual ephemerides are published in several countries for the benefit of astronomers and navigators, one of the most important being the Anglo-American publication *The Astronomical Ephemeris*.

epicentre The point on the surface of the Earth directly above the focus, or site of an earthquake.

epicycle A circle whose centre moves around the circumference of a larger circle (deferent). Early Greek astronomers once thought that the movement of the planets could be explained if they were considered to move around an epicycle which itself moved along a deferent that was centred on the Earth.

epicyclic gear Also called sun-and-planet gear; a gear system in which one or more gear wheels move around the inner or outer circumference of another gear wheel. Epicyclic gears are found in bicycle hub gears and car automatic gearboxes.

epidiascope A kind of projector, which can project not only slides, or transparencies, but also images of opaque objects.

epoxy resin A synthetic resin widely used to make adhesives and paints. For use as an adhesive it is mixed with a curing agent, or hardener, which causes its molecules to cross-link and set hard.

Epsom salts A common mineral form of magnesium sulphate, $MgSO_4.7H_2O$, which is used in medicine as a laxative.

equation In chemistry, a shorthand method of representing a chemical reaction, using symbols to represent the atoms of elements taking part. In mathematics, an equation is a statement that two quantities are equal. It usually involves unknown quantities, or variables. Equations can be classed by the power of the variable. In a linear equation ($x + 1 = 9$) the variable (x) has the power 1; in a quadratic equation ($x^2 - 2x - 6 = 0$), the power 2; and in a cubic equation, the power 3.

equation of state A relationship between the pressure, volume and temperature of a substance, which defines its physical state. The general gas law, $PV = nRT$, is an example.

equator An imaginary line drawn around the Earth midway between the north and south poles. It lies in a plane perpendicular to the Earth's axis. All points on the equator have 0° latitude.

equatorial mounting The usual method of mounting an astronomical telescope. The mounting has one axis (polar axis) pointing towards the celestial pole, and therefore parallel to the Earth's axis. The telescope is carried on another axis (the declination axis) at right-angles to the polar axis. With this mounting, the telescope can readily follow a star across the sky.

equilateral A figure having all sides equal.

equilibrium A state in which a body or a system will remain unless it is disturbed by an outside force. When a body is in equilibrium, the sum of all the forces acting upon it is zero. In chemistry, equilibrium is reached in a reversible chemical reaction when the rate of the forward reaction equals the rate of the reverse reaction.

equinoxes The two times of the year when day and night are of equal length throughout the world. They occur when the Sun lies exactly above the equator, in the spring (spring or vernal equinox) on about March 21 as the Sun is moving north; and in the autumn (autumnal equinox) on about September 23 as the Sun is moving south.

equivalent weight Or gram equivalent; the mount in grams of an element that will combine with or displace 1 g of hydrogen or 8 g of oxygen. It is equal to the element's atomic weight divided by its valency.

Er The chemical symbol for erbium.

erbium (Er) One of the lanthanides, or rare-earth metals (rd 9, mp 1,525°C).

erg A unit of energy in the cgs system, being the work done when a force of 1 dyne acts through a distance of 1 cm. 1 erg = 10^{-7} joule.

ergonomics The study of human beings in relation to their working environment.

Eros An irregularly shaped asteroid, about 35 km (22 miles) long, which sometimes comes within 25 million km (15 million miles) of Earth.

erosion The gradual wearing away of the Earth's surface by natural forces; such as the wind, rain, wave action, glaciers, flowing water and chemicals dissolved in the water.

Equatorial mounting

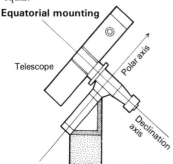

Telescope

Polar axis

Declination axis

Epicyclic gear

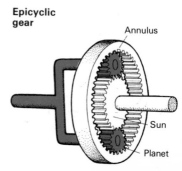

Annulus

Sun

Planet

Es

Es The chemical symbol for einsteinium.

ESA Abbreviation for European Space Agency, the body that coordinates space research in Europe. Its member countries include Belgium, Britain, Denmark, France, Germany (west), Ireland, Italy, the Netherlands, Spain, Sweden and Switzerland. One of its main projects is Spacelab, which is carried into space by the space shuttle.

escalator A moving staircase for carrying passengers. The steps run on an endless belt and fold flat at top and bottom to allow passengers to step on and off more easily.

escapement In a clock or watch, the mechanism that allows the gradual release of energy from the power source (spring, falling weight) and controls the rate of movement of the gear train that turns the hands.

escape velocity Of Earth; the minimum velocity that a body must possess to escape entirely from the Earth's gravity—about 40,000 km/h (25,000 mph). The escape velocity of other planets or moons varies according to their mass and diameter. The escape velocity from the Moon is only about 8,500 km/h (5,300 mph), while that from Jupiter is 220,000 km/h (140,000 mph).

escarpment Or scarp; a steep cliff.

essential oils Volatile oils found in many plants, which give them their characteristic scent. Such oils are often extracted and used in perfumes and cosmetics.

ester A compound formed when an organic acid reacts with an alcohol, water being eliminated. Acetic acid (CH_3COOH) reacts with ethanol (C_2H_5OH) to form the ester ethyl acetate ($CH_3COOC_2H_5$):
$$CH_3COOH + C_2H_5OH \rightarrow CH_3COOC_2H_5 + H_2O$$
Many esters have a pleasant, fruity smell. Fats, oils and waxes are esters.

estrogens Female sex hormones.

etching Cutting into a surface with acid. Printing plates can be made by etching a design in a copper or zinc plate. The metal surface is first coated with a resin, through which the design is then scratched. When treated with acid, the metal is eaten away only where it has been exposed.

ethanal See **acetaldehyde.**

ethane (C_2H_6) A colourless, odourless gas (bp $-88 \cdot 6°C$), which is the second main constituent of natural gas after methane. One of the paraffin hydrocarbons (alkanes), it is the parent compound of the ethyl group, C_2H_5-. It is particularly useful because it can readily be converted by cracking to ethylene.

ESA's Spacelab space station

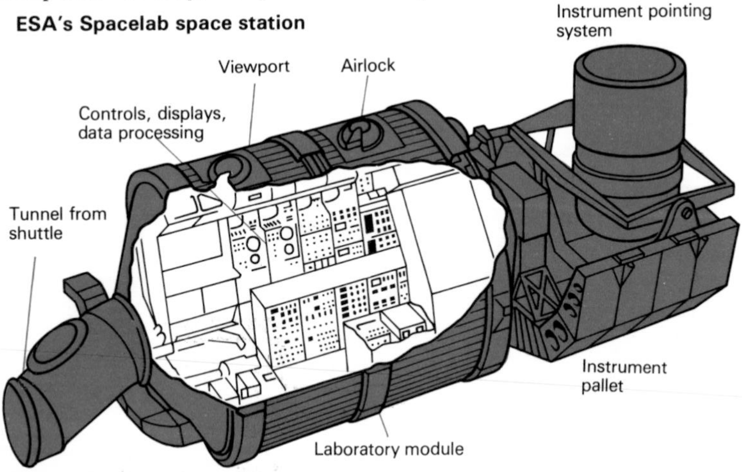

Instrument pointing system

Viewport

Airlock

Controls, displays, data processing

Tunnel from shuttle

Instrument pallet

Laboratory module

evaporites

ethanoic acid See **acetic acid.**

ethanol (C_2H_5OH) Or ethyl alcohol; the best-known alcohol, being the one found in alcoholic drinks such as beers, wines and spirits. It is a highly volatile and flammable liquid (bp $78 \cdot 5°C$), of immense importance in industry as a solvent and as a chemical intermediate. It is made by the fermentation of sugars, followed by distillation.

ethene See **ethylene.**

ether Or aether; an imaginary substance that was once thought to fill all space and to allow the transmission of light and other radiation through space.

ethers Organic compounds of the general formula $R - O - R$, consisting of hydrocarbon radicals linked by an oxygen atom. The most familiar is the anaesthetic known as ether, which is diethyl ether, $C_2H_5OC_2H_5$. It is a volatile, highly flammable liquid (bp $34 \cdot 5°C$) with a distinctive smell and sweetish taste.

ethyl group The hydrocarbon radical $C_2H_5 -$, derived from ethane (C_2H_6).

ethyl acetate ($CH_3COOC_2H_5$) The ester of acetic acid and ethyl alcohol. A sweet-smelling liquid (bp $77°C$), widely used as a solvent.

ethyl alcohol See **ethanol.**

ethylene (C_2H_4) Or ethene; simplest of the olefines (alkenes), which are characterized by the presence of a double bond: $H_2C = CH_2$. A flammable and sweet-tasting gas (bp $- 104°C$), ethylene is very reactive. It is obtained by cracking ethane and during other oil refinery processes. It is an invaluable raw material for the organic chemical industry, eg for the production of ethanol, ether and polyethylene.

ethylene glycol See **glycols.**

ethyne See **acetylene.**

Eu The chemical symbol for europium.

Europa The second largest moon of Jupiter (diameter 3,126 km, 1,942 miles). It has a smooth icy surface, crisscrossed with fracture lines. It orbits Jupiter at a mean distance of 670,000 km (420,000 miles).

europium (Eu) One of the lanthanides, or rare-earth elements (rd $5 \cdot 2$, mp $830°C$).

eutectic A mixture of substances that behaves like a pure substance in melting and solidifying as a whole at a definite temperature. Usually when a liquid mixture of substances is cooled, one component in the mixture separates out first. The eutectic temperature is generally lower than the melting point of either component in the mixture. Solders are alloys having an eutectic composition.

eutrophication The pollution of rivers and lakes with nutrients, such as nitrogenous fertilizers, which promote the growth of algae and plant life. This may lead to serious oxygen starvation and the death of fish.

EVA Abbreviation for extra-vehicular activity—activity outside a spacecraft, or a space 'walk'. Russian cosmonaut Alexei Leonov made the first EVA in 1965, just before astronaut Edward White made the first American EVA.

evaporation The escape of molecules from a liquid surface to form a vapour. It occurs at all temperatures but most rapidly at the liquid's boiling point. Evaporation of water from plants is called transpiration. Solids also evaporate to a limited extent, changing directly into vapour without an intermediate liquid stage—a process called sublimation.

evaporites Sedimentary mineral deposits laid down when salt-laden lakes and seas evaporated. Among the most common evaporites are halite (rock salt), gypsum and anhydrite.

Eutectic point (E) for sodium nitrate solution

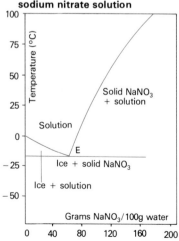

JCB excavator

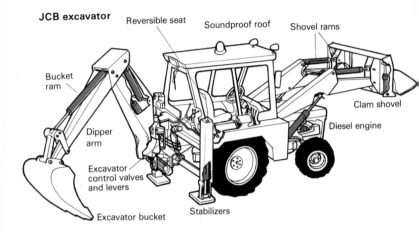

Reversible seat · Soundproof roof · Shovel rams

Bucket ram

Clam shovel

Dipper arm

Diesel engine

Excavator control valves and levers

Excavator bucket

Stabilizers

evening star The name given to the planet Venus when it is prominently visible in the western sky at sunset.

excavator A machine for digging into the ground.

excited state In physics, particles are said to be excited when they absorb energy and transfer to a higher energy level. They may return to their original, ground state by emitting radiation.

exosphere The outermost part of the Earth's atmosphere, beyond the ionosphere, starting at about 500 km altitude.

exothermic reaction One that gives out heat. Normal combustion is an exothermic reaction.

expanding universe A concept in astronomy that considers the universe as a whole to be getting bigger. There is evidence for it in the observation that most galaxies are receding from us, with the farthest ones travelling the fastest.

expansion The increase in dimensions of a body as the external conditions change. Generally materials expand when they are heated (thermal expansion). Each material has a characteristic rate of expansion expressed as a coefficient of expansion, which is the fractional increase in length or volume for unit rise in temperature.

Explorer 1 The first American space satellite, launched from Cape Canaveral by the Juno 1 rocket on January 31, 1958. It made the first significant discovery of the space age—of the Van Allen radiation belts—and remained in orbit until April 1970, having orbited the Earth 58,376 times.

explosives Substances that burn very rapidly, creating large volumes of gases, which cause devastating shock waves as they expand. The original explosive was gunpowder, invented in the 10th century AD by the Chinese. Modern high-explosives require a detonator to set them off (see **nitroglycerin, TNT**).

exponential In mathematics, an exponential function is a relationship of the form $y = a^x$. x is called the exponent. The term exponential growth refers to a very rapid rise in something, such as population; y grows very rapidly as x increases.

exposure meter An instrument used in photography to gauge the optimum time of exposure for a film. One type employs a sensor of cadmium sulphide (CdS), whose resistance changes according to the light striking it. Another type has a sensor such as selenium, which generates a current in proportion to the light striking it.

extender An inert substance, such as chalk, added to paints or plastics.

extragalactic nebulae The former name for galaxies.

extrapolation Calculating the value of a function beyond the range of known values. This can be done graphically, for example, by extending a curve.

extraterrestrial Existing beyond the Earth. Extraterrestrial intelligence refers to beings that might exist elsewhere in the universe.

extra-vehicular activity See **EVA.**

extrusion Forcing a material through a die to give it a uniform cross-section. Metal tubes and plastic pipes are often made by extrusion. Flexible tubes like toothpaste tubes are made by impact extrusion—the material is struck by a sudden impact and flows into shape.

extrusive rocks Or volcanic rocks; rocks formed from magma that has poured out or been ejected on to the Earth's surface. Lava is an extrusive rock.

eye The sense organ by which animals see. The essential parts of the human eye are the cornea, pupil, iris, lens and retina, all of which reside in a roughly spherical eyeball. Light passes through the transparent cornea and through the pupil, whose diameter is varied by the coloured iris. It then enters the lens, which brings it to a focus on the retina. The retina then sends appropriate messages to the brain via the optic nerve. The eyeball is filled with a watery fluid called the aqueous humour and a jelly-like material called vitreous humour.

eyepiece A lens system in a telescope or microscope through which an observer views the image formed by the objective lens. Various eyepieces are often available for these instruments to permit different magnifications.

F

F The chemical symbol for fluorine.

facsimile An exact copy or reproduction of, say, a picture or manuscript. Facsimile machines transmit pictures and other images by wire or radio. They incorporate a scanning device and a photoelectric cell that converts variable light intensities into electric signals for transmission.

factor In mathematics, a number that goes exactly into another without a remainder. The factors of 6 are 3 and 2.

factorial Of a given number, the product of that number and all numbers less than it. So factorial 6, written 6!, equals $6 \times 5 \times 4 \times 3 \times 2 \times 1$ (= 720).

factory system The modern method of manufacturing, in which a production task is split into a sequence of activities (division of labour). Workers are gathered in one place to perform a particular activity in the production sequence, usually with the aid of machinery. Richard Arkwright pioneered the factory system in the cotton industry in the 1770s. Contrast **cottage industry.**

faculae Bright regions of the Sun's surface that are somewhat brighter than average. They often appear near sunspots.

Fahrenheit scale A scale of temperature, named after the German physicist Gabriel Fahrenheit. On this scale, once widely used in English-speaking countries, water freezes at 32° and boils at 212°.

falling star See **meteor.**

fallout Radioactive dust that falls from the atmosphere after a nuclear explosion, often at a large distance from the blast. Two of the most long-lived radioactive isotopes in fallout are caesium-137 and strontium-90, which is hazardous because it can enter the body via milk from cows feeding on contaminated grass.

fanjet See turbofan.

farad A unit of capacitance, being the capacitance of a capacitor when one coulomb of electricity raises its potential by one volt; named after Michael Faraday.

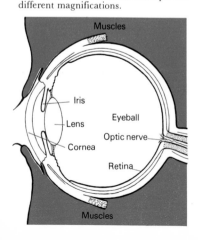

Muscles
Iris
Lens
Eyeball
Optic nerve
Cornea
Retina
Muscles

Faraday

faraday A unit of electricity used in electrolysis, named after Michael Faraday. It is the amount of electricity (in coulombs) that liberates the equivalent weight in grams of a substance at an electrode.

Faraday's laws Of electrolysis: (1) The amount of a substance liberated during electrolysis is proportional to the amount of electricity passed. (2) The relative amounts of substances liberated are in proportion to their equivalent weights.

fast reactor Or breeder reactor; a nuclear reactor that utilizes fission by means of fast neutrons. It uses a fuel of enriched uranium and plutonium and requires no moderator. In present fast-reactor designs liquid sodium is used as a coolant.

fathom A unit used to measure water depth, equalling 6 ft (1·83 m).

fatigue In engineering, a condition under which metals fail under relatively light loads when those loads are applied repeatedly. Structures such as airframes and machines must be carefully designed to avoid cyclic vibrations to prevent premature failure by fatigue.

fats Important animal foods found in animal tissues and in plants, particularly in the seeds. Chemically they are glycerides, esters of fatty acids and glycerol. Fats that are liquid at room temperature are usually termed oils.

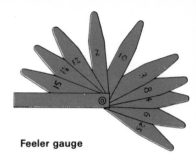

Feeler gauge

fatty acids Carboxylic acids found in fats and oils. The most important are palmitic, stearic acids and oleic acids. Soaps are the sodium or potassium salts of such fatty acids.

fault A break in the Earth's crust caused by the relative movement of the rocks on either side. The surface where the fracture takes place is called the fault plane.

Fe The chemical symbol for iron.

feedback A method of controlling a system, which uses the output of the system to modify the input. A classic case of feedback is an electrical thermostat. Information about the temperature (a measure of the output) is fed back to the switch that turns the electricity (input) on or off. Feedback is an essential element in automated systems.

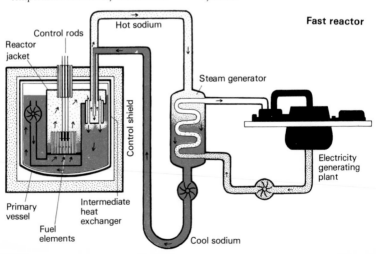

Fast reactor

Control rods
Hot sodium
Reactor jacket
Steam generator
Control shield
Electricity generating plant
Primary vessel
Intermediate heat exchanger
Fuel elements
Cool sodium

feeler gauge A combination of metal leaves of known thickness, used to measure narrow gaps, such as the clearance between the electrodes of a sparking plug.

Fehling's solution A solution used in chemical analysis to test for sugars and other reducing agents. It is a blue solution containing copper sulphate, potassium hydroxide and Rochelle salt. When glucose, for example, is added to it, a red precipitate of cuprous oxide is formed. Such a test is carried out to test for sugar in the urine of diabetics.

feldspar Or felspar; the most abundant mineral group in the Earth's crust, found in igneous rocks. They are mainly sodium, potassium or calcium aluminium silicates. The two main categories are alkali feldspars (sodium-potassium) or orthoclase feldspars (sodium-calcium). They are found in a variety of colours.

fermentation Most commonly, the process in which yeasts or other micro-organisms act on sugar or starch to make alcohol, with carbon dioxide being given off as a waste product.

fermium (Fm) An artificial radioactive element (at no 100).

ferric compounds Those containing trivalent iron; now often written Fe(III).

ferrimagnetism A kind of permanent magnetism displayed typically by magnetic oxide materials, such as the mineral magnetite and iron oxide/ceramic materials called ferrites. Unlike ferromagnetic materials, they conduct electricity very poorly.

ferrites Magnetic iron oxide/ceramic materials containing other metals, such as nickel, cobalt and barium. They possess a kind of magnetism called ferri-magnetism. Computers sometimes have ferrite memory cores.

ferromagnetism The type of strong magnetism displayed by iron, nickel and cobalt.

ferrosilicon A steel containing silicon whose magnetic properties make it suitable for transformer cores.

ferrous compounds Those containing divalent iron; often written Fe(II).

ferrous sulphate ($FeSO_4.7H_2O$) Also called copperas and green vitriol; a common iron salt that forms pale green crystals, used in tanning and dyeing and making ink.

fertilizers Natural or artificial materials added to the soil to increase the fertility. Fertilizers supply essential elements to the soil which plants require for healthy growth. The main elements required are nitrogen, phosphorus and potassium. Nitrogen can be supplied by animal manures, nitrates, and ammonium salts; phosphorus by phosphate rock and superphosphate; potassium by potash minerals.

fibres See **cotton; jute; linen; man-made fibres; silk; synthetic fibres; wool.**

fibreglass Glass formed into fine fibres used for insulation and as reinforcement in plastics (see **GRP**). In fibreglass production glass marbles are melted, and the molten glass is forced through tiny holes in a kind of spinneret, emerging in the form of long filaments. Or the molten glass is drawn through orifices and blown into a kind of glass wool by jets of high-pressure steam.

fibre optics A branch of physics concerned with the transmission of light by means of bundles of glass fibres. The light is transmitted along the fibres by being totally internally reflected from the inner fibre walls. Fibre-optic devices are used in medicine to see inside the body. Optical fibres are also coming into use in telecommunications to carry signals, so replacing copper cables.

field The region around a body where a certain influence is felt, such as electric field, gravitational field and magnetic field. The magnetic field around a magnet can be revealed by means of iron filings, which are placed on a card held over the magnet, as shown here.

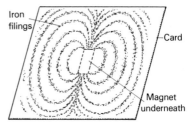

filament A thin thread. In an electric-light bulb the filament is made of tungsten wire. Continuous filament is the endless thread formed during the manufacture of synthetic fibres.

film

film In photography, a flexible strip of transparent cellulose acetate, which holds a light-sensitive coating, or emulsion. Colour films are composed of four different emulsion layers—on the top is one sensitive to blue light, then comes a yellow filter layer, followed by layers sensitive to green light and red light respectively.

filter In general a device that holds something back or removes something. A coloured light filter removes certain wavelengths from the light passing through it. In electronics, a band-pass filter blocks certain signal frequencies. In chemistry, filters are used to remove solids from liquids (see **filtration**).

filtration Removing solids from a fluid by means of a porous material (paper, cloth, wire mesh, gravel) that allows liquid through but holds back the solids. The clear liquid remaining after filtration is called the filtrate. In the chemical industry filtration is often done under pressure in a filter press.

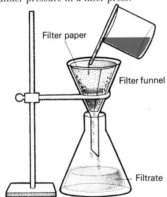

finder A low-power telescope with a wide field of view, mounted on a larger telescope.

fire Rapid combustion accompanied by the production of light (flame) and heat. Man has used fire for at least half a million years. But he has been able to make fire only for about 10,000 years, initially by striking flints against pyrites or by using the bow drill.

fireball See **bolide**.

firedamp An explosive mixture of methane gas and air that often accumulates in coal mines.

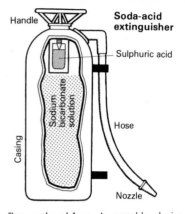

fire extinguisher A portable device used to put out a fire. Soda-acid extinguishers hold water, which is expelled under pressure when sulphuric acid reacts with sodium bicarbonate to produce carbon dioxide gas. Other extinguishers use chlorinated hydrocarbons, such as carbon tetrachloride, but they tend to produce toxic fumes. Dry extinguishers include those containing carbon dioxide, which is stored under pressure as a liquid and released as a gas to blanket the fire. Other dry chemical extinguishers release a powdery cloud of sodium bicarbonate and other chemicals. They generate carbon dioxide when the fire heats them.

fireworks Colourful explosive devices used for entertainment, invented, like the gunpowder they usually contain, by the Chinese about 1,000 years ago. They are coloured by metal salts, such as copper (green), sodium (yellow), strontium (red) and barium (pale green).

first point of Aries One of two points on the celestial sphere where the plane of the celestial equator intersects the plane of the ecliptic. The Sun is at this point on March 21, at the vernal equinox. The first point of Aries is the zero point for the measurement of sidereal time.

first point of Libra One of the two points on the celestial sphere where the plane of the celestial equator intersects the plane of the ecliptic. The Sun is at this point on about September 23, at the autumnal equinox.

Fischer-Tropsch process A method of producing petrol and other liquid hydrocarbons from coal, developed in Germany by Franz Fischer and Hans Tropsch and used there during World War 2.

fission, nuclear The splitting of the nucleus of an atom. Some heavy atoms, notably uranium, undergo fission either spontaneously or when they are bombarded by neutrons. When nuclear fission occurs, other lighter nuclei are produced, along with neutrons or other atomic particles. A certain amount of matter appears to be destroyed during fission, but it has actually been transformed into energy, which is given out as light, heat or radiation (see **nuclear energy**). See also **nuclear reactor**; contrast **fusion, nuclear.**

fixation of nitrogen Extracting nitrogen from the atmosphere to make compounds for use as fertilizers, explosives, and so on. The commonest method is the Haber process.

fixed points On a temperature scale, temperatures that are accurately reproducible, such as the freezing point and boiling point of water.

fixed stars A term used by early astronomers to distinguish the stars, which appear to be fixed to the celestial sphere, from the 'moving stars', or planets.

fixing In photography, treatment of the film after developing to make it insensitive to light. This is done by removing the unchanged silver halides by an agent such as hypo, or sodium thiosulphate.

flame The light given out when something burns.

flame test A simple method of identifying certain elements in qualitative chemical analysis. A sample of the mixture to be analysed is picked up on a platinum wire and held in a Bunsen flame. A vivid crimson will indicate the presence of strontium; orange-yellow, sodium; pale green, barium; lilac, potassium. The flame-test principle is used in a more sophisticated way in the flame photometer.

flammable Able to burn. The term inflammable is often, but incorrectly, used instead. Something that does not burn is correctly termed non-flammable.

flaps On an aeroplane, control surfaces at the trailing edges of the wings inboard of the ailerons. They are moved downwards to provide extra lift when landing.

flares On the Sun, sudden and brief bursts of radiation, usually associated with sunspots. Atomic particles, X-rays and radio waves are emitted, which affect the ionosphere on Earth, leading to brilliant auroras and disruption of long-distance radio communications.

flashbulb A source of artificial light used in photography when natural lighting conditions are inadequate. The bulb contains fine wire of aluminium, magnesium or zirconium and is filled with oxygen. It is held in a flash gun and is electrically ignited by closing of the camera shutter.

flash point Of a flammable liquid, the lowest temperature at which it will produce enough vapour to ignite.

Fleming's rules An aid to remembering the direction of motion, magnetic flux (field) and emf (current) in electrical machines, using the hands. To find directions in the electric generator, use the right hand. Extend the thumb, first finger and second finger at right-angles to one another. The First finger then points in the direction of the Flux, the seCond in the direction of the Current, and the thuMb in the direction of Motion. If you use the left hand, then the directions apply to the electric motor.

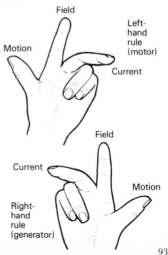

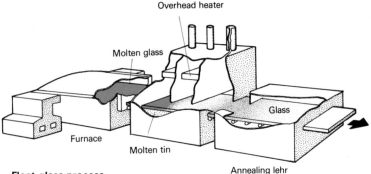

Float-glass process

flint One of the commonest minerals, being an impure form of silica, SiO_2. It is usually dark grey in colour and is commonly found in the form of nodules in chalk and limestone. It is very hard and fractures in sharp conchoidal (shell-like) chips. It was the material pre-historic man used for making tools.

flint glass An alternative name for lead crystal glass.

float chamber The part of a carburettor that ensures a constant head of petrol. A float in the chamber allows in petrol through a needle valve if the petrol level drops, but closes the valve as soon as the chamber is full.

float glass Flat glass formed by allowing molten glass to flow on to a bath of molten tin. It is not distorted like ordinary sheet glass and also has the bright 'fire-finish' that plate glass lacks. The float-glass process was developed in Britain by Pilkington Brothers.

flocculation The coming together of fine particles to form larger ones in the form of soft flakes.

flotation process In metallurgy, a method that uses wetting agents to bring about the separation of ore particles. In the process the ore is finely ground and then mixed with water and suitable wetting agents. Air is then bubbled through the mixture, forming a froth. The ore particles attach themselves to the bubbles and are carried away in the froth. Unwanted dirt (gangue) particles are unaffected and remain behind. Different types of ores can also be separated by flotation.

flow chart A pictorial representation of a process. In computer programming, the flow chart details the sequence of operations that is to be carried out.

flowers of sulphur Fine crystals of sulphur obtained by condensing sulphur vapour.

fluid Any substance that flows readily and takes the shape of its containing vessel. Gases and liquids are both fluids.

fluid flywheel A kind of automatic clutch used in some car automatic transmission systems. It consists essentially of an oil-filled housing containing vaned impellers connected to the shafts between which motion is to be transmitted. When one shaft and impeller turn, the whirling movement of the oil set up causes the other impeller to turn.

fluidics The study of systems and devices that use the flow of gas or liquid to operate control systems.

fluidized bed A system used in chemical processing in which solid particles are suspended in a gas stream. The system then behaves much like a fluid.

fluid mechanics The study of the behaviour and properties of fluids at rest and in motion. Two main branches are aerodynamics, concerned with the flow of gases, particularly air; and hydraulics, concerned with the flow of liquids.

fluorapatite The common phosphate mineral calcium fluophosphate, $Ca_5(PO_4)_3F$, often found as tiny green crystals.

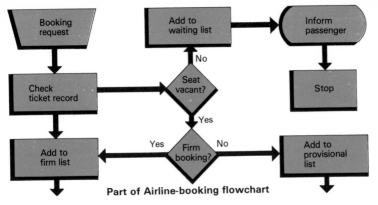

Part of Airline-booking flowchart

fluorescein A deep red organic compound ($C_{20}H_{12}O_5$), whose solutions display vivid fluorescence, even at dilutions of 40 million to 1. It is widely used as a dye and indicator.

fluorescence The property of certain substances to absorb light of one wavelength and radiate it at another wavelength. The fluorescence ceases when the source of illumination is removed (compare **phosphorescence**). Some minerals, including fluorite, fluoresce when they are exposed to invisible ultraviolet light.

fluorescent lamp One that produces light by fluorescence. The lamp takes the form of a long tube, the inside of which is coated with phosphors. The tube is filled with mercury vapour. When electricity is passed through the gas, it emits ultraviolet light, which causes the phosphors to fluoresce and give out visible white light.

fluoridation The practice of adding fluorides to drinking water as a means of reducing tooth decay in children.

fluorides Salts of hydrofluoric acid (HF).

fluorine (F) A choking poisonous gas, yellowish-green in colour, which is the most reactive of all the non-metallic elements (bp $-188 \cdot 1°C$). It is the first of the halogen series of elements. Combined with hydrogen it forms hydrogen fluoride (HF), whose solution is hydrofluoric acid, which gives rise to the fluorides. Hydrofluoric acid is one of the few substances that will attack glass.

fluorite Or fluorspar; the main fluorine mineral (CaF_2), which fluoresces. It may appear in various colours—blue, purple, green, red and yellow—or colourless if pure. Blue John is a well-known purple or blue variety found in Derbyshire.

flutter A distortion of sound in a sound reproduction system caused by irregularities in turntable speed. Wow is similar but of lower frequency.

flux In metallurgy, a substance added to enable a mixture to melt more readily and extract impurities. Limestone is added during iron smelting and forms a slag that absorbs impurities. In soldering, a flux of resin or chloride solution is applied to clean the metals to be joined and prevent their oxidation.

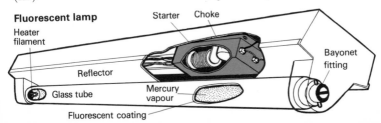

Fluorescent lamp

95

flying boat An amphibious aeroplane that has the underside of its fuselage shaped rather like the hull of a boat and usually has small floats attached to the wing tips. It contrasts with a float plane, which is an amphibian with a conventional fuselage but floats instead of a wheeled undercarriage.

flying saucer A UFO with a saucer shape.

flying shuttle A device invented by John Kay in 1733, which carries the weft yarn through the shed during weaving. Its invention was the first of a series of mechanical improvements in textile making that sparked off the Industrial Revolution.

flywheel A heavy wheel attached to a rotating shaft to smooth out its motion. The most familiar example is the flywheel at the end of the crankshaft in a petrol engine. It smoothes out the motion imparted to it by the separate power strokes of the pistons.

FM Abbreviation for frequency modulation.

Fm The chemical symbol for fermium.

f-number Also called f-stop, or simply stop; a number used in photography to describe the aperture of a camera lens. The f-number is the ratio of the focal length of the lens to its diameter. A small f-number indicates a wide aperture; a large f-number, a narrow aperture.

foam In chemistry, a kind of colloid in which bubbles of gas are dispersed in a liquid.

focal length Of a lens, the distance from the centre of the lens to its principal focus.

focus Of a lens, the point at which light rays form a sharp image. The principal focus of a converging lens is the point at which parallel rays entering the lens meet. The principal focus of a diverging lens is the point from which rays of light which entered the lens parallel appear to come from.

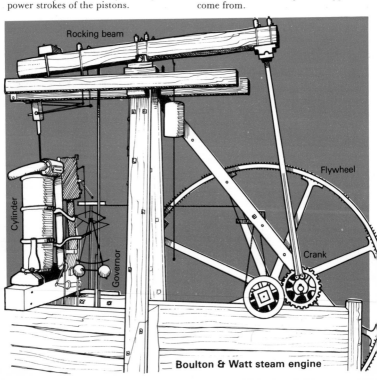

Boulton & Watt steam engine

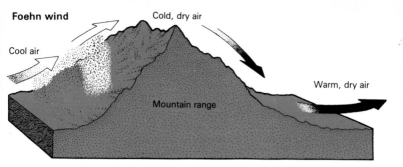

Foehn wind

Cold, dry air

Cool air

Warm, dry air

Mountain range

foehn Or föhn; a warm dry wind blowing down a mountainside, particularly one in the Alps. The wind is dry because the air has deposited its moisture on the other side of the mountain, and it is warm because it becomes compressed on its descent.

fog A low-level cloud, formed of a suspension of water droplets in the atmosphere. Water vapour in the atmosphere forms a fog when it condenses on dust particles in the air. In industrial regions smoke may become trapped in the fog, forming smog.

folding In geology, the wavy structure found in rock strata caused by movements of the Earth's crust.

Fomalhaut (α Piscis Austrini) One of the brightest stars in southern skies (magnitude 1·1) in the constellation Piscis Austrinus. It is widely used in navigation because it is located in an otherwise inconspicuous part of the heavens.

food preservation Means of preventing the spoilage of food so that it can be stored. This may be done by canning—sealing and sterilizing in cans—freezing, drying, or by treatment with salt, sugar or smoke. The various methods aim at preventing chemical spoilage, which is generally caused by oxidation or the action of enzymes; and spoilage by microorganisms, such as yeasts, moulds and bacteria.

fool's gold See **pyrite**.

foot A unit of length in the Imperial measuring system. It equals 12 inches, one-third of a yard, or 0·305 metre.

foot-candle A unit of illumination now little used, being the illumination of a surface 1 foot from a light source of 1 candle. The modern unit is the lux.

foot-pound The Imperial unit of work, defined as the work done when a force of 1 pound acts through a distance of 1 foot.

force An agent that can cause a body to move from a state of rest, alter its motion, or distort it. The magnitude of a force (F) is given by the product of the mass (m) of the body acted upon and the acceleration (a) it produces in that body: $F = ma$. There are physical forces, like a blow; magnetic and electric forces; gravitational forces; and nuclear forces. The modern unit of mechanical force is the newton. Forces are also sometimes expressed in dynes or poundals. Force is a vector quantity—it has direction. See **parallelogram of forces.**

forging Shaping hot metal by hammering or by the more gradual application of pressure. A common forging machine is the drop forge, which shapes a workpiece by sudden blows. A hydraulic press forges by means of gradually increasing pressure.

formaldehyde (HCHO) Also called methanal; a flammable gas with a pungent smell. It is perhaps best known as a 40% solution in water, formalin, which is used to preserve biological specimens and as a disinfectant. Formaldehyde is polymerized with urea, phenol and melamine to make a series of thermosetting resins and plastics.

formalin See **formaldehyde.**

Formica Trade mark for a heat-proof laminated plastic made from formaldehyde resins.

formic acid (HCOOH) Also called methanoic acid; first of the series of carboxylic acids; the acid in ants' stings. It is a pungent-smelling, fuming liquid and a strong acid, used domestically as a defurring agent for kettles.

formula

formula In chemistry, see **chemical formula.** In physics and mathematics, a shorthand method—using symbols—of showing a relationship between various quantities or variables. For example, the area A of a triangle is given by the formula $A = \frac{1}{2} bh$, where b is the length of the base and h is the perpendicular height.

Fortin barometer An accurate mercury barometer, named after Jean Fortin.

Fortran A computer language developed for mainly scientific purposes. It is short for formula translation.

fossil fuels Coal, natural gas and petroleum, which are the hydrocarbon remains of plants and marine organisms that lived on Earth hundreds of millions of years ago.

fossils The remains in the rocks of plants and animals that lived a long time ago. Some fossils, such as shells and coral, are preserved in their original form. Often, however, the fossils have been altered, usually by the action of minerals. Petrified trees and many fossil bones are like this. Some of the most perfect fossils are those of insects trapped in amber—fossil resin.

Magnified microfossils in chalk

Foucault pendulum A long pendulum with a heavy bob used to demonstrate that the Earth rotates, first devised in 1851 by the French physicist Jean Foucault. When the pendulum is set swinging, the plane of swing gradually rotates, because of the Earth spinning.

foundations The base put down to support a structure, such as a bridge, skyscraper or dam. Building foundations often consist of a block, or raft of concrete. In soft ground, pile foundations are used.

foundry A place where metal is cast. Founding is another name for casting.

four-colour process The usual method of colour printing, which employs printing plates of four colours—yellow, cyan, magenta and black. In the process illustrations are first colour separated—split into appropriate proportions of black and the three primary colours—blue, red and green. The printing plates are then made from these separations and printed in ink of the corresponding complementary colour, ie yellow, cyan and magenta.

Fourdrinier machine The usual type of paper-making machine, named after the Fourdrinier brothers who perfected it in England in 1803. In the machine, pulp is spread over a moving wire-mesh belt, from which the water drains. The resulting damp web is then passed around steam-heated rollers and finally rolled smooth by heavy calender rollers.

fourth dimension The dimension of time. Relatively considerations require that any event must be located not only in terms of the three dimensions—location in space relative to the three perpendicular axes—but also in time. See **space-time.**

Fourdrinier paper-making machine

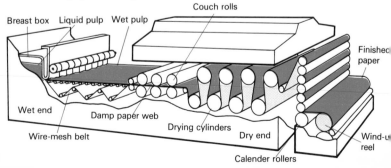

Couch rolls
Breast box Liquid pulp Wet pulp
Finished paper
Wet end
Damp paper web
Wire-mesh belt
Drying cylinders
Dry end
Wind-up reel
Calender rollers

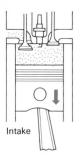

Intake

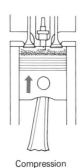

Compression

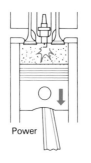

Power

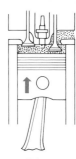

Exhaust

four-stroke cycle Or Otto cycle; the operating cycle of most petrol and diesel engines, based upon the strokes, or up-and-down movements of the pistons. It was first exploited in 1876 by the German engineer Nicolaus August Otto. In the four-stroke petrol engine cycle, petrol-air mixture is drawn into an engine cylinder on the first stroke (down) of the piston. On the second stroke (up) the mixture is compressed. Near the top of this stroke, the mixture is ignited by a spark. The mixture burns and produces gases that drive the piston down on its third, or power stroke. As the piston moves up on its fourth, exhaust stroke, it expels burnt gases from the cylinder.

Fr The chemical symbol for francium.

fractional crystallization A technique for separating a mixture of substances with different solubilities from a solution by repeated crystallization.

fractional distillation Or fractionation; a distillation method that separates liquids with different boiling points, using a fractionating column. Crude oil is separated into numerous parts, or fractions with different boiling points.

fractionating column A vertical tube packed with rings or plates or fitted with bubble trays that is used in fractional distillation. The packing allows distilled vapour to condense and evaporate at various levels throughout the column. Every time this happens the vapour becomes richer in the most volatile liquid, and eventually separation can be achieved.

Francis turbine The common type of water turbine, which has spiral vanes.

francium (Fr) A radioactive alkali metal found only in trace quantities in the Earth's crust. Its most stable isotope has a half-life of only about 20 minutes.

Frasch process A method of mining sulphur in underground deposits. In the process, superheated steam and compressed air are pumped down a set of concentric pipes. The steam melts the sulphur, and then the compressed air forces the molten sulphur to the surface.

Fraunhöfer lines Dark lines in the spectrum of sunlight, named after Joseph von Fraunhöfer, who first studied them. They are absorption lines, caused by the absorption of certain wavelengths by gases in the Sun's outer atmosphere.

free fall The motion of a body in space under the influence of gravity alone. A spacecraft orbiting the Earth is in free fall.

free radical An atom or group of atoms with an unpaired electron that may have an independent existence, even though it is normally found in combination.

freeze drying A method of drying substances by first freezing them and then subjecting them to a vacuum, whereupon the ice formed sublimes. Many foods are now freeze dried, as is blood plasma.

freezing The transformation, or change of state, from liquid to solid. For a certain pure substance at a given pressure freezing takes place always at the same temperature—the freezing point. Freezing points are generally quoted for the substance at standard atmospheric pressure, 760 mm of mercury.

freezing mixture A mixture containing ice and water and certain salts. The salts absorb heat as they dissolve, thus lowering the temperature.

freezing point See **freezing.**

French chalk An alternative name for powdered talc.

Freons A series of halogenated hydrocarbon liquids or gases used as refrigerants. Freon 12, for example, is dichlorodifluoromethane (CCl_2F_2).

frequency The rate of vibration of a wave or the number of waves that pass a given point in a certain time. It is measured in hertz, or the number of cycles (complete vibrations) per second. Frequency is inversely proportional to wavelength, so an increase in frequency means a decrease in wavelength.

frequency modulation (FM) One method of imposing a signal on a radio carrier wave. The signal is made to alter the frequency of the wave. Contrast **amplitude modulation.**

Fresnel lens A lens constructed of a number of concentric rings so as to provide a short focal length. Fresnel lenses are used particularly in lighthouses to produce a narrow concentrated light beam.

friction A force that resists relative motion between two surfaces in contact. Brakes work by friction; and machines must have their moving parts lubricated to reduce friction between them. Static friction is the force needed to start one surface sliding over another. It is greater than kinetic, or sliding friction, which is the force needed to keep the surface sliding. Rolling friction is the friction between rolling bodies, such as a wheel and the road. It is very much less than sliding friction. The coefficient of friction is the force required to push a load over a surface divided by the size of the load.

front In meteorology, the boundary between two moving air masses, one hot, one cold. It is a region of unstable weather, prone to cloud and rain.

frost Ice deposited on objects on the ground when water vapour in the air is cooled below freezing point. It forms by sublimation.

froth flotation See **flotation.**

fructose ($C_6H_{12}O_6$) A simple sugar (monosaccharide) found in fruit and honey.

fuel cell An electric cell in which chemical energy is converted directly into electricity. The most successful type, used to produce electricity on Apollo missions and currently on the space shuttle, employs the reaction between hydrogen and oxygen gases. The gases combine to form water in the presence of a catalyst.

fuel injection Introducing fuel directly into the cylinder in an internal combustion engine. This is always done in diesel engines and is now being increasingly used in petrol engines, replacing the conventional carburettor.

fuels Substances utilized to produce heat energy. The fossil fuels release energy when they are burned. Nuclear fuels release energy by fission.

fulcrum The point about which a lever pivots.

fulgurite A tube-like glassy mineral that was formed as a result of a lightning strike. Fulgurites have been found over 20 m (65 ft) long.

fuller's earth See **diatomaceous earth.**

full Moon See **phases of the Moon.**

fulminates Salts of fulminic acid (HONC), used in detonators for explosives, eg mercury fulminate.

fumaric acid See **cis-trans isomerism.**

fumarole A vent in the ground that gives off fumes, occurring mainly in volcanic regions.

fumigant A poisonous gas or vapour used to kill insects or vermin. Fumigants include hydrogen cyanide, sulphur dioxide, carbon disulphide and halogenated hydrocarbons.

fundamental particles See **elementary particles.**

fundamental unit A unit of measurement based upon a standard, such as the mass of one kilogram, the length of one metre and the length of time of one second. These standards are the fundamental units for the SI system. Other units are derived.

fungicide A chemical used to kill fungus diseases. Bordeaux mixture and lime sulphur have long been used as fungicides, but synthetic organic compounds have now largely replaced them.

furfural ($C_4H_3O.CHO$) An amber liquid used widely as a solvent and as an intermediate in the production of plastics and synthetic fibres and resins.

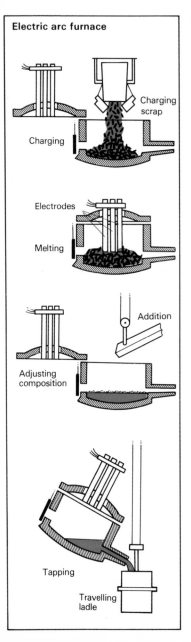

Electric arc furnace

Charging scrap

Charging

Electrodes

Melting

Addition

Adjusting composition

Tapping

Travelling ladle

furnace A structure in which fuel is burned to produce heat. Furnaces are used in conjunction with water-filled boilers to provide hot water (as in home-heating systems) or steam (as in steam turbines). They are also used in metallurgy for smelting ores and refining metals. Industrial furnaces include the blast furnace, for smelting iron ore; the open-hearth furnace, for refining steel; and the electric furnace, for refining high-quality steel.

furring In kettles and pipes, the deposition of calcium and magnesium carbonates when hard water is boiled.

fuse In an electrical circuit, a thin wire that heats up and melts when the current rises above a certain level. By melting and breaking the circuit, it protects other devices in the circuit. In explosive devices, a safety fuse is a cord containing gunpowder used to fire explosives from a distance. When one end is lit, the cord burns slowly towards the explosives, giving time for people setting the charge to get clear.

fusee A stepped drum employed in some chronometers to keep constant tension on the mainspring as it runs down, so improving the accuracy of timekeeping.

fusel oil An oily mixture of different alcohols produced during fermentation, contributing to the flavour of beers and wines.

fusible alloys Alloys with a low melting point. Some, such as Wood's metal, have a melting point below the boiling point of water. Generally, fusible alloys usually contain bismuth, cadmium, tin and lead in various proportions.

fusion Melting. With pure substances fusion occurs at a fixed temperature called the melting point. Fusion is accompanied by the absorption of energy—the latent heat of fusion.

fusion, nuclear The joining together of light atoms to form heavier ones, a process accompanied by the release of an enormous amount of energy. The fusion of heavy hydrogen, such as deuterium and tritium, into helium in the Sun and stars releases the energy that keeps them shining. It is the process exploited by man to make the hydrogen bomb. Attempts are being made with tokamaks and other devices to harness fusion energy for power generation.

G

g The acceleration due to gravity, being approximately 981 cm per second per second ($32 \cdot 2$ ft per second per second).

Ga The chemical symbol for gallium.

gabbro A common dense igneous rock rather like basalt but made up of coarser grains of plagioclase feldspar, pryoxene and olivine.

gadolinium (Gd) A rare-earth element (rd $7 \cdot 9$, mp $1,320°C$) occasionally used in nuclear reactor control rods.

gain In electronics, an increase in signal power.

galaxies Once called extragalactic nebulae; systems of stars. Stars are not sprinkled evenly throughout space but gathered into widely separated irregular, elliptical or spiral-shaped galaxies. Our own galaxy, or the Galaxy, is of the spiral type and is called the Milky Way. It is an average-sized galaxy, about 100,000 light-years across and contains approximately 100,000 million stars. Galaxies also contain a considerable amount of interstellar dust and gas, which may be concentrated into visible nebulae. Some stars are grouped together in open clusters and globular clusters.

Galaxy, the See **Milky Way**.

galena Or lead glance; the main ore of lead, being lead sulphide (PbS). It forms distinctive cubic crystals, and often contains valuable traces of silver.

Galilean telescope A simple refracting telescope, as invented by Galileo, consisting of a convex object lens and a concave eyepiece. It produces an erect image. The system is no longer used in telescopes, but is still employed in opera glasses.

gallic acid A substance found in the galls that grow on trees, particularly oaks. It is used to make writing ink and photographic developers. Its chemical name is trihydroxybenzoic acid, $C_6H_2(OH)_3COOH$.

gallium (Ga) A rare metallic element (rd $5 \cdot 9$, mp $29 \cdot 8°C$) with the second lowest melting point of all the solid elements. It resembles aluminium in properties (Mendeleyev predicted its existence and called it eka-aluminium). Gallium ansenic'? is a useful semiconductor material.

gallon A measure of liquid capacity in Imperial units, equal to 8 pints, or $4 \cdot 55$ litres. 1 British gallon = $1 \cdot 2$ US gallons.

galvanizing Coating metal, particularly iron and steel, with a protective layer of zinc. The zinc layer provides physical and also electrochemical (galvanic) protection. It forms with iron an electric cell in which it is the anode. And when the layer is broken, the zinc corrodes away rather than the iron.

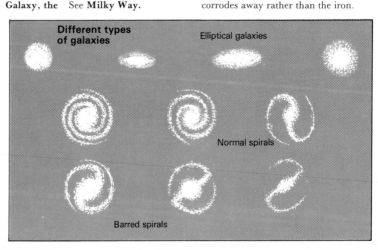

Different types of galaxies

Elliptical galaxies

Normal spirals

Barred spirals

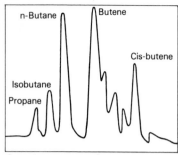

Labels on the trace, from left to right: n-Butane, Butene, Cis-butene, Isobutane, Propane

Gas chromatography trace

galvanometer A sensitive instrument used to measure electric current. One common instrument is the D'Arsonval galvanometer, which consists of a coil of fine wire suspended between the poles of a permanent magnet. When current passes through the coil, a slight magnetic field is set up, which interacts with the existing field and causes the coil to move. A needle is attached to the coil and indicates the strength of the current on a graduated scale.

gamma-rays Electromagnetic radiation of very short wavelength given out by radioactive isotopes and during nuclear fission. They are very penetrating and constitute a major hazard when handling nuclear materials.

gangue The unwanted material mined with ores, which usually needs to be removed (eg by flotation) before smelting.

Ganymede The largest moon (diameter 5,276 km, 3,278 miles) of Jupiter, and indeed the largest in the whole solar system. Bigger than the planet Mercury, Ganymede has a low density and undoubtedly contains substantial quantities of water and ice. Its surface has dark and light regions, both of which are peppered with bright icy craters.

garnet A group of common silicate minerals, whose crystals can make attractive gemstones. Garnets are usually deep red, but may also be black, green and transparent. They are hard (Mohs hardness about 7) and are widely used as abrasives. Garnets are double silicates of the general formula $3MO.N_2O_33SiO_2$, where M = calcium, magnesium, iron or manganese, and N is usually aluminium.

gas One of the states of matter. In gases the molecules are far apart and have little interaction. Gases fill completely any vessel in which they are confined. They are readily compressible. Their behaviour can be described by the gas laws. A perfect gas follows these laws well, though a real gas deviates from them at high pressures and low temperatures. A gas is described as a vapour if the temperature is below its critical temperature.

gas chromatography A method of separating and analysing minute traces of volatile substances from a mixture. It is the most sensitive method of chromatography, in which the mixture is vaporized and then flushed through a packed column by an inert gas such as nitrogen. The packing has a liquid coating, through which the different components in the mixture migrate at different rates, thus achieving separation.

gas equation An equation that combines Boyle's law and Charles's law connecting the pressure (P) of a gas with its volume (V) and the absolute temperature (T): $PV = nRT$, where R is a constant known as the gas constant and n is the number of moles of gas.

gas, fuel See **coal gas; natural gas; producer gas; water gas.**

gasket A flat sheet of asbestos, copper, cork or even treated paper for making a gas-tight joint. The head gasket in a car engine ensures a tight seal between the cylinder head and block.

gas laws Laws that describe the behaviour of a gas in relation to its volume, temperature and pressure. They include Boyle's law and Charles's law, which are combined together to give the gas equation. The laws hold well for a perfect gas but less well for real gases.

gas mask Or respirator; a face mask to protect the wearer from poisonous gases. Air enters the mask through a filter of activated charcoal, which absorbs the poison.

gasohol A mixture of petrol (gasoline) with alcohol, now being marketed in some countries.

gas oil Or diesel oil; a petroleum fraction boiling at about 300°C, used mainly for fuelling diesel engines.

gasoline See **petrol.**

Gas turbine (turboprop)

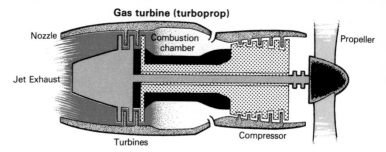

gasometer Or gasholder; an inverted cylindrical container in which gas is collected and stored above water. Gasometers are a prominent feature of gasworks.

gas turbine An internal combustion engine that burns fuel to produce hot gases, which spin a turbine to produce power. The jet engine is a form of gas turbine. In a typical gas turbine, air is taken in, compressed by a compressor and fed to a combustion chamber. Fuel is introduced and ignited in the compressed air, and the hot gases exhaust through the turbine. Power is taken from the turbine to drive the compressor.

gauge An instrument for measuring a quantity, such as pressure, temperature or water level.

gauge, railway The distance between the two rails. The standard gauge, introduced by the railway pioneer George Stephenson, is $1 \cdot 4$ metres (4 ft 8 ½ ins). About 60 % of all railway track is of this gauge. Most of the rest is narrow gauge, or less than the standard.

gauss A unit used to indicate the density of a magnetic field, or magnetic flux density. It is named after the German mathematician Karl Gauss. The Earth's magnetic field at the surface is about $0 \cdot 5$ gauss. That of Jupiter is 30 times stronger.

Gay-Lussac's law Of gaseous volumes; a law first advanced by the French chemist Joseph Gay-Lussac in 1808, which states that when gases combine to form a gaseous product, the volumes of reacting gases and of the product are in simple proportion. For example, hydrogen and oxygen combine to form water vapour in the ratio by volume of $2:1:2$ ($2H_2 + O_2 \rightarrow 2H_2O$).

Gd The chemical symbol for gadolinium.

Ge The chemical symbol for germanium.

gear A device in a machine that transmits motion between one shaft and another. It usually consists of a toothed wheel which engages with another toothed wheel, thus preventing slip. The teeth may be cut straight, as in spur gears, or at an angle, as in helical gears.

gearbox Part of a machine, such as a car, in which gears are used to change the speeds of rotation of various shafts. A large gearwheel with many teeth will drive a small gear with few teeth faster. A car gearbox contains a number of gearwheels of different sizes mounted on two parallel shafts. Movement of the gear lever causes certain pairs of meshing gearwheels to be locked on the driving and driven shafts.

gear pump See pump.

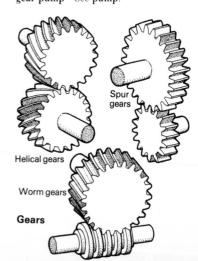

Spur gears

Helical gears

Worm gears

Gears

gegenschein Or counterglow; a faint glow in the night sky sometimes observed in the opposite direction from the Sun, caused by the reflection of sunlight from dust particles in space.

Geiger counter Or Geiger-Müller counter; a device for detecting and counting charged particles and radiation, developed by the German physicist Hans Geiger. It consists essentially of a metal cylinder (cathode) with a wire (anode) running along the middle, filled with gas at low pressure. A high voltage is applied between the cylinder and wire so that it is near discharge potential. When charged particles enter the tube, they ionize the gas and allow electricity to flow momentarily. This generates a voltage pulse, which can trigger a counting mechanism or make a noise in a loudspeaker.

Geissler tube A glass tube containing gas at low pressure, used to demonstrate the effect of electric discharge through gases. It was invented by Heinrich Geissler in 1858.

gel A type of colloid that is solid or semisolid. It consists of solid particles dispersed in a liquid, which has become highly viscous. A jelly is a gel.

gelatin A protein obtained from animal skins and bones. It is an easily digested food and is most widely used by the food industry. When dissolved in hot water and allowed to cool, it forms a gel.

gelignite A form of dynamite containing nitrocellulose and potassium nitrate as well as nitroglycerin.

gem Or gemstone; a precious stone used in jewellery, prized for its beauty. The most precious gems are diamonds, emeralds, sapphires and rubies, which are exceptionally beautiful when expertly cut, very durable, and extremely rare. Like most gems, they are mineral crystals. The weight of gems is measured in carats.

Gemini The Twins; a zodiacal constellation in the northern hemisphere whose twin bright stars are Castor and Pollux.

Gemini project A series of 10 two-man missions into space between 1965 and 1966 which the Americans carried out as a prelude to the Apollo project.

generator, electric See **alternator**; **dynamo**; **electric generator.**

genetic engineering Altering the genetic make-up of living things. Biochemists can now change the genes in simple organisms, such as bacteria, and are using the technique to produce hormones such as insulin.

geocentric Centred on the Earth. The ancient concept of the universe was geocentric.

geochemistry See **geology.**

geode A hollow stone or rock cavity, often lined with well-formed crystals. Some of the finest crystal specimens have come from geodes.

geodesic dome A dome-shaped structure formed of triangular struts, devised by R. Buckminster Fuller in the US.

Geodesic dome

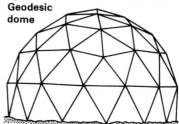

geodesy Large-scale measurement of the Earth's surface, taking into account its curvature.

geography The study of the Earth as a whole—its physical features, climate, countries, plant and animal life and population.

geology The study of the composition and structure of the Earth, its surface and the changes that take place thereon. Two major branches of geology are geophysics and geochemistry, which are concerned with the physical and chemical processes that take place in the Earth's crust.

geometric isomerism See **cis-trans isomerism.**

geometry The branch of mathematics concerned with the size and shape of plane and solid figures.

geophysics See **geology.**

geothermal energy Heat energy extracted from deep inside the Earth. In some volcanic regions and elsewhere steam and hot water are piped from geysers and underground hot-water reservoirs to provide power (as in California) or district heating (as in Iceland).

German silver

German silver An alternative name for nickel silver.

germanium (Ge) A rare semimetallic element (rd 5·3, mp 958°C) widely used as a semiconductor. Before its discovery in 1886, its existence had earlier been predicted by Mendeleyev, who called it eka-silicon.

getter A substance used in vacuum tubes to remove traces of gases remaining. Barium and magnesium are often used as getters.

G-forces The forces astronauts and pilots experience when their craft are accelerating rapidly. Space shuttle astronauts experience about 3G on take-off—three times the normal pull of gravity.

geyser A kind of hot spring that periodically shoots steam and water into the air. Some geysers are very regular. The famous 'Old Faithful' in Yellowstone National Park in the US erupts for about 4 minutes every 65–6 minutes.

giant stars Highly luminous stars of low density and large size, typically about 20 times bigger than the Sun. See **Hertzsprung-Russell diagram; red giant; supergiant.**

gibbous The shape of the Moon when between quarter and full phases. The closer planets Mercury, Venus and Mars can also appear gibbous.

giga- (G) One thousand million, as in GeV—one thousand million electron-volts.

gilding Decorating metal, plaster, glass, and so on with gold leaf or powder.

gimbals A means of suspending an object so that it can maintain its position despite the movement of the supporting frame. A gyroscope, for example, is mounted in gimbals.

glacier A moving mass of ice, which is a major agent of erosion. Valleys that once had glaciers flowing through them have a typical U-shape.

glass One of Man's most useful materials, which is transparent, easy to shape, cheap to make, and resistant to practically all chemical attack. Glass looks as if it is crystalline, but it is not. It is amorphous—lacking structure. It is actually a supercooled liquid with very high viscosity. Common glass is soda-lime glass, made from sand, soda ash and lime. Different types of glass are obtained by varying the recipe (see **borosilicate glass; lead-crystal glass**). Various methods are employed to produce flat glass (see **float glass; plate glass**).

glasses See **spectacles.**

glass fibre See **fibreglass.**

glasspaper An abrasive paper made by gluing particles of glass on paper or cloth.

glass wool A form of fibreglass blanket, used for insulation or as a filter medium.

glassine A thin, dense form of paper highly resistant to grease and air.

Glauber's salt A form of hydrated sodium sulphate ($Na_2SO_4.10H_2O$) that has laxative properties.

glaze A glassy coating applied to pottery to make it watertight and for decoration.

glider An unpowered aeroplane that glides on air currents. Sailplanes, or soaring planes, are gliders with very large wings specially designed to make the best use of normal air currents and upward-rising thermals.

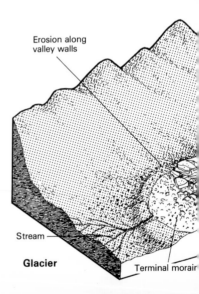

Erosion along valley walls

Stream

Glacier

Terminal moraine

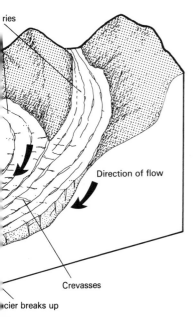

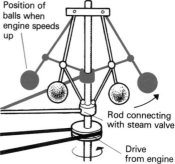

globular cluster A cluster containing up to 100,000 stars, found near the centre of galaxies. They are mainly old (population II) stars.

glucose Or dextrose; a simple sugar ($C_6H_{12}O_6$) found in grapes and other fruits, and honey. Most sugar circulates in the body as glucose.

glue An adhesive substance obtained from animal skin, bone and tissue, or from fish. It consists mainly of the protein, collagen.

gluten The protein in wheat that enables bread made from flour to rise.

glyceride An ester of an organic acid and glycerol.

glycerin See **glycerol.**

glycerol Or glycerin, a colourless, viscous liquid with a sweet taste. A trihydric alcohol ($HOCH_2CHOHCH_2OH$), it has widespread use in medicine and industry. Fats and oils are its esters (glycerides) with fatty acids. When they are treated with sodium hydroxide or potassium hydroxide during soap-making, glycerol is produced as a by-product. Glycerol is used to make the explosive nitroglycerin.

glycols Or diols; organic alcohols with two hydroxyl groups; a dihydric alcohol. The best known is the antifreeze, ethylene glycol ($HOCH_2CH_2OH$).

gneiss A medium-to-coarse grained metamorphic rock that displays prominent banding, but shows little tendency to cleave along the bands.

gnomon A simple device that uses the position or length of a shadow to indicate the time; or the arm on a sundial, which casts a shadow.

gold (Au) One of the most precious metals (rd 19·3, mp 1,063°C) of attractive pale yellow colour. Often found native, it has a high density and is resistant to most chemicals. Ordinary acids do not attack it, only aqua regia will dissolve it.

Position of balls when engine speeds up

Rod connecting with steam valve

Drive from engine

governor A device for regulating the speed of a machine which works by feedback. James Watt invented the steam-engine governor, in 1788, which works by means of rotating balls (see diagram). When the engine rotates, the balls rotate, and they are pivoted so that they move outwards and upwards as the engine speed increases. As they do so, they close the valve allowing steam into the engine, and the engine speed consequently drops. The reverse happens if the engine speed drops. So for a particular governor setting the engine runs at a more or less constant speed.

graduated scale Or graduation; a scale on an instrument divided up into suitable units.

Graham's law Of diffusion; the rate of diffusion (R) of a gas is inversely proportional to the square root of its density (d): $R \propto 1/\sqrt{d}$. It is named after the British chemist Thomas Graham.

ries

Direction of flow

Crevasses

cier breaks up

grain

grain (gr) An Imperial unit of weight. 7,000 grains = 1 pound; 1 grain = 0·0648 g.

gram (g) Or gramme; the fundamental unit of mass and weight in the cgs system. 1 g = 0·035 ounce. 453·6 g = 1 pound.

gram-equivalent The equivalent weight in grams.

gram-molecule See **mole**.

gramophone Or record player; a device for reproducing sound from a record, or disc, invented (as the phonograph) by the prolific American inventor Thomas Edison. The essential parts are a turntable for spinning the record at the correct speed; a pick-up for picking up the vibrations from the record grooves and translating them into equivalent electrical currents; and an amplifier for strengthening these currents for feeding to a loudspeaker. Most gramophones are now stereophonic, feeding signals to two speakers.

granite One of the commonest igneous rocks, formed when magma cooled slowly beneath the surface. Its crystals, typically of feldspar, quartz and mica, are easily visible.

granulation A constantly changing pattern of bright patches on the Sun's surface, caused by hot gas welling up from the Sun's interior.

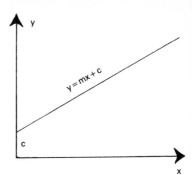

$y = mx + c$

graph A diagram that relates two variable quantities, plotted usually with right-angled, or Cartesian axes. Plots on the graph may be straight-lined or curved. Simple lines and curves can be described by equations.

graphite One form in which carbon occurs naturally, as soft black flakes. Unusually for a non-metallic substance, it conducts electricity and is used for making electrodes for batteries and furnaces. It is used as a moderator in some nuclear reactors. Contrast **diamond.**

graptolite A common fossil found in sedimentary rocks, between about 250–550 million years old. They often have the appearance of a sawblade.

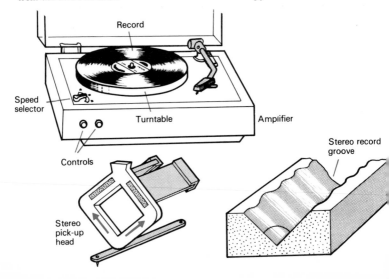

Record

Speed selector

Turntable

Amplifier

Controls

Stereo pick-up head

Stereo record groove

graticule A network of lines. In a microscope and telescope eyepiece they act as a scale or frame of reference. On a map the graticule is the network of lines that show latitude and longitude.

graver See **burin.**

gravimetric analysis A form of quantitative chemical analysis in which the substance to be estimated is converted to a compound of known composition, which can be weighed.

gravitation One of the fundamental forces in the universe, the attraction between any two masses. We call the Earth's attraction gravity. Isaac Newton put forward his law of gravitation in 1687, which states that every particle of matter attracts every other particle of matter with a force (F) proportional to the product of their masses (M_1, M_2) and inversely proportional to the square of their distance (R) apart: $F = GM_1M_2/R^2$, where G is the gravitational constant.

graviton A hypothetical quantum of gravitational energy, analogous to the photon of electromagnetic energy.

gravity The Earth's gravitational attraction. The force of gravity is responsible for a body's weight on Earth. The weight of a body is equal to the product of its mass and the acceleration due to gravity.

gravure A printing method that makes use of plates that hold ink in recesses; an intaglio process.

grease A thick, oily substance used for lubrication. In the past, greases were obtained from animal fats (eg goose grease), but most greases today are obtained from petroleum.

great circle A circle on the surface of a sphere (particularly the Earth), whose plane cuts through the centre. Great circle routes by sea or air are the shortest routes between any two points on the Earth's surface.

greenhouse effect An atmospheric condition in which carbon dioxide prevents heat escaping from a planet. Venus has a thick atmosphere where this effect is marked. There are fears that increasing concentrations of carbon dioxide in the Earth's atmosphere might produce a greenhouse effect that could have catastrophic consequences, such as melting the polar ice caps.

green vitriol See **ferrous sulphate.**

Greenwich Mean Time (GMT) Time along the Greenwich meridian, which is taken as a time standard for the whole world.

Greenwich meridian The prime meridian of the Earth, 0° longitude, which passes through Greenwich in London. All other meridians of longitude are measured in degrees east or west of Greenwich.

Gregorian calendar See **calendar.**

grid In the power industry, a network of electricity transmission lines that distribute power to a region or a country.

grid reference A means of identifying the location of a place on a map by reference to a grid of horizontal and vertical lines.

Grignard reagents Organic compounds containing magnesium that are widely used in chemical synthesis. They are made by reacting magnesium with an organic halide, such as ethyl bromide. They have the general formula RMgX, where R is an alkyl or phenyl group and X is a halogen.

grinding A machining operation in which an abrasive wheel or belt is used to sharpen tools or shape metal.

ground See **earth.**

ground-effect machine A machine such as a hovercraft, which relies on interaction with the ground to achieve lift.

ground state The lowest and most stable energy state of a particle.

ground waves Radio waves that travel directly from transmitter to receiver, without being reflected from the sky.

group See **periodic table.**

group relationship The chemical relationship that exists between elements in the same vertical group of the periodic table. It is the consequence of a similarity in electronic structure.

grout In constructional engineering, a fluid cement mixture injected into rocks to increase their strength.

GRP An abbreviation for glass-reinforced plastic, a moulding material now widely used in the construction of boat hulls, for example.

guano A useful natural fertilizer made up of the droppings of birds or bats. It is high in nitrogen and phosphorus. The largest deposits of bird guano are found on Peruvian islands.

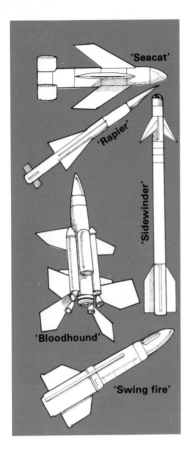

'Seacat'

'Rapier'

'Sidewinder'

'Bloodhound'

'Swing fire'

gun cotton A type of nitrocellulose used as an explosive.

gunmetal An alloy of copper (88%), tin (10%) and zinc (2%) that is very durable and corrosion resistant. It was formerly used to cast cannon.

gunpowder Or black powder; the original explosive substance, invented about 1,000 years ago by the Chinese. It is a mixture of about 75% saltpetre (potassium nitrate), 15% powdered charcoal and 10% sulphur.

gutta-percha A kind of natural plastic, made by processing the latex from certain Malaysian trees. It is soft, pliable and water-resistant. It was once widely used for electrical insulation, especially for marine cables.

gypsum One of the commonest minerals, being hydrated calcium sulphate, $CaSO_4.2H_2O$. It is often chalky in appearance and texture, but also occurs in the form of silky crystals as selenite, or satin spar; and as fine-grained alabaster. It loses part of its water of crystallization when heated to become plaster of Paris.

gyrocompass A compass that uses the constant directional properties of a spinning gyroscope to indicate direction. Unlike a magnetic compass, a gyrocompass is unaffected by local magnetic fields. Once set spinning and pointing due north, the gyrocompass will continue to do so.

gyroscope A device containing a spinning wheel, or rotor, mounted in gimbals so that it is unaffected when the supporting frame is moved. It is the inertia of the rotor that gives the gyroscope its fascinating properties.

guided missiles Projectiles that are guided to their target after launching. Some ground-to-ground anti-tank missiles and torpedoes are guided by signals sent through a wire. Many ground-to-air and air-to-air guided missiles contain homing devices, such as heat-seeking sensors, which enable them to home in on their targets. Others, such as the Cruise missile, have a built-in inertial guidance system which is less subject to countermeasures.

gums Sticky substances exuded by plants, often used as adhesives. A widely used gum is gum arabic, obtained from acacia trees.

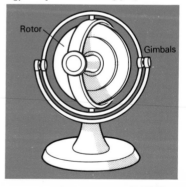

Rotor

Gimbals

H

H The chemical symbol for hydrogen.

Haber process Or Haber-Bosch process; the main method of manufacturing ammonia, devised by Fritz Haber in 1908. In the process nitrogen (from liquid nitrogen) and hydrogen (from methane) are passed over a heated iron catalyst (temperature about 500°C, pressure over 200 atmospheres), when the following reaction takes place:

$$N_2 + 3H_2 \rightleftharpoons 2NH_3$$

Hadar (β Centauri) the second brightest star (magnitude $0 \cdot 6$) in the constellation Centaurus and tenth brightest in the heavens. It lies about 390 light-years away.

hadron An elementary particle like the proton, neutron and meson, which takes part in the strong interaction in the atomic nucleus.

haematite A major iron oxide (Fe_2O_3) ore, so called because of its usual deep blood-red colour.

hafnium (Hf) A comparatively rare transition metal (rd $13 \cdot 3$, mp $2,000°C$), whose most important use is in nuclear reactor control rods because it readily absorbs neutrons.

hahnium (Ha) An artificial radioactive element (at no 105) whose most stable isotope has a half-life of only about 3 seconds.

hailstones Hard balls of ice that fall usually during a thunderstorm. They have a characteristic layered structure formed as they travel up and down in the thundercloud before they fall.

Hale telescope A 508-cm (200-inch) diameter reflecting telescope at Palomar Observatory in California, which was, from its completion in 1948 until the early 1970s, the largest reflector in the world. It is named after the American astronomer George Ellery Hale.

half-life A measure of a substance's radioactivity. It is the time taken for half a given sample of radioactive material to decay, that is, for half its nuclei to change into different nuclei. It can range from a fraction of a second for astatine to over 4,500 million years for uranium.

half-tone process A method of making printing plates to reproduce pictures. The picture is photographed through a screen, which breaks up the image into patterns of tiny dots. Shades of grey are represented by different densities of black dots.

halides Compounds containing halogen elements, particularly the salts of the halogen acids, such as hydrofluoric and hydrochloric acids.

halite Or rock salt; a common mineral form of sodium chloride (NaCl), found throughout the world in massive underground deposits.

Hall-Héroult process The electrolytic method by which aluminium is extracted industrially. It was developed independently and at the same time (1886) in the US by Charles M. Hall and in France by Paul Héroult, who coincidentally were both born and died in the same years (1864–1914).

Halley's comet

Halley's comet The most famous of all comets, first sighted as far back as 240 BC, and depicted on the Bayeux Tapestry at its 1066 appearance. Its 76–7 year period was first established by English astronomer Edmond Halley. It appeared in 1910, and was next detected in 1982 as it began closing for its 1986 return.

halo A luminous ring occasionally observed around the Sun or Moon, caused by the refraction and reflection of light by ice crystals in thin cloud.

halogens The highly reactive elements that occupy Group 7 of the periodic table, comprising fluorine, chlorine, bromine and iodine. (The unstable astatine is also a halogen, but is only found in negligibly small amounts.)

hardness Or Mohs. hardness; a measure of the hardness of a mineral on a scale originally devised by Friedrich Mohs in the 19th century.

hard water Water containing dissolved minerals that prevents soap lathering during washing. Compounds of calcium and magnesium are the main culprits, combining with soap to form an insoluble scum. Temporary hardness is caused by the presence of bicarbonates, but this can be removed by boiling the water, whereupon the insoluble carbonates are deposited (they form the fur in kettles). Permanent hardness, caused by calcium and magnesium sulphates, cannot be removed by boiling and has to be removed by other means (see **water softening**).

hardware The equipment or machines involved in computing; contrast **software**.

harmonics Of a wave motion such as sound; related waves that are superimposed on the main one, or fundamental. A violin string, for example, vibrates as a whole to produce a main note—the fundamental. But superimposed on it are a number of others caused by the string vibrating at frequencies that are fractions of the fundamental.

harvest Moon The full Moon occurring close to the autumnal equinox when it rises for several nights at about the same time and should be bright enough for harvesting to continue after sunset.

H-bomb See **hydrogen bomb**.

He The chemical symbol for helium.

head In hydraulics, the height of water above a certain point, which provides potential energy.

heat A form of energy that is a reflection of the kinetic energy possessed by atoms and molecules. It is measured in joules, calories or Btus.

heat exchanger A device in which heat from one fluid is passed to aother. Most domestic hot-water tanks are heat exchangers, the water being heated by pipes through which hot water from the boiler is circulated.

heat shield The coating of the outside of a manned spacecraft that dissipates the heat developed during re-entry. On early spacecraft, such as Apollo, the heat shield worked by ablation—boiling away. The heat shield on the spece shuttle is made up mainly of silica tiles that re-radiate the heat gradually.

heat transfer The transfer of heat, which takes place from a hotter to a cooler body by means of conduction, convection or radiation, or combinations of them.

heat treatment The controlled heating and casting of metals to produce more favourable properties. See **annealing; quenching; tempering**.

Harmonics

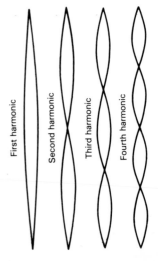

First harmonic Second harmonic Third harmonic Fourth harmonic

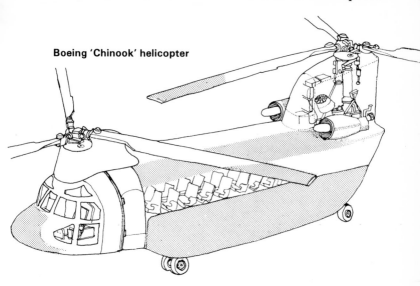

Boeing 'Chinook' helicopter

Heaviside layer Or Heaviside-Kennelly layer; a region of the Earth's ionosphere that reflects radio waves and makes possible long-distance radio communications.

heavy chemicals Those produced in very large quantities by the chemical industry, such as sulphuric acid, caustic soda, sodium carbonate and superphosphate.

heavy hydrogen Isotopes of hydrogen containing one or two neutrons as well as a proton; see **deuterium; tritium.**

heavy spar See **barytes.**

heavy water Or deuterium oxide (D_2O); water composed of the heavy hydrogen isotope, deuterium, and oxygen. It is used as a moderator in some nuclear reactors.

hectare A unit of area in the metric system, equivalent to $2 \cdot 47$ acres; 1 acre = $0 \cdot 4$ hectares.

hecto- A prefix meaning 100, as in hectogram, 100 g.

Heisenberg uncertainty principle In nuclear physics, it is impossible to determine both the position and the momentum of a particle accurately at the same time because of the wave nature of matter. It was first stated by the German physicist Werner Heisenberg in 1927.

helicopter A versatile VTOL aircraft that derives propulsion and lift from a set of horizontally rotating blades. The rotor blades have an aerofoil cross-section, and their pitch is altered to achieve a change in thrust for movement in different directions. Helicopters with a single main rotor also have a small vertical rotor at the tail to counteract the tendency of the body of the craft to spin round. Twin-rotor helicopters have their rotors spinning in opposite directions.

heliocentric Centred on the Sun; the solar system is heliocentric.

heliograph A former method of long-distance communication that used mirrors to flash signals with reflected sunlight.

helium (He) The next lightest gas after hydrogen and one of the noble gases. It has the lowest boiling point of any element, $-269 \cdot 7°C$ or $3 \cdot 5$ K. It is obtained mainly from natural gas.

helix A spiral curve that lies on the surface of a cone or cylinder. DNA has a double-helix structure.

hemisphere Half of a sphere; particularly half of the sphere of the Earth, such as the northern hemisphere, north of the equator; and the southern hemisphere, south of the equator.

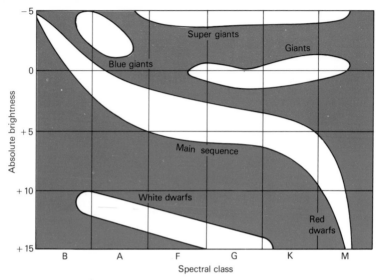

Absolute brightness

Spectral class

hemp   A tall plant whose stems contain fibres for making coarse fabrics and twines. Hemp seeds are a source of the narcotic drug hashish and marijuana.

henry   A unit of inductance, being the inductance of a circuit in which electric current changing at a rate of 1 amp per second induces an emf of 1 volt.

Henry's law   Of solubility; at a given temperature, the mass of gas that dissolves in a given volume of liquid is directly proportional to its pressure.

herbicide   A substance that kills plants; a weedkiller. Selective herbicides can kill some plants without harming others.

heroin   A powerful painkilling drug derived from opium. It is an alkaloid, and its regular use quickly leads to addiction.

hertz   A measure of frequency; number of cycles per second.

Hertzsprung-Russell diagram   A diagram that shows the relationship between the absolute brightness of a star and its spectral class or temperature. When data are plotted on the diagram, the stars fall into distinct categories (see diagram).

Hess's law   No matter in how many stages a chemical reaction may be carried out, the total heat absorbed or evolved is the same.

heterocyclic compounds   Mostly organic compounds containing rings that include atoms other than carbon. Pyridine is an example:

$$\begin{array}{ccc} & \text{H} & \\ & \text{C} & \\ \text{HC} & & \text{CH} \\ | & & \| \\ \text{HC} & & \text{CH} \\ & \text{N} & \end{array}$$

heterogeneous   Of non-uniform composition.

hexagon   A plane figure with six sides; a regular hexagon has equal sides and equal interior angles.

Hf   The chemical symbol for hafnium.

Hg   The chemical symbol for mercury, from the Latin hydrargyrus.

high fidelity (hi-fi)   The attributes of a high-quality sound reproduction system, which reproduces sounds faithfully with little distortion. The standard hi-fi set-up includes electronically matched turntable, tape deck and tuner, with an amplifier (and often a preamplifier) and two or more loudspeakers, each with two or more individual cones for reproduction of a particular frequency range.

high tension   High voltage.

high explosive   A powerful explosive, such as dynamite and TNT, which needs to be detonated.

histamine An organic substance that is released into the bloodstream as a result of an allergy attack. Antihistamines are drugs prescribed to prevent the release of histamine in, for example, hay-fever attacks.

histogram A simple graphical way of presenting statistics in which quantities are represented by vertical bars.

Ho The chemical symbol for holmium.

hole In electronics, the absence of an electron in a semiconductor. The progressive filling up of holes by electrons but creating other holes, can be regarded as equivalent to the movement of holes with a positive charge.

holmium (Ho) One of the lanthanides, or rare-earth elements (rd 8·8, mp 1,500°C).

Holocene Epoch The geological time span we are in at present, which has lasted for about 10,000 years, marked by a generally warm climate.

holography A kind of three-dimensional photography using neither camera nor lenses but beams of laser light. A hologram holds the image in the form of interference patterns. These patterns are made on a piece of film when part of a split laser beam reflected from an object interferes with the other part, which illuminates the film directly. The image is recreated by shining laser light through the hologram.

homogeneous Of uniform composition.

homologous series A series of compounds, each of which differs by a simple structural unit. They are most common in organic chemistry. The paraffin hydrocarbons, or alkanes, form a homologous series, in which the structural unit is CH_2:
CH_4, C_2H_6, C_3H_8, C_4H_{10}, etc.

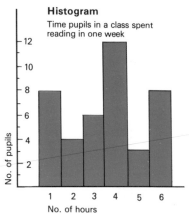

Histogram
Time pupils in a class spent reading in one week

No. of pupils (vertical axis)
No. of hours (horizontal axis)

homopolymer A polymer formed from the same monomer. Polyethylene is a homopolymer, formed from ethylene monomer. Compare **copolymer**.

honing A machining process for finishing cylinder bores to high accuracy, using an abrasive rotating head.

Hooke's law Of elasticity; named after English physicist Robert Hooke. The amount an elastic body deforms (strain) is proportional to the deforming force (stress). And when the stress is removed, the body regains its original dimensions. Hooke's law is true of metals and other materials only for small stresses and strains. When the stress rises above a certain limit—the elastic limit—permanent deformation results.

horizon The imaginary line where the sky seems to meet the land or sea. To an average-sized person on the beach, the horizon out to sea is nearly 4 km (2½ miles) away.

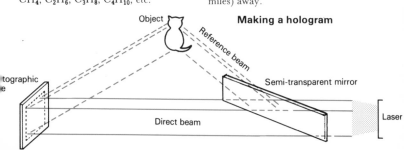

Making a hologram

Object
Reference beam
Photographic plate
Semi-transparent mirror
Direct beam
Laser

hormones

hormones Chemicals produced by glands in the body that control metabolism and all kinds of body activities. Most are steroids or proteins. Some of the main hormones are insulin, adrenalin, cortisone and the sex hormones testosterone and progesterone.

hornblende A series of common minerals consisting largely of calcium, magnesium and aluminium silicates.

horology The science of measuring time, including the design and construction of clocks and watches.

horoscope In astrology, a drawing that shows the relative positions of the Sun, Moon and planets at the time a person is born, from which astrologers try to make predictions about the future.

Horsehead nebula A dark nebula in the constellation Orion that looks like a horse's head; about 1,000 light-years away.

horsepower A common unit of power, first adopted by James Watt as equivalent to the power of a working horse. It is a rate of working of 30,000 foot-pounds a minute. 1 hp = 746 watts. The output of an engine is usually given in brake horsepower (bhp).

hovercraft Or air-cushion vehicle; an amphibious craft that glides over a surface on a 'cushion' of compressed air. The air is compressed by a fan and delivered through inward pointing jets underneath the craft. To reduce the rate of escape of the air, many hovercraft have flexible 'skirts' around their base. They are generally powered and steered by means of backward facing propellers.

hp See **horsepower.**

Hubble's constant In astronomy, the ratio between the distances of the external galaxies and the velocity at which they are apparently receding.

humidity The amount of moisture in the atmosphere. The amount of moisture in the air, compared with the amount required to saturate it is termed the relative humidity.

hundredweight (cwt) a unit of weight in Imperial units, being 112 lb (50·8 kg).

hurricane An intense cyclone that occurs in the Caribbean region. Winds spiral violently around the hurricane at speeds of up to 300 km/h (200 mph). In the centre is a calm region, the eye, about 30 km (20 miles) across.

Hyades A famous open cluster of stars in the constellation Taurus, around the prominent red star Aldebaran.

hydrate A chemical compound containing water of crystallization. Ordinary copper sulphate is a hydrate: $CuSO_4.5H_2O$.

hydraulic press A machine in which power is applied by means of hydraulic (liquid) pressure. The English engineer Joseph Bramah patented the hydraulic press in 1795. It works by applying a small force on a liquid, by means of a small piston, and then the liquid transmits the pressure to a large piston, which applies a magnified force.

hydraulics A branch of engineering concerned with utilizing liquids, especially water. It is concerned particularly with the way they flow and transmit pressure. Many devices, such as the hydraulic press and the hydraulic brakes of cars, rely on Pascal's law.

Hovercraft

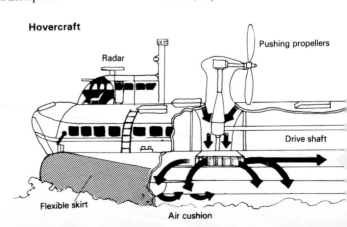

Radar · Pushing propellers · Drive shaft · Flexible skirt · Air cushion

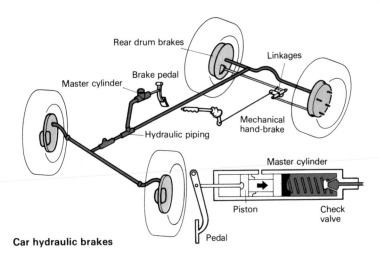

Car hydraulic brakes

hydrazine A fuming liquid that is a powerful reducing agent. It smells rather like ammonia, from which it can be derived, and its methyl derivatives (such as UDMH) make excellent rocket fuels.

hydride A compound of hydrogen with another element, such as sodium hydride, NaH, and silane, SiH_4.

hydriodic acid A strong acid formed when hydrogen iodide (HI) gas dissolves in water; the parent of the iodides.

hydrobromic acid A strong acid formed when hydrogen bromide (HBr) gas dissolves in water; the parent of the bromides.

hydrocarbons A very large group of organic compounds made up of hydrogen and carbon only. Natural gas and petroleum are made up almost entirely of hydrocarbons. The two main types of hydrocarbons are the aliphatics, whose molecules are made up of chains of carbon atoms; and the aromatics, whose molecules contain one or more benzene rings.

hydrochloric acid Once called muriatic acid; one of the three main mineral acids (with sulphuric and nitric), being a solution in water of hydrogen chloride (HCl) gas. It is the parent of the chlorides.

hydrocyanic acid See **prussic acid**.

hydrodynamics The science of liquids in motion, particularly water.

hydroelectric power (HEP) The production of electricity using the energy of flowing water. In a typical hydroelectric power plant, water trapped in a reservoir behind a dam is channelled through water turbines to which electricity generators are coupled. The biggest hydroelectric power station in the world is at Krasnoyarsk in Siberia. It has a proven output of over 6,000 megawatts. See also **tidal power**.

hydrofluoric acid A strong acid formed when hydrogen fluoride (HF) gas dissolves in water. It is the only acid that attacks glass.

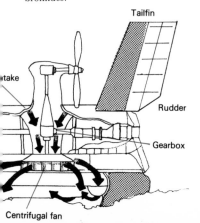

117

hydrofoil

hydrofoil An underwater wing (compare **aerofoil**) attached to the hull of boats to make them lift out of the water when they travel at speed. By lifting out of the water, hydrofoil craft are subject to minimum water resistance and can thus travel at high speeds—over 120 km/h (75 mph).

hydrogen (H) The lightest and structurally the simplest of the chemical elements (at no 1). It is the most abundant element in the universe, making up most of the mass of stars and found scattered between the stars in nebulae. It is the nuclear fusion of hydrogen that keeps the stars shining. On Earth, hydrogen is most commonly found combined with oxygen as water (H_2O). Hydrogen is flammable and makes an excellent fuel and, when liquid (at $-253°C$), a rocket propellant.

hydrogenation A chemical reaction in which hydrogen is added to an unsaturated organic compound in the presence of a catalyst, usually nickel. Hydrogenation is utilized in petroleum refining and in making magarines.

hydrogen bomb Or H-bomb; a nuclear bomb that exploits the fusion of heavy hydrogen into helium as an energy source. In an H-bomb, a small atomic bomb is used as a trigger to produce high enough temperatures for the fusion of the hydrogen (as deuterium and tritium) to occur. H-bombs typically have the explosive force of several megatons (million tons) of TNT.

hydrogen bond A weak type of intermolecular bonding found in the water molecule and elsewhere. It forms between hydrogen atoms and electronegative atoms such as oxygen. It results, for example, in hydrogen-bonded compounds being less volatile than they should. Theoretically, water should be a gas at room temperature, like the chemically related hydrogen sulphide.

hydrogen bromide See **hydrobromic acid.**

hydrogen chloride (HCl) A colourless gas with a pungent smell, which fumes in moist air. It dissolves readily in water to form hydrochloric acid.

hydrogen cyanide See **prussic acid.**

hydrogen fluoride See **hydrofluoric acid.**

hydrogen iodide See **hydriodic acid.**

hydrogen ions (H^+) The ions present in acid solutions that give them their characteristic properties. The pH of a solution is defined as the log_{10} of the reciprocal of the hydrogen-ion concentration. A hydrogen ion is a hydrogen atom without its single electron; in other words, it is a proton.

hydrogen peroxide (H_2O_2) A viscous liquid that is a powerful oxidizing agent. It decomposes on heating into water and oxygen. It is commonly found as a water solution (eg for domestic bleaching).

hydrogen sulphide (H_2S) A poisonous gas with a disgusting smell of bad eggs. It is flammable and burns in air with a blue flame. Readily prepared by the action of acid on iron pyrites, it is useful in the laboratory as a reagent in chemical analysis.

hydrolysis A decomposition, sometimes reversible, brought about by water, as in the general equation: $AB + H_2O \rightarrow BH + AOH$. Acids, bases, salts and organic compounds such as esters may be hydrolysed. In the body the hydrolysis of carbohydrates, fats and proteins forms the basis of digestion, the processed being catalysed by enzymes.

hydrometer A device for measuring the density of liquids, consisting usually of a weighted glass bulb with a long stem, on which a scale is marked. It floats higher or lower according to whether the liquid has a high or low density.

hydrophilic, hydrophobic Water-loving, water-hating. Terms used, for example, to describe the action of detergent molecules.

hydrophone An underwater microphone used, for example, in sonar.

hydroponics The growing of plants in salt solution rather than soil.

hydroquinone Or quinol; an organic reducing agent, $C_6H_4(OH)_2$, widely used in photographic developing.

hydrosphere The total waters in and on the Earth's crust, including the oceans, rivers and lakes, ground water, and water vapour in the atmosphere, together with water locked in glaciers and the polar ice caps. It is estimated that it amounts to about 1,500 million cubic kilometres (360 million cubic miles).

hydrostatics The branch of fluid mechanics concerned with the forces in liquids at rest.

The world's most advanced hydrofoil craft, the Boeing Jetfoil, passing under London's Tower Bridge.

hydroxides

hydroxides Inorganic compounds containing the hydroxyl group or ion, OH^-. Metal hydroxides are bases and, if soluble, are termed alkalis.

hydroxyl group The group —OH, which is present in inorganic metal hydroxides and in organic alcohols and acids. Hydroxides dissolve to form hydroxyl ions (OH^-), which give solutions their alkaline characteristics.

hygrometer An instrument for measuring the humidity of the air. The common hair hygrometer uses the principle that human hair lengthens when it absorbs moisture. The psychrometer is another common type of hygrometer.

hygroscopic compound One that tends to absorb moisture from the air but does not dissolve. Compare **deliquescence**.

hyperbola An open curve that is one type of conic section.

hyperdermic syringe A device used to administer drugs under the skin. It was invented by the French doctor Charles Pravaz in 1853.

hypergolic propellants Those that ignite spontaneously when they are mixed, eg hydrazine and nitrogen tetroxide.

hypermetropia See longsightedness.

hyperons A group of short-lived elementary atomic particles that have a greater mass than the neutron. They decay into protons or neutrons.

hypersonic flight Flight at speeds over about five times the speed of sound.

hypo The compound sodium thiosulphate, $Na_2S_2O_3.5H_2O$, which is widely used in photography as a fixing agent.

hypochlorites Salts (usually of sodium and potassium) of hypochlorous acid, $HClO$, which are powerful oxidizing and bleaching agents.

hypotenuse The longest side of a right-angled triangle.

hypothesis A theory advanced to explain observed facts; a provisional explanation.

hypsometer An instrument for determining the boiling point of water, and hence to find the external atmospheric pressure and thus altitude.

hysteresis A phenomenon that occurs in a magnetic material when it is being magnetized, in which the degree of magnetization lags behind the field that is causing it.

I

I The chemical symbol for iodine.

ice Water in its frozen state, which forms at $0°C$. Unusually for a substance, ice is less dense than liquid water, a result of hydrogen bonding.

ice ages Periods of the Earth's history when large areas were covered in vast sheets of ice. Over the past million years there have been four major ice ages (sometimes called together the Ice Age). Between them the ice retreated and warm interglacial periods occurred. It is thought that we are presently in a warmer interglacial period.

iceberg A huge floating mass of ice, formed typically when a glacier reaches the sea and breaks apart. Most of the mass (about 85%) of an iceberg is hidden beneath the surface, leaving only a deceptively small tip showing.

Iceland spar A clear form of calcite ($CaCO_3$) that displays the phenomena of double refraction.

ideal gas See **perfect gas**.

igneous rocks The 'fire-formed' rocks, which resulted from the solidification of magma either on the surface (extrusive) or underneath (intrusive). Basalt is an extrusive, or volcanic, igneous rock; granite an intrusive, or plutonic igneous rock.

ignition coil A simple transformer used in the ignition system of a car to produce a high voltage for the sparking plugs. Low-voltage battery current is passed through the primary coil, which has few windings, and is periodically interrupted by the contact-breaker. This induces a high voltage in the many windings of the secondary coil.

ilmenite A black heavy mineral; the main ore of titanium. It is a mixed iron and titanium oxide, $FeTiO_3$.

ILS See **instrument landing system**.

image In optics, the representation of an object or scene formed in an optical instrument. A real image can be made visible on a screen, while a virtual image cannot, even though it can be seen by the eye.

immiscible liquids Liquids that do not mix, such as oil and water.

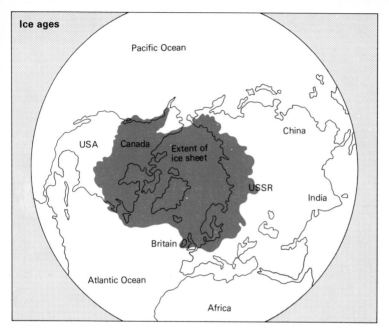

Ice ages

Pacific Ocean

China

USA　Canada　Extent of
ice sheet

USSR

India

Britain

Atlantic Ocean

Africa

impedance In electricity, the total resistance to the flow of electric current, including both electrical resistance and reactance.

Imperial units A system of units once widely used in English-speaking countries, which includes the inch, foot, yard, mile, pound and gallon. It has now been largely replaced with units based on the metric system.

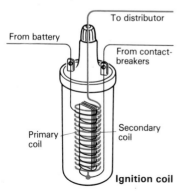

To distributor

From battery

From contact-
breakers

Primary
coil

Secondary
coil

Ignition coil

impermeable Not allowing water or other liquids to flow through.

implosion A bursting or collapse inwards; the opposite of an explosion.

impulse In physics, a force that acts for a short time; equal in magnitude to the change in momentum it produces.

In The chemical symbol for indium.

incandescence The light emitted when a body is heated to a high temperature.

incandescent lamp See **electric-light bulb.**

inch (in) A unit of length in Imperial units. $1 \text{ in} = 2 \cdot 54 \text{ cm}$; $1 \text{ cm} = 0 \cdot 39 \text{ in}$.

Indian ink Or Chinese ink; a dense black ink containing fine particles of carbon.

indicator A substance used in volumetric analysis to mark, or indicate, the end point of a reaction by changing colour. Litmus, methyl orange and phenolphthalein are widely used indicators for simple acid-base titrations. See **universal indicator.**

indigo A blue dye used since ancient times, once obtained from the indigo plant, but now made synthetically. Its formula is $C_{16}H_{10}N_2O_2$.

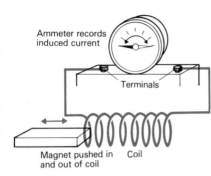

Ammeter records induced current

Terminals

Magnet pushed in and out of coil   Coil

indium (In) A rare soft metal (rd 7·3, mp 156·6°C), named after the indigo colour some of its compounds make when burned in a flame.

induction coil See **ignition coil.**

induction, electromagnetic The generation of an emf in an electrical circuit in a changing magnetic field. The emf is proportional to the rate at which the field is changing. The phenomenon of electromagnetic induction, discovered independently by Michael Faraday and Joseph Henry, is the basis of the electric generator and motor.

induction motor An alternator in which the current in the armature is induced, rather than supplied.

Industrial Revolution A period of history marked by the widespread introduction of machines in manufacturing and the beginnings of the factory system. In Britain the Industrial Revolution began in the early 1700s; it happened at different times elsewhere.

inert Lacking in chemical activity.

inert gases See **noble gases.**

inertia The property by which a body resists any disturbance of its state of rest or existing motion. Newton's first law of motion is often called the law of inertia.

inertial guidance A method of navigation, used for example by submarines and missiles, which relies on data from accelerometers and gyroscopes along three axes to determine direction and distance travelled. The data are fed into a computer, which works out a position.

infinitesimal Smaller than any quantity you can imagine, but greater than zero.

infinity A number larger than anyone can imagine, symbol ∞.

inflammable Able to burn; the correct term is flammable. Something that does not burn should be termed nonflammable.

infrared radiation Electromagnetic radiation, whose wavelength is longer than visible light and lies beyond the red end of the spectrum. We cannot see it, but we can feel it as heat.

inhibitor A negative catalyst.

ingot A metal casting produced after smelting of convenient size for subsequent processing.

injection moulding A common method of shaping plastics by squirting molten plastic into a water-cooled mould.

ink A solution or paste used for writing or printing. Writing inks may be made from galls, iron salts and dyes. Printing inks are made from dyes or pigments in an oil or varnish base.

inorganic chemistry The branch of chemistry that studies all elements and their compounds except most carbon compounds. Carbon compounds are the province of organic chemistry.

insecticide A substance that kills insects. Some, such as pyrethrum and derris, are derived from plants. But most are now synthetic organic compounds, such as DDT and dieldrin, many of which have the drawback of being toxic to higher life.

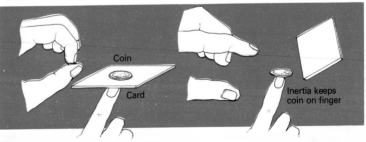

Coin

Card

Inertia keeps coin on finger

insoluble Cannot be dissolved.

instrument landing system (ILS) A radio guidance system that helps aeroplanes land in conditions of bad visibility. It uses two radio transmitters (localizer and glide slope), which transmit narrow beams that indicate the correct glide path. Instruments on the plane lock on these beams and indicate to the pilot how he should manoeuvre to hit the glide path.

insulator A substance that prevents the flow of heat or electricity. Cork, fibreglass and asbestos are good heat insulators; rubber, plastics and glass are good electrical insulators.

insulin A hormone that controls the way the body uses sugar. It is a protein, produced in the pancreas. Lack of insulin causes the disease diabetes.

intaglio process One in which a design is cut into a substance. Gravure is an intaglio printing process.

integers The set of all whole numbers (positive and negative) and zero.

integrated circuit A complete electronic circuit formed in a semiconductor crystal (usually silicon). The various components (transistors, diodes, etc) are formed in situ by 'doping' the appropriate areas with chemicals. The process of large scale integration (LSI)—incorporating many components in a small area—made possible the silicon chip.

integration In mathematics, an operation performed in the calculus, which for example gives a method of finding the area under a curve.

Intelsat The name of a series of communications satellites financed by more than 100 members of the International Telecommunications Satellite Organization. Several Intelsat IV and V satellites are in stationary orbit over the Atlantic, Indian and Pacific Oceans.

interaction In nuclear physics, a force that exists between atomic particles. The electromagnetic interaction is responsible for the attractive force between electrically charged or magnetized bodies. The strong interaction is the force that binds the protons and neutrons together in the nucleus. The weak interaction, some 10^{21} times weaker, also exists between elementary particles and gives rise to such phenomena as radioactive beta decay.

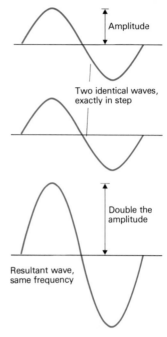

Amplitude

Two identical waves, exactly in step

Double the amplitude

Resultant wave, same frequency

interference Of waves; the phenomenon that results when two waves interact. If two identical waves are exactly in step, then the amplitude of the resulting interference wave is doubled. Conversely, if the two waves are 180° out of step, then they cancel each other out. In sound, interference can produce rhythmic beats. In light, it gives rise to alternate light and dark interference fringes. See also **Newton's rings.**

interferometer An accurate instrument that uses interference effects for measurement. There are opical, acoustical and radio interferometers. Radio interferometry is the basis of operation of many radio telescopes.

intermediate chemical A product of the chemical industry that requires further processing into a finished product.

internal combustion engine (ICE) A heat engine in which fuel is burned in an enclosed space to produce gases that provide power as they expand. The petrol, diesel and jet engines are familiar internal combustion engines.

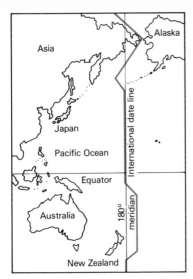

Map labels: Alaska, Asia, Japan, Pacific Ocean, Equator, Australia, New Zealand, International date line, 180° meridian

international date line An imaginary line joining the north and south poles at the 180° meridian, which is taken as the starting point for the new calendar day. There is a 24-hour time difference between one side of the line and the other.

interstellar space The space between the stars, which is not entirely empty but contains certain amounts of dust and gases. Where this interstellar matter is relatively thick, we see it as a nebula.

invar An iron alloy that expands very little when heated, used for the construction of precision instruments. Invented by the Swiss physicist Charles Guillaume, it contains 36% nickel and traces of other metals.

inverse square law A law in which the effect of something (light intensity, gravity, electrical charge) varies inversely as the square of its distance away. For example, the light intensity from a source 20 metres away will be only one-quarter that when the source is 10 metres away. The distance is doubled, therefore, the effect is reduced to a ¼:

$$\frac{1}{2^2} = \frac{1}{4}$$

inversion In chemistry, the conversion by hydrolysis of cane sugar ($C_{12}H_{22}O_{11}$) into a mixture of glucose and fructose (both $C_6H_{12}O_6$). Optically the two solutions differ. The former rotates the plane of polarized light to the right (is dextrorotatory), while the later rotates it to the left (is laevorotatory).

inversion, temperature A freak condition in the atmosphere that results in a cool air layer being trapped beneath a layer of warm air. It can aggravate pollution by trapping smoke and moisture.

invisible ink Colourless ink that can be treated or developed to make it visible. Potassium ferrocyanide solution is suitable. It is colourless, but turns into Prussian blue when 'developed' in a solution of a ferric salt.

Io The third largest moon of Jupiter, which the Voyager space probes found was volcanically active. Some 3,630 km (2,255 miles) in diameter, it is a vivid orange-red colour.

iodides The salts of hydriodic acid (HI), the best known of which is silver iodide, used in photographic emulsions.

iodine (I) One of the halogen elements (rd 4·9, mp 113·6°C). It exists as shiny black crystals, which sublime to give a pungent violet vapour. Most iodine is extracted from Chile saltpetre, which contains traces of sodium iodate, $NaIO_3$; and seaweed. Tincture of iodine is an antiseptic solution of iodine in alcohol.

iodoform A yellow crystalline organic compound (CHI_3), used as an antiseptic.

ion An electrically charged atom or group. In general, metals and hydrogen form positive ions (or cations), while non-metals form negative ions (or anions). They do so by respectively losing or gaining electrons. Salts such as sodium chloride are made up of ions, the bond being termed an ionic, or electrovalent, bond. When in solution, such substances ionize, or form ions, which are capable of carrying electric current. They enable solutions to conduct electricity. Ions can be formed in gases by electrical discharge.

ion exchange A process in which certain ions in a solution are replaced by others, for the purposes of separation, analysis, or more commonly water-softening. Various synthetic resins or materials called zeolites are used to promote ion exchange. See **water-softening.**

ionic bond See **electrovalent bond.**

ionization The process of forming ions.

ionosphere The part of the Earth's upper atmosphere, above about 80 km (50 miles) altitude, where the atmospheric gases are substantially ionized. It has the property of reflecting radio waves, making possible long-distance radio communications. Auroras occur in the ionosphere. Jupiter and Saturn are other planets known to have an extensive ionosphere.

Ir The chemical symbol for iridium.

iridescence A rainbow-like play of colours displayed by certain objects, such as soap bubbles and opals, caused by interference effects.

iridium (Ir) A rare metal (rd 22·4, mp 2,443°C) of the platinum group that is exceptionally hard, dense and chemically resistant.

iris See **eye.**

iris diaphragm An adjustable aperture, often of overlapping metal leaves, which controls the amount of light entering an optical system, such as a camera.

iron (Fe) The most important of all metals (rd 7·9, mp 1,539°C) because it can be alloyed with traces of carbon to form steel. It is a transition metal, many of whose salts are coloured. It has two oxidation states, or valencies, of 2 and 3. These give rise respectively to ferrous, or Fe(II) compounds; and ferric, or Fe(III) compounds. Another important property of iron is that it is magnetic. It occurs widely in the Earth's crust as minerals—iron oxides such as haematite and magnetite; carbonates such as siderite; and sulphides such as iron pyrites, or pyrite. The oxides and carbonates are valuable as ores. See also **cast iron; pig iron; wrought iron.**

iron oxides Compounds of iron and oxygen. They include ferrous, or iron(II) oxide, FeO; ferric, or iron(III) oxide, Fe_2O_3; and ferrosoferric oxide, Fe_3O_4, in which both oxidation states occur. See **haematite; limonite; magnetite.**

irradiation Exposure to radiation. Irradiation is carried out in nuclear reactors to produce artificial radioisotopes. It is also used as a means of pest control and sterilization, of foodstuffs and equipment.

isinglass A gelatin-like substance made from the swimming bladders of fish, used as an adhesive and also as a clarifying agent in wines and beers.

isobars Lines on a weather map joining points of equal atmospheric pressure.

isocyanides Organic nitrogen compounds containing the $-N \equiv C$ group, noted for their repulsive smell.

isomerism In chemistry, a condition when two or more compounds have the same chemical formula but different structures and properties. The hydrocarbon butane (C_4H_{10}) has two isomers: normal (n-) butane, whose structure is:

$$
\begin{array}{c}
\text{H}\quad\text{H}\quad\text{H}\quad\text{H} \\
|\quad\ |\quad\ |\quad\ | \\
\text{H}-\text{C}-\text{C}-\text{C}-\text{C}-\text{H} \\
|\quad\ |\quad\ |\quad\ | \\
\text{H}\quad\text{H}\quad\text{H}\quad\text{H}
\end{array}
$$

and isobutane (2-methylpropane) whose structure is:

$$
\begin{array}{c}
\text{H}\quad\text{H}\quad\text{H} \\
|\quad\ |\quad\ | \\
\text{H}-\text{C}-\text{C}-\text{C}-\text{H} \\
|\quad\ |\quad\ | \\
\text{H}\quad\ |\quad\text{H} \\
\\
\text{H}-\text{C}-\text{H} \\
| \\
\text{H}
\end{array}
$$

This kind of isomerism is called structural isomerism. Other types include cis-trans isomerism and optical isomerism.

isomorphous Compounds are described as isomorphous when they crystallize in the same form. The alums are isomorphous.

isoprene (C_5H_8) An unsaturated liquid hydrocarbon obtained from petroleum, used to make synthetic rubber. Natural rubber is itself made up of isoprene units.

isotherms Lines drawn on a weather map joining points at the same temperature.

isotopes Atoms of the same element that contain the same number of protons but different numbers of neutrons. They are chemically identical and differ only in mass. Most elements exist in nature as a mixture of isotopes. For example, tin has no fewer than eight stable isotopes. See also **radioisotopes.**

isotropic Having uniform properties throughout.

J

jack A lifting device that employs a screw thread or hydraulic action.

jacquard loom A loom whose action is guided by a series of punched cards. Developed in 1801 by Joseph-Marie Jacquard in France, the jacquard mechanism was an early milestone in automation.

jade A hard tough coloured mineral (usually white or green), widely used for carving. There are two main types of jade—nephrite and the more prized jadeite.

jasper A variety of the silica mineral chalcedony (SiO_2). It is hard and takes a high polish and is valued for jewellery and ornaments. It can be found in many colours, but chiefly reddish, the colour being due to the presence of haematite.

JET An abbreviation of Joint European Torus, a machine built by European atomic scientists to conduct experiments in nuclear fusion. Located at Culham, near Abingdon, it is a type of tokamak.

jet A very hard black ('jet-black') variety of lignite coal, used in jewellery.

jet engine An internal combustion engine that burns kerosene and achieves its thrust by jet propulsion. A type of gas turbine, it is the main engine used in modern aircraft. It is simpler in construction, uses cheaper fuel and is more reliable than the petrol piston engine and can propel aircraft at much higher speeds. The three main types of jet engines are the turbojet, turbofan and the turboprop. In the turbojet and turbofan all the propulsive thrust comes from jet propulsion, while in the turboprop part of the thrust is provided by propeller. British airman Frank Whittle patented the jet engine in 1930, but the first jet-powered plane was the German Heinkel-178, which flew in 1939. See also **pulse jet; ramjet.**

jet propulsion A method of propulsion in which a body is propelled forwards by means of the discharge backwards of a stream, or jet of fluid. The forward thrust is the result of reaction to the backward jet. This follows from the third of Newton's laws of motion.

jet streams Narrow and fast-flowing air currents that surge through the Earth's atmosphere at a height of about 10,000 metres (in the stratosphere). They flow eastwards and can reach speeds of over 300 km/h (200 mph). Similar jet streams are found in the atmospheres of Jupiter and Saturn, where wind speeds in excess of 1,800 km/h (1,100 mph) have been recorded.

joule A unit of work or energy, defined as the work done when a force of 1 newton moves over a distance of 1 metre. It is equal to 10^7 ergs, and 0.74 foot-pounds. In electrical units 1 joule is the energy released in 1 second by a current of 1 ampere flowing through a resistance of 1 ohm. It is named after the English physicist James Prescott Joule.

Concorde's Rolls-Royce Olympus jet engine

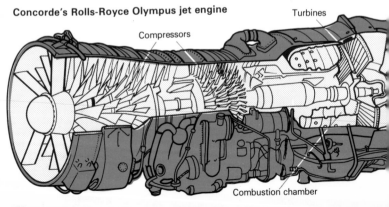

Turbines

Compressors

Combustion chamber

Joule-Thomson effect The change of temperature (usually cooling) that occurs when gases are suddenly expanded. This effect is utilized to liquefy gases, such as air.

Julian calendar The calendar devised by Julius Caesar in 46 BC, in which every fourth year was deemed a leap year. It was corrected in 1582 by Pope Gregory XIII, giving the present Gregorian calendar, which has fewer leap years (see **leap year**).

Jupiter The largest planet in the solar system, 143,000 km (88,700 miles) in diameter and more than twice as massive as all the other planets put together. It lies some 780 million km (480,000,000 miles) from the Sun, around which it travels in 11·9 Earth years, while it rotates on its axis every 9 hours 50 minutes. It has at least 16 satellites, of which the four largest—Ganymede, Callisto, Io and Europa—can be clearly seen with binoculars. The surface of Jupiter is marked with parallel belts and zones, which are the result of the parallel banding of the clouds due to Jupiter's swift rotation. A very prominent feature is a large red spot, which is a giant oval storm centre some 28,000 km (17,000 miles) across. Jupiter also has a faint ring system.

Jurassic Period A period of geological time between about 190 and 136 million years ago.

jute A coarse natural fibre obtained from the stems of jute plants, which grow in Asia.

K The chemical symbol for potassium, from the Latin, kalium.

kaleidoscope An optical toy invented by the Scottish physicist David Brewster in 1817. It consists essentially of a tube containing two plane mirrors inclined at an angle to each other (usually at 30°) and touching along one edge. At one end is a viewing window, and at the other a double window containing coloured glass pieces or beads, which can be rotated. Multiple reflection of the glass or beads takes place in the mirrors, producing beautiful symmetrical patterns.

kaolin Or china clay; a soft, pure white clay used for making fine pottery. Decomposed feldspar, it is an aluminium silicate, $Al_2O_3.SiO_2.2H_2O$.

kapok A soft, shiny cotton-like natural fibre obtained from the seed pod of kapok trees, which are native to the tropics. Very light and water-resistant, it is used for filling cushions, mattresses and life-jackets.

Kekulé structure A method of depicting the structure of benzene as a six-carbon ring devised by the German chemist August Kekulé. The Kekulé structure is:

which may be represented as:

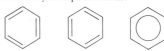

The latter is used in this book because it reflects current ideas on the structure of the benzene molecule.

kelvin The unit of temperature of the absolute temperature scale, symbol K. The freezing point of water is 273·16K (not °K).

Kelvin scale See **absolute temperature**.

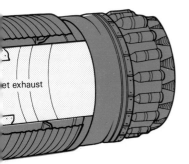

et exhaust

Kepler's laws

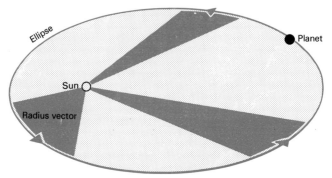

Kepler's laws Of planetary motion; advanced by Johannes Kepler in the early 1600s. Law 1: The planets orbit the Sun in ellipses, with the Sun at one focus. Law 2: The radius vector (see diagram), an imaginary line joining the planet to the Sun, sweeps out equal areas in equal times. Law 3: The square of the time it takes the planet to orbit the Sun (the planet's 'year') is directly proportional to the cube of the planet's mean distance from the Sun.

keratin A protein that is the main constituent of wool, hair, nails and feathers.

kernite A mineral found together with borax and, like borax, a valuable source of boron. It is hydrated sodium tetraborate, $Na_2B_4O_7.4H_2O$.

kerosene Or paraffin oil; one of the most useful fractions (boiling point range $150°–300°C$) obtained from petroleum refining. It consists of liquid hydrocarbons, and is a valuable fuel for jet engines and also for domestic heating.

ketones A series of organic compounds containing the group $>C = O$, the simplest of which is acetone, or dimethyl ketone $(CH_3)_2CO$.

kidney ore A variety of haematite that occurs as rounded, kidney-like masses.

kieselguhr A variety of diatomaceous earth, which is very porous and is used as an absorbent medium for nitroglycerin in the manufacture of dynamite.

kiln An oven or furnace used for drying, or firing pottery and other ceramics.

kilo- A prefix meaning one thousand, as in kilometre—1,000 metres.

kilogram (kg) The fundamental unit of mass in the SI system. It equals 1,000 g, and $2·2$ lb; 1 lb = $0·45$ kg.

kilometre (km) A unit of length in the metric system, equal to 1,000 metres. 1 km = $0·62$ miles; 1 mile = $1·61$ km.

kilowatt (kW) A unit of power, equal to 1,000 watts.

kilowatt-hour (kWh) A unit of electrical energy, being the energy released when 1 kilowatt is used for 1 hour.

kimberlite Also called blue ground; a dark, heavy igneous rock in which diamonds can be found. Its most common mineral is olivine. It is found particularly as pipes and dykes in the Kimberley district of South Africa.

kinematics A branch of mechanics concerned solely with motion and not with the agencies that cause the motion. It deals with such things as time, speed and distance, not with mass or forces.

kinetic energy The energy a body possesses because of its motion. The kinetic energy of a body with mass m moving at a speed v is $\frac{1}{2}mv^2$.

kinetic theory of gases A theory that describes and explains the basic nature and behaviour of gases. According to the kinetic theory, gases are composed of molecules, which are in constant random motion and which are far apart relative to their size. The kinetic energy of the molecules is a measure of the heat of the gas and increases—the molecules move faster—as the temperature rises. The molecules exert no appreciable force on one another, and if they collide they do so elastically—with no energy loss. The gas laws fit in with this theory, which holds well for a perfect gas though less well for a real gas.

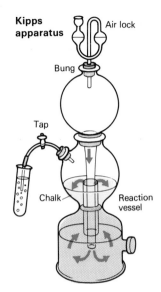

Kipps apparatus

Air lock

Bung

Tap

Chalk

Reaction vessel

kurchatovium

kite In mathematics, a quadrilateral that has two pairs of equal adjacent sides. The diagonals of a kite intersect at right-angles.

Klein bottle An interesting kind of vessel that has only one surface. It represents a special topological surface.

klystron An electronic device used to produce or amplify microwaves. Klystrons are used, for example, in radar equipment.

knocking Or pinking; tapping noises that occur inside the cylinder of a petrol engine. It is caused by premature ignition of the petrol-air mixture prior to sparking.

knock-rating See **octane rating**.

knot A unit of speed used by ships and aircraft, being one nautical mile per hour. The British nautical mile equals 6,080 ft ($1 \cdot 15$ miles), while the international nautical mile is fractionally shorter (6,076 ft, 1,852 metres).

Kr The chemical symbol for krypton.

krypton (Kr) One of the inert, or noble gases, which is about three times heavier than air. It is present in trace amounts in the atmosphere. It is used in high-wattage electric-light bulbs and in electronic flash tubes.

kurchatovium An alternative name for rutherfordium.

Kipp's apparatus A common laboratory apparatus designed for producing gas. It consists of a bulbous funnel inside a double-bulbed vessel (see diagram). To make hydrogen sulphide, for example, iron pyrites is placed in the middle bulb and acid poured into the funnel. When the gas tap is opened, the acid flows into the bottom bulb and then up into the middle one. There it reacts with the pyrites to produce hydrogen sulphide gas, which leaves via the tap. When the tap is closed, gas pressure forces the acid out of the middle vessel and the reaction stops.

Kirchoff's laws Of electric circuits; in any network of wires the sum of all the currents going into a junction is the same as the sum of all the currents going out (junction theorem). In a closed circuit, or loop, the sum of the emfs is equal to the sum of the voltages across each of the resistances (loop equation).

kite A simple flying device consisting essentially of a fabric or paper-covered framework, which obtains lift in much the same way as an aeroplane wing. Two main types are the ages-old, diamond-shaped eddy kite; and the box kite, invented by the Australian Lawrence Hargrave in 1893.

Simple eddy kite

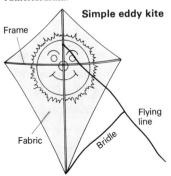

Frame

Flying line

Fabric

Bridle

La

L

La The chemical symbol for lanthanum.

laboratory A place equipped with scientific apparatus where scientists conduct experiments, carry out analysis, and test products and processes.

lac See **shellac**.

lacquer A type of varnish, which is often clear, but may also be coloured or opaque. Traditional lacquers are made from natural products, such as lac, but modern lacquers are made from synthetic resins, such as polyurethanes and nitrocellulose.

lactic acid The acid that gives sour milk its taste, formed by the bacterial fermentation of lactose. Of chemical formula $CH_3CH(OH)COOH$, lactic acid exists in several isomeric forms, two of which display optical activity.

lactose The sugar in milk. It is a disaccharide, $C_{12}H_{22}O_{11}$.

laevorotatory See **optical activity**.

lagging The insulation around pipes and boilers to prevent heat loss or, as in domestic water systems, attack by frost.

Lagrangian points In astronomy, gravitationally stable points in space for a body influenced by the gravitational effects of two larger bodies.

lakes In dyeing, an insoluble colouring substance formed when a soluble dye combines with a mordant. Lakes are used as pigments in paints, varnishes, cosmetics and printing inks.

lambert A unit of luminance (brightness) named after the German physicist Johann Lambert, being the brightness of a perfect diffuser of light that gives out one lumen per sq cm.

laminar flow Also called streamline flow and viscous flow: continuous steady flow in a fluid. Contrast **turbulent flow.**

laminating A method of constructing materials from a number of thin layers. Heatproof surfaces are made from plastic laminates. One type of safety glass consists of a sandwich of toughened glass with plastic film inside. Plywood is a kind of laminate.

lampblack A form of carbon, as given off by an oil lamp; a kind of soot.

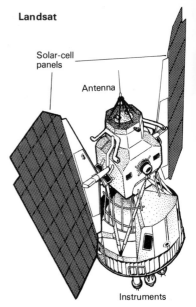

Landsat

Solar-cell panels

Antenna

Instruments

Landsat An American Earth-survey satellite; formerly known as ERIS (Earth Resources Technology Satellite). The first of the series was launched in 1972. It photographs the Earth in green, red and infrared light and provides 'false colour' pictures which can reveal a wealth of data of surface rocks, mineral deposits, plant life, and so on.

lanolin A soft yellowish grease obtained from sheep's wool, used in ointments and cosmetics.

lanthanides The series of closely similar elements that begins with lanthanum, which are generally known as the rare earths.

lanthanum (La) The first member of the rare-earth metals (rd 6·2, mp 920°C), used in the alloy Misch metal and as a getter in vacuum technology.

lapilli Tiny pieces of lava formed from magma ejected during a volcanic eruption, which may have a layered structure rather like hailstones.

lapis lazuli A deep blue rock used for ornaments and as a source of the pigment ultramarine. It is a metamorphic rock often found with crystalline limestone.

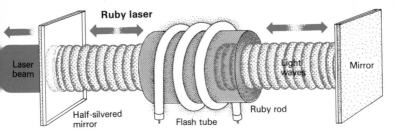

Ruby laser

Laser beam

Half-silvered mirror

Flash tube

Ruby rod

Light waves

Mirror

lapping A kind of polishing operation carried out with a piece of soft metal, such as lead, charged with a very fine polishing powder.

Large Magellanic Cloud See **Magellanic Clouds.**

laser An electronic device that produces an intense and very pure, parallel beam of light known as coherent light. Laser light can be focused to form a powerful cutting and welding tool. It can be used in telecommunications to carry signals along optical fibres. Surgeons and eye specialists also sometimes use laser instruments. The main types of lasers are ruby crystal lasers, gas lasers and semiconductor lasers. The ruby laser can produces only pulses of laser light; the other two types can produce continuous beams. A widely used gas laser uses carbon dioxide. A common semiconductor laser uses gallium arsenide crystals. The word 'laser' stands for 'light amplification by stimulated emission of radiation'. See also **holography.**

latent heat The amount of heat evolved or absorbed by a substance when it undergoes a change of state, which occurs without a change in temperature. The heat change involved when a solid melts or a liquid freezes is called the latent heat of fusion. The heat change involved when a liquid boils or a vapour condenses is called the latent heat of vaporization.

latent image The invisible image that forms in photographic film after exposure due to chemical changes in the silver salts in the film emulsion. The latent image is changed into a visible image during developing.

latex A milky juice found in certain plants and trees, particularly the rubber tree. It consists of a colloidal suspension of resins and fats in a watery solution.

lathe The most important machine tool used in engineering workshops, on which the machining process of turning is carried out. The lathe has facility to grip and rotate a workpiece and bring a variety of cutting tools to bear on it.

Chuck

Cross-slide

Tailstock

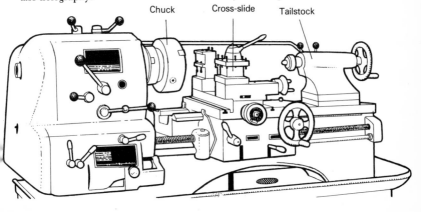

latitude and longitude

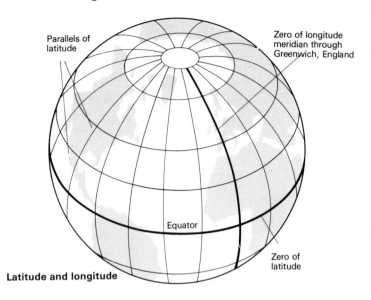

Parallels of latitude

Zero of longitude meridian through Greenwich, England

Equator

Zero of latitude

Latitude and longitude

latitude and longitude The standard system of coordinates for locating the position of a point on the Earth's surface. Lines of latitude, or parallels, are drawn parallel with the equator; lines of longitude, or meridians, are drawn at right-angles to them, linking the north and south poles. The latitude and longitude are measured in degrees of angle. Latitude is measured in degrees north or south of the equator, which has latitude 0°. At the north pole, the latitude is 90° North; at the south pole, 90° South. The starting point for longitude is the meridian passing through Greenwich. This prime meridian has longitude 0°. Longitudes are measured 0°–180° East of the prime meridian, or 0°–180° West of it.

lattice A regular network of points in space. In a crystal lattice the atoms are arranged in a regular three-dimensional network.

laughing gas The anaesthetic gas nitrous oxide (N_2O), which may cause a patient to laugh involuntarily.

lava The molten rock, or magma, that pours out of a volcano; and also the rock formed when it solidifies. When lava cools quickly, it may form volcanic glass (obsidian), which is amorphous, or non-crystalline. If it cools more slowly, basalt forms, in which tiny crystals can just be distinguished.

lawrencium (Lr) An artificial radio-active element (at no 103) obtained by nuclear bombardment.

LCD An abbreviation for 'liquid crystal display', one method of forming the digits on the face of a digital watch. The display contains a thin layer of liquid crystal, a liquid that twists the plane of polarization of light passing through it when it is electrically altered. This phenomenon is used to prevent light being reflected from certain segments of the display, so forming parts of the digits.

LD process See **basic-oxygen process.**

leaching In geology, a process in which water percolates through the surface soil and dissolves mineral salts from it. In metallurgy, leaching means treating ores with, for example, an acid to dissolve the minerals they contain.

lead (Pb) A soft, bluish-grey heavy metal (rd 11·3, mp 327·3°C) that resists corrosion. It is used mostly in alloys, such as pewter, type metal and solder. Another major use is in lead-acid car batteries. The main lead ore is galena, lead sulphide (PbS).

Lead forms a variety of oxides: PbO (litharge), PbO_2, Pb_3O_4 (red lead). In these oxides lead occurs in the divalent state—plumbous, or lead(II)—or the tetravalent state— plumbic, or lead(IV). Litharge and red lead are well-known pigments, as is white lead, or lead carbonate ($PbCO_3$). Lead compounds must, however, be used with care because many are poisonous.

lead-acid battery The battery used in cars. It is an accumulator, made up of secondary cells, which can be recharged with electricity after they have discharged. The battery, when charged, has negative plates of lead dioxide (PbO_2) and positive plates of lead. The electrolyte is sulphuric acid. During discharge both plates are altered to lead sulphate, but revert back to their original state under charge. The standard battery has six sets of cells, with a total output of 12 volts.

lead-chamber process One method of producing sulphuric acid from sulphur dioxide (SO_2). The gas is mixed with air and nitrogen oxides and passed through several lead-lined chambers, whereupon it changes into sulphur trioxide. This is absorbed in water to form sulphuric acid. It has been largely superseded by the contact process.

lead glass Also called lead crystal; a dense glass containing lead oxide, which has a diamond-like sparkle when expertly cut.

lead tetraethyl Or tetraethyl lead, $Pb(C_2H_5)_4$; an organic lead compound added to petrol to prevent premature ignition of the fuel mixture. It is a source of pollution.

leap year A year containing 366 days instead of the usual 365, by the addition of February 29. An extra day is needed in the present calendar because the solar year is actually $365\frac{1}{4}$ days long. Leap years are years that are divisible by 4 (1984, 1988, 1992, etc). Century years, however (1600, 1700, 1800, etc), are only leap years when they can be divided by 400. So the year 2000 will be a leap year, whereas 1900 was not.

leather A material made by tanning animal hide or skin. Tanning preserves it and makes it flexible and water-resistant. Chrome leather is leather made using chromium salts for tanning.

Leblanc process The first important industrial chemical process, devised by the French chemist Nicolas Leblanc in 1791. The Leblanc process, which converts salt to soda (sodium carbonate), was later superseded by the ammonia-soda process.

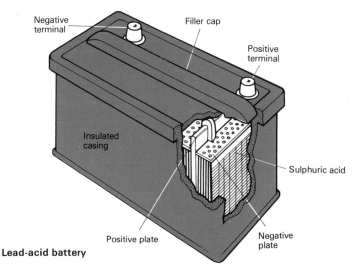

Lead-acid battery

Le Chatalier's principle

Le Chatalier's principle When a system in equilibrium is subject to a change in conditions, it will adjust itself so as to try to oppose that change. This was first stated in 1888 by the French chemist Henri-Louis Le Chatalier.

Leclanché cell The forerunner of the modern dry cell, devised by the French chemist Georges Leclanché, who developed it in 1866. It has electrodes of carbon/manganese dioxide (+) and zinc (–), and an electrolyte of ammonium chloride solution. It has an emf of about 1·5 volts.

LED An abbreviation for 'light-emitting diode', one method of forming the digits in digital displays in watches and other apparatus. The digits are formed of dots of gallium arsenide phosphide or other material that glows when conducting electricity.

lens A piece of glass or other transparent material, which has at least one curved surface. It is used in optical instruments to form magnified or reduced images and to focus images on the eye, a screen, piece of film and so on. A lens has this focusing ability because light passing through it is refracted (bent) at the curved surfaces. Convex, or converging lenses tend to bring light rays together; concave, or diverging lenses tend to spread them out.

Lenz's law When a current is induced in a conductor, because of relative motion between the conductor and a magnetic field, the induced field will tend to oppose the motion. It is named after the Russian physicist Heinrich Lenz, who first discovered it in 1834.

Leo The Lion; a distinctive zodiacal constellation lying between Cancer and Virgo, whose main star is the brilliant (magnitude 1·3) Regulus.

Leonids A meteor shower occurring between about November 14 and 20, which appears to come from the constellation Leo.

lepton A class of elementary particles, which includes the electron, muon and neutrino.

letterpress A method of printing from a raised surface. Contrast **gravure** and **offset-litho.**

level See **spirit level.**

lever One of the simplest machines, which consists of a rigid beam that pivots about a fixed point, or fulcrum. There are three classes of levers, distinguished by where the effort is applied and from where the load is lifted. A first-class lever has effort and load on opposite sides of the fulcrum. A second-class lever has the load between the effort and the fulcrum. A third-class lever has the effort between the load and fulcrum. If in a lever the effort arm (distance from effort to fulcrum) is longer than the load arm, then a mechanical advantage is gained. Less effort than the load is required, but the effort travels farther. The law of equilibrium, or law of the lever, discovered by Archimedes, states that effort × length of effort arm = load × length of load arm.

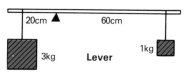

Lever

Leyden jar The earliest type of capacitor (electrical condenser), so named because it was first devised in Leyden (Leiden) in 1746.

Li The chemical symbol for lithium.

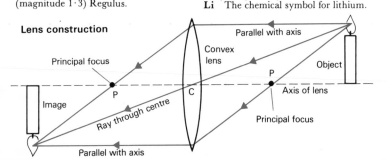

Lens construction

Parallel with axis

Principal focus

Convex lens

Object

Image

P

C

P

Axis of lens

Ray through centre

Principal focus

Parallel with axis

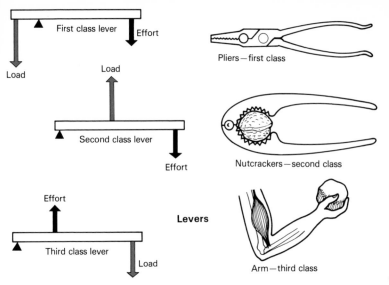

First class lever — Effort, Load

Load — Second class lever — Effort

Effort — Third class lever — Load

Levers

Pliers—first class

Nutcrackers—second class

Arm—third class

Libra The Scales; a zodiacal constellation of little astronomical interest, lying between Scorpio and Virgo.

libration In astronomy, a regular nodding motion or oscillation of the Moon's face up and down and from side to side which enables us to see nearly 60% of its surface. It results from irregularities in the Moon's orbital motion.

Liebig condenser The common type of condenser in the chemical laboratory, named after the German chemist Baron von Liebig. See **condenser.**

lie detector Also called a polygraph; an instrument that magnifies and records minute changes in a person's blood pressure, respiration, pulse rate and other body functions while he is being questioned. When a person lies, the body functions may alter in a characteristic way, which the lie detector can perceive.

lift Also called elevator; a machine that carries people and goods up and down a vertical shaft. The American engineer Elisha Graves Otis invented the first safety lift in 1857.

lift, aerodynamic The upward force experienced by an aeroplane wing due to its aerofoil shape.

light A type of electromagnetic radiation to which our eyes are sensitive. White light, which we receive from the Sun, is a mixture of light of different wavelengths. We can split up white light into its component wavelengths by passing it through a prism. We see the spread of the wavelengths as a spread of colours—the colours of the spectrum. The wavelengths vary from about 4×10^{-7} metre (violet) to about 7×10^{-7} metre (red). The study of light is known as optics. In a vacuum light travels at a velocity of 299,793 km (186,282 miles) per second.

light-fast A term applied to dyes and pigments which relates to their resistance to fading.

light meter See **exposure meter.**

lightning A brilliant flash of light that accompanies the discharge of atmospheric electricity between clouds or between clouds and the ground. Electric potentials of hundreds of millions of volts can build up in thunder-clouds. When the voltage is sufficient to overcome the insulation of the air, discharge occurs as lightning. Fork lightning follows a zigzag path to the ground. Ball lightning consists of what looks like balls of fire. Sheet lightning is a diffuse flash.

lightning conductor A metal rod placed high on a building and connected by a conductor to the ground. If lightning strikes, the discharge passes harmlessly through the conductor to the ground rather than destructively through the structure of the building. It was invented by the American scientist and statesman Benjamin Franklin in the 1850s.

light-year The distance light travels in a year, $9 \cdot 46 \times 10^{12}$ km ($5 \cdot 88 \times 10^{12}$ miles). It is a unit astronomers use to express the enormous distances in space. It has now been largely superseded by the parsec, which is equal to $3 \cdot 26$ light-years.

lignin One of the chief constituents of wood (up to about 30%), which imparts strength and rigidity. It is a complex carbohydrate.

lignite Or brown coal; a low-grade brownish-black coal containing up to about 75% carbon when dry, but over 50% water as mined.

lime The name given to certain products derived from calcium carbonate or limestone. Quicklime is calcium oxide, CaO. Slaked lime is calcium hydroxide, $Ca(OH)_2$. Lime-water, a solution of slaked lime in water, is a useful alkali.

limelight A light used in the early theatre, which consisted of calcium, heated to incandescence by an oxyhydrogen flame. Limelight was brilliant and soft. The device was invented by Thomas Drummond in 1816.

limestone One of the commonest sedimentary rocks, made up of impure calcium carbonate, which occurs as calcite and aragonite. It may have been formed by precipitation from lime-laden waters, or from the accumulated deposits of shells and corals. It often contains fossils. Acidic rainwater slowly dissolves the calcium carbonate in lime-stone, and limestone regions are invariably riddled with caves, which often contain stalagmites and stalactites.

Linde process A method of liquefying air on a large scale, developed by Carl von Linde in 1895. It relies on the effect that when compressed air is suddenly expanded, it cools.

linear accelerator Or linac; see **accelerator, particle.**

linear induction motor A type of electric motor in which the rotor (rotating part) and stator (stationary part) are 'opened out' straight. The 'rotor' travels in a straight line when it moves (rather than rotate, as it normally does). The motor can thus provide a useful means of propulsion, for example, for 'hover' and 'maglev' trains.

linen One of our commonest fabrics, made from the long fibres found in the stalks of flax plants. It is strong and absorbs water readily.

lines of force Imaginary lines within a magnetic or electric field which show the direction·of that field at any point. In a magnetic field, for example, the lines can be plotted from the way iron filings align themselves in the field.

linoleum A cheap floor covering made from linseed oil, resins and gums, mixed with powdered wood or cork and coloured with pigments. It has a felt or hessian backing.

Linotype 'slug'

linotype machine A typesetting machine that casts type in whole lines. It assembles individual moulds of the type characters into a line and then injects hot metal into them to form a 'slug' of type. Compare **monotype.**

linseed oil A useful oil obtained from the seeds of the flax plant. It oxidizes and forms a tough skin on exposure to the air, and because of this is widely used in paints and varnishes. It is also used to make linoleum.

liquefaction Changing a gas into a liquid.

liquefied petroleum gas (LPG) Or bottled gas; a mixture of propane, butane or other gases obtained from natural gas and liquefied under pressure. When bottled, it is widely used as a portable heat source.

liquid One of the three main states of matter, intermediate between a solid and a gas. The molecules in a liquid are relatively free to move but still influence one another. A liquid takes the shape of any container it is put in, but has a

Offset litho

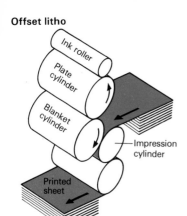

Ink roller

Plate cylinder

Blanket cylinder

Impression cylinder

Printed sheet

definite volume and is virtually incompressible. It also finds its own level. It has a kind of surface 'skin' caused by the net downward attraction on the surface molecules. We call this phenomenon surface tension.

liquid air Air that has been liquefied by being cooled to temperatures below about – 196°C. A pale blue liquid, it consists mainly of liquid oxygen and liquid nitrogen.

liquid crystal See **LCD.**

liquid hydrogen Hydrogen cooled below its boiling point (– 253°C). It is an excellent fuel, used particularly as a propellant in rockets such as Ariane and in the space shuttle.

liquid oxygen Oxygen cooled below its boiling point (– 183°C). One of its major uses is as a rocket propellant.

litharge Lead monoxide (PbO), a yellow compound used in making glass and paints.

lithium (Li) The third element in the periodic table and the lightest solid element (rd 0·53, mp 180°C). It is the first of the alkali metals.

lithography A method of printing from a flat surface, originally from stone. In modern printing practice the surface is zinc or aluminium. The printing plate is treated photographically so that it carries an image, which attracts greasy printing ink. The remainder of the plate can be wetted to repel the ink. When paper is pressed against the plate, only the inked image is transferred. These days litho printing is done by the offset method. Between the printing plate and paper is a rubber-covered roller, which transfers the inked image from the plate to the paper.

litmus A mixture of coloured dyes extracted from certain lichens, used as an indicator in chemical analysis. It is a different colour in acidic (red) and alkaline (blue) solutions.

litre The main unit of volume in the metric system, equal to 1 cubic decametre, or 1000 millilitres (or cc). 1 litre = 0·22 gallon. 1 gallon = 4·55 litres.

lock A device for securing, usually, a door of some kind. Locks have been used for over 4000 years. The Romans developed the warded lock, which contains wards, or obstacles that the key must pass to turn. In the 1770s Robert Barron in England developed the forerunner of the modern mortise lock, whose bolt is trapped by levers. The American Linus Yale Jr invented the pin-tumbler cylinder lock in 1865. This lock has a number of pins and drivers (see diagram). Only the right key will raise all the pins to the correct heights so that the cylinder can turn.

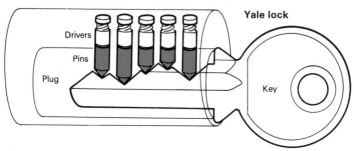

Yale lock

Drivers

Pins

Plug

Key

locomotive

locomotive A vehicle used to haul trains. Cornishman Richard Trevithick built the first steam locomotive in 1804 for the tramway at Pen-y-darran in South Wales. But not until George Stephenson designed 'Rocket' (1829) was the locomotive a reliable machine. From then on the railways developed rapidly worldwide. Today steam locomotives have been superseded practically everywhere by electric and diesel locomotives. Electric locomotives may pick up their power from a live 'third rail' alongside the track or, via a pantograph, from an overhead power line. Diesel locomotives are powered by diesel engines, and one of three types of power transmission may be employed—mechanical, electric or hydraulic. Diesel-electric transmission is most common. Gas-turbine locomotives are now coming into more widespread use. They include the fast French Turbotrains à Grande Vitesse (TGVs).

lodestone A magnetic variety of the iron oxide mineral magnetite. 'Lodestone' means 'guiding' stone because when a piece is suspended, it acts like a compass.

logarithms Sets of numbers used to aid calculations, based upon the exponent of a number, or power to which it is raised. If $a = b^n$, n is the logarithm of a to the base b. The base of common logarithms is 10, and we write a logarithm to the base 10 as $\log_{10}$. The other widely used type of logarithms are Naperian, or natural logarithms, which have the base e (about $2 \cdot 72$).

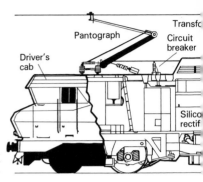

long-chain molecule The typical molecule of a polymer or plastic, which consists of thousands and even millions of repeated units (monomers) linked together in a long chain. See also **polymerization.**

longitude See **latitude and longitude.**

longsightedness Also known as hypermetropia; a common type of defective vision in which a person cannot focus properly on nearby objects. His eyes bring the light rays to a focus behind the retina. The condition is remedied by the use of spectacles with convex lenses.

loom A machine for weaving cloth, whose ancestry dates back at least 7000 years. On a loom one set of threads (the warp) is stretched lengthwise over a frame. Then devices called heddles raise and lower certain of the threads to make a gap, or shed, through which another

Principle of loom

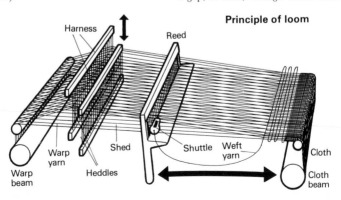

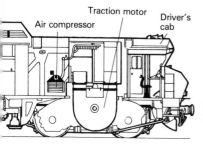

Electric locomotive

Overhead conductor wire

Traction motor

Driver's cab

Air compressor

thread (the weft) is passed in a shuttle to make a line of weave. A comb-like reed then beats, or firms, the line of weave before the heddles are reversed, and the shuttle carries the weft thread through to make another line of weave. Modern shuttleless looms use different methods of passing the weft thread through the shed, using needles, compressed air and even water jets.

loran An electronic system of navigation that uses beams of pulses from two radio transmitters. By comparing times of arrival of the pulses a navigator can work out his position.

loudspeaker A device that converts electrical signals into sound waves. It works in the opposite way to a microphone. The commonest type is the moving-coil speaker, illustrated here. A coil of wire is attached to the apex of a large paper cone and is located between the poles of a permanent magnet. When an electrical current passes through the coil, it sets up a magnetic field. This field interacts with that of the magnet and causes the coil to be attracted or repelled. This vibrates the cone, which causes sound to be emitted. In hi-fi equipment, several separate speakers are included in the speaker unit, each sensitive to a particular frequency range. The high-frequency one is called a tweeter; the low-frequency one, a woofer.

LPG See **liquefied petroleum gas.**

Lr The chemical symbol for lawrencium.

LSD A powerful alkaloid drug that causes hallucinations. It is lysergic acid diethylamide ($C_{20}H_{25}N_3O$).

Lu The chemical symbol for lutetium.

lubricant A substance that reduces friction between moving parts in a machine. The commonest lubricant is oil derived from petroleum. For high-temperature duty, synthetic oils based on silicones are used. Where oils would be too thin, greases are used, which are oils thickened with soap or fat. Some solids are occasionally used for lubrication, including graphite and talc. Some plastics are self-lubricating.

lucite See **PMMA.**

lumen The SI unit of luminous flux (or amount of light), being the amount of light emitted within unit solid angle by a point source having an intensity of 1 candela.

luminescence The emission of light from a body by other means than by heat. See **bioluminescence; fluorescence; phosphorescence.**

luminosity In astronomy, a star's absolute brightness, usually expressed in magnitude.

Luna Or Lunik; a series of probes sent by Russia to the Moon. Luna 2 crashlanded on the Moon in 1959. Luna 9 made the first successful soft-landing in 1966 and relayed the first close-up pictures of the lunar soil. In 1970 Luna 16 soft-landed, scooped up a sample of soil and returned it to Earth; and Luna 17 soft-landed the first wheeled vehicle on the Moon, Lunokhod.

lunabase The name often given to the material making up the Moon's seas, or maria. It is dark volcanic rock, similar to the basalts on Earth.

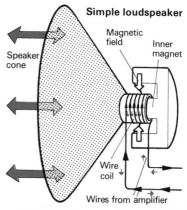

Simple loudspeaker

Magnetic field · Inner magnet · Speaker cone · Wire coil · Wires from amplifier

lunar orbital rendezvous

lunar orbital rendezvous The technique used in the Apollo project to travel to and return from the Moon. It involved the use of a separate Moon-landing craft (lunar module), which separated from the parent craft in orbit and returned to it after the landing.

Lunar Orbiter A series of American probes to the Moon, which mapped the Moon's surface in 1966 and 1967 as a prelude to the successful Apollo landings.

lunation A synodic month—the time between one new Moon and the next, equalling 29 days 12 hr 44 min.

Lunokhod 1 The first wheeled vehicle that travelled on the Moon. It was carried there by the Russian probe Luna 17 in November 1970. During a 322-day life span, it travelled more than 3 km (2 miles).

lustre The characteristic surface appearance of a mineral. Galena has a metallic lustre; quartz, a vitreous lustre; sulphur, a resinous lustre; asbestos, a silky lustre; talc, a pearly lustre.

lutetium (Lu) One of the rare-earth metals, or lanthanides (rd 9·8, mp 1700°C).

lux The SI unit of illuminance, being 1 lumen per sq metre.

lyophilic, lyophobic Terms applied to colloids, meaning solvent-loving, solvent-hating. When the solvent is water, the appropriate terms are hydrophilic, hydrophobic.

M

macadam A road surface consisting of layers of graded stone chippings, devised by the Scottish engineer John Loudon MacAdam in the early 1800s. In modern road-building tar is included as a binder in the surface layer; hence the term tarmac.

machine A device for performing useful work. It may modify and transmit forces and motion or convert one kind of energy into another. A heat engine is a machine that converts mechanical energy into electrical energy.

machines, simple The three simplest machines are the lever, inclined plane and wheel-and-axle, from which three more simple machines are derived—the pulley, wedge and screw.

machine tool A machine that shapes metal by cutting or grinding. The most important is the engineering lathe, which shapes by turning. Other machine tools perform drilling, boring, planing, shaping and milling operations. Many machine tools now work automatically under numerical control. American inventor Eli Whitney pioneered the use of machine tools in the 1790s to make interchangeable parts for muskets.

Lunokhod lunar vehicle

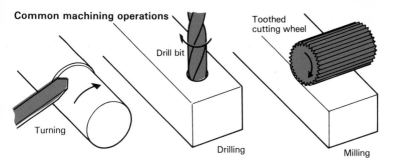

Common machining operations

Drill bit

Toothed cutting wheel

Turning

Drilling

Milling

Mach number A number used to decribe the speed of aircraft. It is the ratio of an aircraft's speed to the local speed of sound. So Mach 1 equals the speed of sound; Mach 2, twice the speed of sound; and so on. The Mach number varies with height above sea level, as the air density falls; and with ambient temperature. It is named after the Czech physicist Ernst Mach.

mackerel sky A cloud formation consisting of groups of small dappled clouds that can be likened to the markings on a mackerel. Its proper name is cirrocumulus.

macromolecules The very large molecules of polymers.

mafic minerals Dark minerals in igneous and metamorphic rocks, such as olivine, amphibole and mica.

Magellanic Clouds The two galaxies nearest to our own. The Large Magellanic Cloud (LMC) and Small Magellanic Cloud (SMC) lie respectively 170,000 and 205,000 light-years away in the southern hemisphere. They are irregular in shape.

maglev An abbreviation for 'magnetic levitation', a method used to suspend high-speed trains above a track and thus reduce friction. Maglev depends on the principle of induced magnetic repulsion.

magma Molten material found in the Earth's interior, which forms igneous rocks when it solidifies within the Earth's crust (intrusive rocks) or on the Earth's surface (extrusive rocks). It contains water vapour and volatile gases as well as rock-forming material.

magnesite A mineral ore of magnesium, being magnesium carbonate, $MgCO_3$. It is an excellent refractory material, used to line furnaces.

magnesium (Mg) One of the alkaline-earth metals (rd $1\cdot7$, mp $650°C$). Lightest of the structural metals, it is used with aluminium in aircraft alloys. It is obtained by the electrolysis of molten magnesium chloride, obtained from seawater or by processing other magnesium minerals. It burns in air with a brilliant white light. Among its many important compounds are the refractory magnesia, (magnesium oxide, MgO) and the antacid milk of magnesia, which is a suspension in water of magnesium hydroxide, $Mg(OH)_2$. See also **Epsom salts.**

magnet See **magnetism.**

magnetic field The region around a magnet or a wire carrying an electric current in which magnetic forces act. These can be detected, for example, by a compass needle. The presence of a magnetic field around a magnet can be indicated by placing a card on top and sprinkling iron filings on it. The Earth has a strong magnetic field (see **magnetism, Earth's**); so have the planets Jupiter and Saturn. See also **magnetosphere.**

magnetic storm A disturbance in the Earth's magnetic field and ionosphere brought about by the eruption of sunspots and flares on the Sun's surface. The solar flares emit streams of protons and electrons, which interact with the Earth's upper atmosphere, causing brilliant auroras and upsetting radio reception.

magnetic tape Plastic tape coated with magnetic particles, usually finely powdered iron or chromium oxide. It is used in audio and video tape recording, holding a magnetic 'pattern' of sound or vision signals.

magnetism

magnetism The phenomenon associated with the devices we call magnets, typified by the magnetic mineral magnetite, which is able to attract pieces of iron. The Earth itself behaves as if it were a huge magnet (see **magnetism, Earth's**). Magnetism is also produced when an electric current passes through a conductor. Electricity and magnetism are closely related and are studied together in the branch of physics called electromagnetism. In a simple bar magnet the magnetism appears to be concentrated near the ends, which will point north and south if the magnet is suspended because of the presence of the Earth's magnetic field. The end, or pole, pointing north we term the north pole; the south-pointing pole, we term the south pole. When two like poles (two 'souths' or two 'norths') are brought together, they repel one another. When two unlike poles are brought together, they attract one another. See also **diamagnetism; ferromagnetism; paramagnetism.**

magnetism, Earth's Or geomagnetism; the magnetism displayed by the Earth, which is believed to result from the powerful electrical currents generated in the Earth's liquid core as the Earth spins. The magnetic axis of the

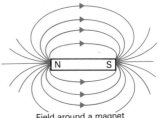

Field around a magnet

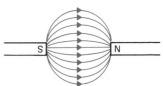

Unlike poles attract

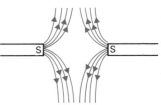

Like poles repel

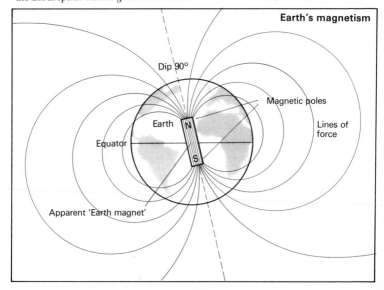

Earth's magnetism

Dip 90°

Magnetic poles

Earth

Lines of force

Equator

Apparent 'Earth magnet'

Earth does not coincide with the geographic axis. A compass needle points to magnetic north, the angle between this and true north being called the magnetic variation or declination.

magnetite Magnetic iron oxide (Fe_3O_4), the lodestone ('guiding stone') of the ancients, who used it as a compass. It is one of the main iron ores.

magneto A simple electrical generator, used in the ignition system of most motorcycles. It consists essentially of a rotating magnet, which induces an alternating current in a coil.

magnetohydrodynamics See **MHD.**

magnetometer An instrument that measures the strength of a magnetic field, used in geological surveys.

magnetomotive force Or MMF, also called magnetic potential. It is the equivalent in magnetism of the electric potential, or emf, in electricity.

magnetosphere The region in space around a body, such as the Earth or Jupiter, where its magnetic influence is felt. The shape of the magnetosphere is not spherical because of interaction with the solar wind. This results in the magnetosphere being flattened in the direction of the Sun, and elongated into a tail away from the Sun.

magnetron A vacuum tube used as an oscillator to produce microwaves for use in radar.

magnifying glass A convex, or converging, lens used to magnify objects. When the glass is held close to the object (closer than the focal length), an upright virtual image is formed. When held further away, a real but inverted image is formed. By focusing the Sun's rays, a magnifying glass can be used as a burning glass to make fire.

magnitude In astronomy, a measure of a star's brightness. The apparent magnitude is the brightness as it appears to an observer. The absolute magnitude, or luminosity, is the brightness the star would have at a distance of 10 parsecs ($32 \cdot 6$ light-years).

magnox An early type of British nuclear reactor (used in Calder Hall, for example).

mains electricity Or the mains; the electricity that flows into the home via the national grid.

main sequence The main series of stars on the Hertzsprung-Russell diagram.

malachite A vivid green mineral, which is one of the major ores of copper. It is basic copper carbonate, $Cu_2CO_3(OH)_2$.

malleability A property of a metal that allows it to be hammered into thin sheets.

maltose Or malt sugar; the sugar in malt (see **brewing**). It is a disaccharide, $C_{12}H_{22}O_{11}$.

manganese (Mn) A hard, brittle transition metal (rd $7 \cdot 4$, mp $1250°C$), whose main use is in steelmaking. It is added to steel as a ferromanganese alloy and improves the quality of the metal by combining with oxygen and sulphur. Manganese exhibits several valencies (oxidation states) in its compounds. In manganese chloride ($MnCl_2$) its valency is 2; in manganese dioxide (MnO_2) it is 4; in potassium manganate (K_2MnO_4) it is 6; and in potassium permanganate ($KMnO_4$) it is 7.

manganese nodules Small rounded lumps of minerals found on the deep ocean floor, particularly in the Pacific Ocean. They contain a high proportion of manganese, together with cobalt, copper, nickel and iron.

man-made fibres Textile fibres produced by man using either natural materials, such as cellulose, or synthetic materials. Fibres made from cellulose include acetate and rayon. Synthetic fibres include nylon, polyester and acrylic fibres. See also **fibreglass.**

manometer A simple device for measuring gas pressure. It often consists of a U-tube containing liquid, to one limb of which is attached the gas supply. From the difference between the liquid levels in the two limbs, the gas pressure can be determined.

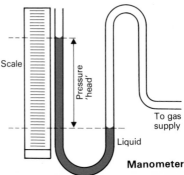

Manometer

mantle

mantle The part of the Earth that lies beneath the crust. It extends from about 30 km (20 miles) from the surface to a depth of about 3000 km (2000 miles), where the core begins. It is thought to consist of silicate rock such as olivine. Plastic flow of the upper mantle is one possible explanation of continental drift.

marble A common metamorphic rock formed when limestone or dolomite is subjected to heat and pressure in the Earth's crust. It is often off-white in colour with attractive coloured streaks.

mares' tails The popular name for the long wispy cirrus clouds, which appear high in the sky.

margarine A manufactured butter substitute made from animal fats and vegetable oils, together with milk products, emulsifiers, vitamins, colouring matter and preservatives. The French chemist Hippolyte Mège-Mouries invented margarine in 1869.

maria Or seas; the name given to the dark regions of the Moon (singular 'mare'). They are flat plains formed from vast lava flows.

marijuana Also called cannabis and hashish; an intoxicating drug obtained from the Indian hemp plant, *Cannabis sativa*.

Mariner probes A series of American space probes to Venus, Mars and Mercury. Mariner 2 to Venus was the first successful interplanetary probe in 1962. Mariner 9 mapped Mars in detail from orbit, beginning in 1971. Mariner 10 took the first close-up pictures of Mercury in 1974/5.

Mars The Red Planet; the fourth planet in the solar system going out from the Sun. Its diameter is 6790 km (4220 miles) at the equator. It lies on average about 228 million km (142 million miles) from the Sun and can approach within 56 million km (35 million miles) of Earth. It turns on its axis every 24·4 hours and circles the Sun in 687 days. Mars has a slight atmosphere of carbon dioxide gas. Its polar caps are mainly of water ice. The surface is orange-red in colour and is strewn with small rocks. Among its main features are a vast canyon (Valles Marineris) 5000 km (3000 miles) long and an extinct volcano (Olympus Mons) some 30 km (20 miles) high. Mars has two tiny satellites, Phobos and Deimos.

mascons Regions of high density (mass concentrations) that exist beneath the maria, or 'seas', on the Moon.

maser A device that emits beams of pure microwave radiation. The word is an acronym for 'microwave amplification by stimulated emission of radiation'. It is the microwave equivalent of the optical laser.

mass Generally, the amount of matter in a body; more accurately, a measure of a body's inertia. The mass of a body is different from its weight. Weight results from the gravitational attraction on the mass of the body and thus varies according to the local strength of gravity. In orbit, for example, weight is zero, but the mass remains the same. The standard unit of mass is the kilogram; or in cgs units, the gram.

mass number The number of nucleons—protons and neutrons—in the nucleus of an atom.

mass production The modern manufacturing method of producing large volumes of goods at low unit cost. It is based on the specialist division of labour and the widespread use of machinery. An essential feature of many mass-production processes is the moving assembly line. The introduction of industrial robots is further increasing the efficiency of mass production.

mass spectroscopy A technique used, for example, in chemical analysis to separate atoms and molecules according to their masses. A substance is first vaporized and then ionized. The ions are accelerated by an electric field and then deflected by a magnetic field. They are deflected by varying amounts, depending on their mass and charge and form a so-called mass spectrum, which can be photographed (in a mass spectrograph) or detected electrically (in a mass spectrometer). Particular atoms can then be identified.

match A splinter of wood or strip of cardboard tipped with a substance that ignites when it is struck on a suitable surface. The modern safety match is tipped with a mixture containing antimony sulphide and potassium chlorate. It will ignite only if it is struck against a striking surface containing red phosphorus. 'Strike anywhere' matches usually contain phosphorus sesquisulphide in the tip.

mathematics The study of quantities and numbers; often called the 'queen of the sciences'. Major branches of mathematics include arithmetic, algebra, calculus, geometry, trigonometry and statistics.

matter The stuff of which every material object is made up. All forms of matter occupy space and have inertia, a property measured by their mass. Matter can exist in three normal states on Earth as a solid, liquid or gas. In the stars matter exists as a plasma. Ordinary matter is made up of molecules consisting of atoms bonded together. The atoms are made up of other tiny particles. In nuclear reactions a certain amount of matter is converted into energy (see **Einstein's equation**); matter can be considered a particular form of energy. See also **antimatter.**

mauve Or mauveine; the first synthetic dye, prepared from coal tar by William Perkin in 1856. It is an aniline dye.

maximum/minimum thermometer A thermometer that records maximum and minimum temperatures over a period.

maxwell The unit of magnetic flux in the cgs system, now replaced by the SI unit, the weber.

Md The chemical symbol for mendelevium.

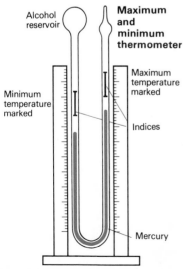

Alcohol reservoir

Maximum and minimum thermometer

Minimum temperature marked

Maximum temperature marked

Indices

Mercury

mean The simple arithmetical average. The mean of a set of marks, for example, is the sum of the marks divided by the number in the set. Compare **median** and **mode.**

mechanical advantage A measure of the effectiveness of a machine. It is the ratio of the force exerted by the machine to the effort, or force applied to the machine.

mechanical engineering The branch of engineering connected with the application of mechanical power. Mechanical engineers, for example, design, develop and manufacture engines and turbines for power production.

memory In electronics, the part of a computer that stores information (data). A computer has two kinds of memory. The ROM (read only memory) permanently stores data essential for the computer to work. The RAM (random access memory) temporarily stores data for each computation.

mechanics The branch of science concerned with the action of forces on bodies. It includes dynamics, kinematics and statics.

mechanization The widespread application of machines, which began during the Industrial Revolution. See also **automation.**

median A kind of average, being the middle value of a series of values when arranged in increasing order. In the series of numbers 2, 5, 7, 12 and 14, 7 is the median.

mega- A prefix meaning one million, as in megawatt (MW), 1 million watts.

megaton A unit used to measure the power of nuclear weapons. A 1-megaton weapon has the explosive violence of 1 million tons of TNT.

melting point The temperature at which a solid changes state into a liquid, when solid and liquid are in equilibrium. For a particular substance, it varies with external pressure. Melting points are generally quoted at standard atmospheric pressure (760 mm mercury).

mendelevium (Md) An artificial radioactive element (at no 101) of the actinide series.

meniscus The curved surface of a liquid in a container, which may be concave (as with water) or convex (as with mercury).

Mercury spacecraft

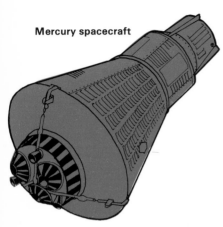

menthol An organic compound related to camphor, noted for its minty taste and smell. It is a white solid (mp 42°C), which produces a cooling sensation when it is rubbed on the skin or when its vapour is inhaled. Its chemical formula is $C_{10}H_{20}O$.

Mercury The planet nearest to the Sun, which lies on average about 58 million km (36 million miles) away. It takes 59 days to spin once on its axis and 88 days to circle once around the Sun. It is a rocky body, 4850 km (3015 miles) in diameter, and looks remarkably like the Moon. It is heavily cratered and has no atmosphere. It is not an easy planet to see because, being so near the Sun, it never rises far above the horizon.

mercury (Hg) Also called quicksilver; one of the few elements (rd 13·55, mp − 38·9°C, bp 356·6°C) and the only metal that is liquid at ordinary temperatures. Its main ore is cinnabar, mercury sulphide (HgS). It is widely used in thermometers and barometers. Its alloys (known as amalgams) with silver and gold are used to fill teeth. As vapour it is used in fluorescent tubes and some street lamps. Mercury exhibits two valencies, or oxidation states—mercurous, or mercury(I); and mercuric, or mercury(II). Calomel, or mercurous chloride (Hg_2Cl_2), is used in medicine. Most mercury compounds are toxic.

Mercury spacecraft The first American manned spacecraft, the type in which pioneer astronaut John Glenn rode into orbit in 1962. Two suborbital and four orbital flights took place, the last in May 1963.

meridian On the Earth, a great circle through the poles perpendicular to the equator. On the celestial sphere, a great circle that passes through the celestial poles, the zenith and the nadir. See **Greenwich meridian.**

meson A kind of elementary particle found in cosmic rays and produced during nuclear bombardment. Mesons belong to a class of particles called hadrons, which have a mass intermediate between leptons (such as electrons) and baryons (such as protons). They include π mesons (pions) and $\varkappa$ mesons (kaons).

mesosphere A region in the Earth's upper atmosphere between about 50 and 80 km (30 and 50 miles), immediately above the stratosphere.

Mesozoic Era A span of geological time between about 225 million and 65 million years ago, covering the Triassic, Jurassic and Cretaceous Periods. Often called the 'age of reptiles', it saw the rise and fall of the dinosaurs.

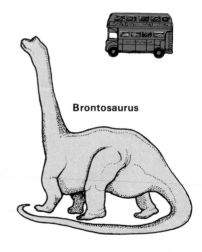

Brontosaurus

Messier number In astronomy, the number given to certain nebulae and star clusters in a catalogue compiled by the French astronomer Charles Messier in 1784. Messier (M) numbers are still often used for identification—M1 is the Crab Nebula.

metaldehyde A polymerized form of acetaldehyde. A white flammable solid, it is used as a fuel in portable heaters.

metals Something like three-quarters of all elements are metals. In general metals have a high density and are hard and shiny. They conduct heat and electricity well. They can be drawn into thin wire (are ductile) and be hammered into thin sheets (are malleable). They are solids with a crystalline structure. In chemical combination, they tend to lose electrons and form positive ions. Not all metals possess all these properties, but most metals possess most of them. One metal is liquid at room temperature—mercury.

metallic bond The kind of chemical bonding found in a metal, in which the atoms are densely packed in a crystalline lattice. Within the lattice the electrons are loosely held and are able to move from atom to atom, allowing the phenomenon of electrical conductivity.

metalloid An element intermediate between a metal and a non-metal, which displays certain properties of both. The metalloids include antimony, arsenic, boron, germanium, silicon and tellurium. Most are semiconductors.

metallurgy The science and technology that deals with extracting metals from their ores and preparing them for use. It is concerned with mineral dressing, smelting and refining operations, and shaping processes such as casting, rolling, forging and extrusion.

metamorphic rock One of the three main types of rocks. Metamorphic rocks are formed when heat and pressures in the Earth's crust alter existing rocks. Shale, for example, undergoes metamorphism and becomes slate.

meteor A bright streak often seen in the night sky caused by a small piece of stony or metallic matter from outer space burning up as a result of friction with the Earth's atmosphere. Meteor showers (or swarms) are seen at certain times of the year when the Earth passes through the debris from ancient comets.

meteor crater See **Arizona Meteor Crater.**

meteorite A lump of stone or metal from outer space that has survived a fiery passage through the Earth's atmosphere. The stony variety, called aerolites, are composed of silicates; while the metal variety, called siderites, are composed of iron/nickel alloys.

meteorology The study of the atmosphere and the weather, particularly with a view to weather forecasting.

methanal See **formaldehyde.**

methane A flammable gas, the main constituent of natural gas. It is found in coal mines in firedamp and is produced in marshes (as marsh gas) by the decomposition under water of vegetable matter. Methane is the first member of the hydrocarbon series of paraffins, or alkanes, having a chemical formula CH_4.

methanoic acid See **formic acid.**

methanol Or methyl alcohol, CH_3OH; the first member of a series of alcohols derived from the alkanes. It can be produced by distilling wood, and is often called wood alcohol. But it is now mainly synthesized from carbon monoxide and hydrogen. It is an invaluable industrial solvent and raw material for chemical synthesis.

methylated spirits A mixture of ethanol with methanol and pyridine and a violet dye; used as a fuel.

methylbenzene See **toluene.**

methyl group The group CH_3-, one of the commonest hydrocarbon groups, or radicals, in organic chemistry, derived from the alkane methane (CH_4). Among common methyl compounds are methanol (CH_3OH) and toluene ($C_6H_5CH_3$).

methyl orange A common laboratory indicator that is red in acid solution and yellow in alkaline solution.

Metonic cycle A cycle of 19 years (or 235 synodic months) after which time the phases of the Moon are repeated on the same days of the months. It was discovered by the Greek astronomer Meton in the 5th century BC.

metre The standard unit of length in the metric system, defined as the length equal to $1,650,763 \cdot 73$ wavelengths of the orange-red line in the spectrum of the isotope krypton-86 in a vacuum. 1 metre (m) = $39 \cdot 37$ inches or $1 \cdot 09$ yards.

metric system

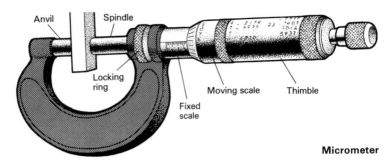

Anvil Spindle

Locking ring

Moving scale Thimble

Fixed scale

metric system A system of weights and measures based on a decimal system of units. It was introduced by the National Assembly in Revolutionary France in 1791. For years it has been used worldwide for scientific measurements and for general purposes in most countries. Some English-speaking countries, however, still adhere to the Imperial system of units.

Mg The chemical symbol for magnesium.

MHD An abbreviation for magneto-hydrodynamics. MHD is a field of study relating to the behaviour of electrically conducting fluids in the presence of electromagnetic fields.

mho Or reciprocal ohm; a unit of electrical conductance, now properly called the siemens.

mica A common silicate mineral noted for its plate-like, flaky structure, which enables it to cleave readily into thin sheets. Among the various types of mica are the common white muscovite and the black biotite.

micro- A prefix (symbol μ) meaning one-millionth when used with a unit, as in microfarad (μF), one-millionth of a farad. When used generally, the prefix means 'very small' as in microscope, microelectronics, etc.

microfilm Film containing greatly reduced images of pictures, documents, and so on. Microfiche is a piece of film attached to a file card on which images are reduced to the size of a small dot.

micrometer A simple measuring device used in the eyepiece of optical instruments such as telescopes and microscopes. It may consist of a scaled grid, or graticule, or moving hairlines moved by a precision screw.

micrometer gauge An instrument for measuring lengths or diameters with great accuracy. It works on the principle that a screw advances a precise amount when it is turned.

micron A unit of measurement (μ) used for small dimensions (as in microscopy) equal to one-millionth of a metre. In SI units it has been replaced by the micrometre (μm).

microphone A device that converts sound waves into variable electrical signals for transmission by wire or radio wave. The essential features of a microphone are a diaphragm, which vibrates when sound waves hit it, and a device to translate these vibrations into a variable electric current. The device may be a piezoelectric crystal, a moving coil or a ribbon suspended in a magnetic field; a capacitor; or, in the case of a telephone microphone, carbon granules.

microprocessor A silicon chip that is self-contained and can function as the central processing unit of a computer. It is often called the 'computer on a chip'.

microscope An instrument for producing enlarged images of tiny objects. A magnifying glass is a simple microscope. The compound microscope consists of two lenses, whose distance apart can be altered to provide focusing. The lower, objective lens forms an enlarged image of the object, which is then viewed and magnified further by the eyepiece lens. The practical limit to magnification with this, the optical microscope is about 2000X. Much greater magnification is provided by the electron microscope.

microtome A machine for cutting very thin slices of specimens for viewing under the microscope.

microwave oven An oven in which food is cooked by being irradiated with microwaves. The microwave radiation penetrates the food and causes its molecules to align themselves continually in different directions. This agitation uniformly and rapidly heats up the food—a chicken roasts in 15 minutes, a potato bakes in 4.

microwaves A form of electromagnetic radiation, with wavelengths ranging from about 1 mm to 30 cm. Microwaves are used in radar and, domestically, in the microwave oven.

mild steel The commonest kind of steel, which contains up to about 0·25% carbon.

mile A standard unit of distance in most English-speaking countries, equal to 1760 yards or 5280 feet. 1 mile = 1·61 km; 1 km = 0·62 miles. The British nautical mile equals 6080 feet. The international nautical mile is 1852 metres (6076) ft.

milk of magnesia A water emulsion of magnesium hydroxide, $Mg(OH)_2$, used as an antacid and laxative.

Compound microscope

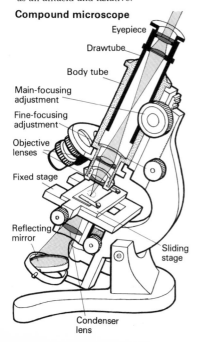

- Eyepiece
- Drawtube
- Body tube
- Main-focusing adjustment
- Fine-focusing adjustment
- Objective lenses
- Fixed stage
- Reflecting mirror
- Sliding stage
- Condenser lens

milk sugar See **lactose.**

Milky Way The diffuse milk-like band that spans the heavens which, when observed through a telescope, resolves into a dense star field. The Milky Way is also the name of the galaxy to which our Sun and all the other stars we can see in the sky belong. The Milky Way in the night sky is a view of a cross-section of our galaxy from the inside.

milli- A prefix meaning one-thousandth, as in millimetre (mm), one-thousandth of a metre.

millibar A unit of pressure used in meteorology, see **bar.**

milling A machining operation carried out with a rotating wheel edged with cutting teeth. On a milling machine, the workpiece moves back and forth beneath the cutter.

mineral A compound of inorganic origin found in the Earth's crust. Every mineral has a definite composition and physical characteristics. Most are crystalline in form. Rocks are made up of one or more minerals. Minerals that are mined and processed to produce metals are called ores. In a more general sense, minerals can be considered substances extracted from the ground by mining, and thus include the fossil fuels—petroleum, coal and natural gas. Among characteristic mineral properties are hardness, lustre and density.

mineral dressing Preparing mineral ores for processing. It includes such things as crushing, separation of unwanted material, and concentration of the ore by such means as flotation.

mining The extraction of coal and minerals from the ground either on the surface (opencast or openpit mining) or underground. The extraction of rock and gravel on the surface is usually termed quarrying. Strip mining is a form of surface mining in which huge power excavators strip off a thin surface layer of earth above the mineral deposit. In hydraulic mining water jets are used to break up and wash away deposits (of clay and gold-bearing gravels, for example).

minor planet See **asteroid.**

Miocene Epoch A period of geological time, extending from about 26 million–7 million years ago, during which higher primates like the apes evolved.

Mira Ceti

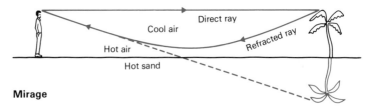

Mirage

Mira Ceti Or o Ceti; the first variable star to be discovered, in the constellation Cetus. It usually varies in magnitude between about 8 and 3, but may exceptionally reach 1st magnitude. Its period is about 330 days.

mirage An optical illusion commonly experienced where a surface is very hot, as in the desert or on roads in the summer. It makes an object, such as a tree, appear to be reflected in a pool of water. Light rays travelling downwards from the top of the tree are gradually refracted upwards as they pass through layers of increasingly warm air and enter the eye. To the eye, it seems that the rays originated below the surface.

mirror A polished surface that reflects light in a regular manner, forming visible images. It usually consists of a sheet of glass silvered on the back. The surface may be flat, giving a plane mirror, or it may be curved, as in a spherical or parabolic mirror. Convex spherical mirrors curve outwards and produce a smaller upright and virtual image. Concave mirrors curve inwards and produce either an inverted real image or an upright virtual image depending on where the object is located.

Misch metal An alloy used for lighter flints, containing cerium and other rare-earth metals.

miscibility The tendency of liquids to mix together. Some liquids (such as alcohol and water) are completely miscible in all proportions. Some (such as water and phenol) are only partly miscible; and others (such as oil and water) are completely immiscible.

mistral A cold dry northerly wind that sweeps down from the Alps along the Rhône Valley in southern France. It blows for up to 100 days a year.

MKS units A system of units based on the metre, kilogram and second, which forms the basis of the SI unit system.

Mn The chemical symbol for manganese.

Mo The chemical symbol for molybdenum.

mode One kind of average used in mathematics. In a set of numbers, the mode is the number that occurs most frequently. In the set 6, 9, 5, 6, 3, 6 and 7, the mode is 6.

moderator In a nuclear reactor, a substance that slows down the neutrons produced during fission to a speed at which they are more likely to bring about fission in other atoms. Reactors with a moderator (such as graphite and heavy water) are termed thermal reactors.

modulation In electronics, varying the characteristics of a radio wave (the carrier wave) by variable signals, representing (in radio transmission) sounds or (in television transmissions) pictures. See **AM; FM.**

modulus A constant or factor of proportionality, as in Young's modulus, for example.

mohair The smooth, lustrous hair of the Angora goat, which is exceptionally resilient and hardwearing.

Mohs hardness A measure of the hardness of a mineral on a scale devised by the German mineralogist Friedrich Mohs. On the scale (see table), which goes from 1 to 10 in order of increasing hardness, a mineral scratches the minerals of lower number and is scratched by minerals of higher number.

1 Talc	6 Orthoclase
2 Gypsum	7 Quartz
3 Calcite	8 Topaz
4 Fluorite	9 Corundum
5 Apatite	10 Diamond

molality The concentration of a solution expressed as the number of moles of solute per kilogram of solvent.

molarity The concentration of a solution expressed as the number of moles of solute dissolved in one litre of solution.

mole A unit of quantity of chemical substance, being the amount (in grams) of a substance that contains as many chemical units (atoms or molecules) as there are in 12 grams of the carbon-12 isotope. It is numerically equal to the molecular weight (relative molecular mass) of the substance in grams. The number of molecules in 1 mole of anything is $6 \cdot 02 \times 10^{23}$—the Avogadro number.

molecular weight Or relative molecular mass; the number of times the molecule of a substance is heavier than one-twelfth of an atom of the carbon-12 isotope. It is the sum of the atomic weights (relative atomic masses) of all the atoms in the molecule.

molecule The smallest part of a substance that can have an independent existence and retain the typical properties of that substance.

molybdenite The main mineral source of molybdenum, being molybdenum disulphide (MoS_2). It has a flaky structure similar to that of graphite, and likewise can be used as a lubricant.

molybdenum (Mo) A transition metal (rd $10 \cdot 2$, mp $2620°C$), used mainly in alloy steels.

moments The moment of a force, or torque, is the tendency of the force to rotate the body to which it is applied. The moment of a force about a fixed point is numerically equal to the product of the magnitude of that force and the perpendicular distance from the line of action of that force to the point. The moment of inertia is a measure of the rotational inertia of a body.

momentum The product of the mass and velocity of a body. It is a vector quantity, having direction. A rotating body has angular momentum.

monazite A complex phosphate mineral containing cerium, lanthanum, traces of other rare-earths, and thorium.

Mond process A method of refining nickel devised by the German-born chemist Ludwig Mond. It involves treating inpure nickel with carbon monoxide (CO) to form the volatile nickel carbonyl, $Ni(CO)_4$, which readily decomposes to yield pure nickel.

Monel A copper-nickel alloy of excellent chemical and corrosion resistance, containing typically about 66% nickel, 30% copper and traces of iron, silicon and manganese.

monobasic acid An acid having one replaceable hydrogen atom, giving rise to one series of salts. Hydrochloric acid is monobasic.

monochromatic light Light of one wavelength, or colour.

monoclinic system See **crystal**.

monomer Small molecules, usually of organic compounds, that can react with one another or with other small molecules to form long-chain molecules (polymers). Ethylene is a monomer, which will react to give the polymer polyethylene (polythene).

monorail A railway that uses only a single rail rather than a twin-rail track. There are two basic types—the suspension monorail, in which the passenger car is suspended by an arm from the track; and the straddle type in which the car straddles the track. The most successful monorail is the Wupperthal Schwebebahn in Germany, introduced in 1901.

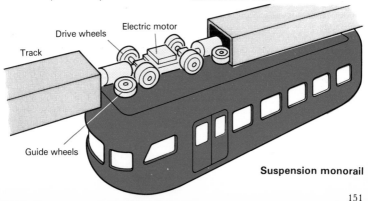

Electric motor
Drive wheels
Track
Guide wheels

Suspension monorail

monosaccharides

monosaccharides Simple sugars, such as pentoses ($C_5H_{10}O_5$) and hexoses ($C_6H_{12}O_6$), which form the 'building blocks' for more complex carbohydrates. Glucose and fructose are monosaccharides.

monosodium glutamate A widely used food flavouring, being the monosodium salt of glutamic acid, which in an amino acid. It is obtained from gluten or soya-bean protein.

monotropy A type of allotropy in which only one allotrope is stable under all conditions.

monotype A method of typesetting in which lines of type are assembled from individually cast characters. Compare **linotype.**

monsoons Seasonal winds that blow from land to sea during the winter and from sea to land during the summer. They are caused by the uneven heating and cooling of land and sea and are the large-scale equivalent of land and seas breezes. The summer monsoons are wet, the winter monsoons dry. The best-known monsoons occur in the tropical regions of South-East Asia.

month A period of time based upon the movement of the Moon around the Earth. The sidereal month of $27 \cdot 3$ days is the time it takes the Moon to rotate on its axis, with respect to the stars. The synodic months of $29 \cdot 5$ days is the time between similar phases of the Moon (eg two full Moons). Calendar months have 30 or 31 days, except for February, which has 28 days normally but one extra day in each leap year.

Moon The Earth's only satellite and nearest neighbour in space, lying on average some 384,000 km (239,000 miles) away. With a diameter of only 3476 km (2160 miles) and a relative density of $3 \cdot 3$, it has a mass 1/81 that of the Earth. Its surface gravity is only one-sixth that of the Earth. It is an airless and waterless world, which suffers extremes of temperature and where no life can exist. The Moon had a captive rotation, which means that it keeps the same side facing towards us as it revolves around the Earth. Two main surface features are visible from the Earth—dark plains called maria (seas) and rugged highlands, which appear bright. The highland regions are heavily cratered. Between 1969 and 1972, 12 American Apollo astronauts explored the Moon on foot and returned with 385 kg of soil rocks and core samples that are still being analysed. The Moon rocks are of two main types similar to basalts and breccias on Earth. They contain the same elements as terrestrial rocks, but are richer in certain elements such as chromium, titanium and zirconium. See also **phases of the Moon.**

moon A satellite.

moraine A mass of debris that accumulates around a glacier as it erodes the surrounding rocks and is deposited when the glacier retreats.

mordant A chemical compound used in dyeing to fix dyes to the fabric fibres. Alums are common mordants.

morning star The planet Venus when it appears in the eastern sky before sunrise.

morphine The most important alkaloid in opium, used as a painkilling drug, formula $C_{17}H_{19}NO_3$. A powerful narcotic, it is extremely habit-forming, quickly leading to addiction.

Morse code A system of signals devised by the American inventor Samuel Morse for sending messages by his telegraph. It uses a code of dots and dashes to represent letters, figures and punctuation (see table).

mortar A building material made from cement, lime and sand, used in bricklaying.

mother of pearl Also called nacre; an iridescent material that lines the shells of certain molluscs, such as the pearl oyster and abalone. It consists of thin layers of calcium carbonate.

International Morse Code					
A ·—	B —···	C —·—·	D —··	E ·	F ··—·
G ——·	H ····	I ··	J ·———	K —·—	L ·—··
M ——	N —·	O ———	P ·——·	Q ——·—	R ·—·
S ···	T —	U ··—	V ···—	W ·——	X —··—
Y —·——	Z ——··	1 ·————	2 ··———	3 ···——	4 ····—
5 ·····	6 —····	7 ——···	8 ———··	9 ————·	0 —————

Close-up of the Moon's Alpine Valley taken by an American Lunar Orbiter probe from a distance of about 250 km.

motion pictures Films shown on a screen which give the impression of continuous motion. Actually what happens is that a series of still photographs taken in rapid succession (at a rate of 24 per second) are projected on the screen at the same rate. Because of the persistence of vision, the eye merges succeeding pictures together to give an impression of motion. See also **ciné camera; projector.**

motor A device that changes one kind of energy into mechanical power, such as an internal combustion engine.

moving-coil instrument One that uses a fine wire coil suspended in a magnetic field, such as a moving-coil ammeter. When current flows through the coil, it sets up a magnetic field. This interacts with the external magnetic field, causing the coil to twist and the instrument needle to register.

multiple proportions, law of When two elements combine with one another to form more than one compound, the weights of one element which combine with a fixed weight of the other are in simple proportion. For example, carbon combines with oxygen to form carbon monoxide and carbon dioxide. It is found that 12 grams of carbon combine with 16 and 32 grams of oxygen respectively to form these compounds, the ratio of the weights of oxygen being 1:2. The law was put forward by John Dalton in 1804 and provided convincing evidence in favour of his atomic theory.

multiple star A star that consists of three or more separate stars linked together by gravity. To the naked eye it appears as a single star but resolves in the telescope and spectroscope into separate components. Mizar, the middle star in the handle of the Plough, consists of six components.

multistage rocket A rocket that consists of a number of separate rockets joined together. All space launching vehicles have multiple rocket stages, usually three. Each stage gives the stages above it a boost before separating. In this way an advantageous overall power-to-weight ratio can be achieved.

muon Or mu-meson; a kind of elementary atomic particle classed now, not as a meson, but as a lepton.

muscovite See **mica.**

myopia See **shortsightedness.**

N

N The chemical symbol for nitrogen.

Na The chemical symbol for sodium, from the Latin 'natrium'.

nadir In astronomy, the point on the celestial sphere directly below an observer; ie on the other side of the Earth.

nano- A prefix meaning one-thousand-millionth, as in nanosecond (10^{-9} second).

napalm A kind of jelly petrol used in flame throwers and incendiary devices. The gel-forming substance is a soap.

naphtha A volatile mixture of liquid hydrocarbons, used as solvents and for making into petrol. It is obtained from petroleum, coal tar and shale oil.

naphthalene A cyclic hydrocarbon ($C_{10}H_8$) with a characteristic smell, used to make dyes and moth-balls.

narcotic A drug that dulls the senses, relieves pain and induces lethargy and sleep. Morphine, opium, cocaine and heroin are among dangerous narcotics that are highly addictive.

NASA The abbreviation for the National Aeronautics and Space Administration, the American agency responsible for the nation's programme for space exploration; founded in 1958.

nascent hydrogen Hydrogen being produced by a chemical reaction, in which form it is a much stronger reducing agent than usual.

native elements Elements that are found pure naturally in the Earth's crust rather than in the form of compounds with other elements. The native elements include the non-metals carbon (as graphite and diamond) and sulphur, and the metals platinum, gold, silver and copper.

natural gas A mixture of gaseous hydrocarbons found in the Earth's crust, resulting from the ages-long decay of marine organisms. The main gases in the mixture are methane and ethane, together with other alkanes, including butane and propane. The last two are removed and liquefied to form bottled gas, or LPG (liquefied petroleum gas).

nautical mile See **mile.**

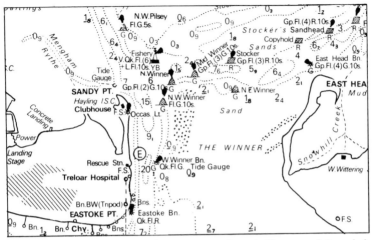

navigation The science of directing a craft or vehicle from place to place. Ancient mariners navigated by dead reckoning and celestial navigation, methods still used today. In dead reckoning, a navigator plots direction and distance from a known position. In celestial navigation, a navigator observes the position of heavenly bodies at a certain time with a sextant and by referring to tables can determine his position. In addition these days there are electronic aids to navigation such as radio direction finders and loran. Submarines may navigate by means of an inertial guidance system.

Nb The chemical symbol for niobium.

Nd The chemical symbol for neodymium.

Ne The chemical symbol for neon.

nebula A cloud of gas and dust between the stars. Bright nebulae emit light themselves or reflect light from nearby stars. Dark nebulae block the light from stars behind them. A planetary nebula is a spherical shell of gas around a central star, formed of matter given off by the star when it exploded as a nova. The term 'extragalactic nebulae' was once used to refer to galaxies.

nebular hypothesis A theory about the origin of the solar system, which suggests that it condensed from a nebula. As the nebula contracted, it began to rotate and formed rings of gas from which the planets and their moons were formed. The core of the nebula became the Sun.

negative, photographic A piece of film that has been developed to show a visible image. It is called a negative because it is light where the scene viewed was dark, and dark where the scene viewed was light.

neodymium (Nd) One of the rare-earth metals (rd $7 \cdot 0$, mp $1024°C$), used in electronics and for making lasers.

neon (Ne) One of the inert, noble gases found in tiny concentrations in the air. It is obtained by fractional distillation of liquid air. It is most widely used for filling discharge tubes (neon lights), which give off a vivid orange-red light.

neoprene One of the first and best synthetic rubbers. Discovered in 1931, it is made by polymerizing chloroprene, which is in turn made from acetylene and hydrogen chloride. It is more resistant to oil, solvents and heat than natural rubber.

Neptune The most distant planet but one (Pluto) in the solar system, circling on average about 4497 million km (2794 million miles) from the Sun. With a diameter of about 50,000 km (31,000 miles) it is the third largest planet. It takes about 15 hr 48 min to rotate once on its axis, and 165 years to circle the Sun. It is a mainly gaseous planet of hydrogen, helium and probably methane. It has two known satellites, Triton and Nereid.

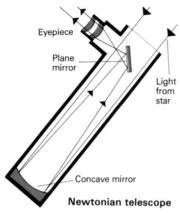

Eyepiece

Plane mirror

Light from star

Concave mirror

Newtonian telescope

neptunium (Np) The first of the artificial elements (at no 93), following uranium in the periodic table and made by bombarding uranium with neutrons. It is radioactive.

nerve gas A poisonous gas that effects the nervous system, usually an organic phosphorus compound.

Nessler's reagent A solution of potassium mercuric iodide ($KHgI_3$) in potassium hydroxide solution, used in chemical analysis as a test for ammonia. When ammonia is bubbled through it, a brownish precipitate is formed.

neutralization A reaction in which an acid reacts with a base to form a neutral solution.

neutrino An elusive elementary particle produced in nuclear reactions. It has neither mass nor charge, but has spin and always travels at the velocity of light. It is very penetrating but has scarcely any effect on the matter it passes through.

neutron A stable elementary particle found in the nucleus of every atom except ordinary hydrogen. It has fractionally greater mass than a proton. It is so called because it is electrically neutral. Because it is electrically neutral, it is very penetrating and can bring about nuclear transmutations, including nuclear fission.

neutron star A hypothetical star of extremely high density and small dimensions (tens of kilometres) made up of solid neutrons. Pulsars could be stars of this type.

new Moon See **phases of the Moon.**

newton The SI unit of force, named after Isaac Newton, being the force that gives a mass of 1 kilogram an acceleration of 1 metre per second per second. 1 newton (N) = 100,000 dynes or 0·225 pound.

Newtonian telescope A reflecting telescope designed by Isaac Newton in 1668. The principle is still widely used. In a Newtonian reflector, the light is gathered by a parabolic mirror and reflected back up the telescope tube to a plane mirror angled at 45°. This mirror deflects the light to the side of the tube, where an eyepiece is located.

Newton's laws of motion describe the relationship between the forces acting on a body and the motion they produce. The first law, the law of inertia, states that a body remains in a state of rest or in uniform motion in a straight line unless it is acted on by an external force. The second law states that the rate of change of velocity—that is, the acceleration (a)—produced by a force is proportional to that force (F) and inversely proportional to the mass (m) of the body: $a = F/m$ or $F = ma$. The third law states that to every action there is an equal and opposite reaction.

Newton's rings A series of alternately bright and dark circles or other patterns that can be seen when a convex piece of class rests on a flat glass plate. The rings are caused by interference between light reflected from the top of the flat surface and bottom of the convex surface.

NGC The abbreviation for 'New General Catalogue of Nebulae and Clusters of Stars'. Nebulae, star clusters and galaxies are commonly identified by an NGC number, the number in the Catalogue which was compiled originally in 1888 by J. L. E. Dreyer. The Andromeda nebula is NGC224.

Ni The chemical symbol for nickel.

niacin Also called nicotinic acid; one of the B complex vitamins found in lean meat, whole grain cereals and green vegetables. Deficiency of niacin causes pellagra.

nickel (Ni) One of our most important metals (rd 8·9, mp 1453°C), which is hard and corrosion resistant. It is widely used in alloys, such as cupronickel (with copper), invar (with iron) and nichrome and stainless steel (with chromium and iron). It is also used for nickel plating and as a catalyst. It is chemically very similar to iron and cobalt, and like them is strongly ferromagnetic. Its major ore is the iron-nickel sulphide, pentlandite.

nickel silver Also called German silver; an alloy containing copper, nickel and zinc, used as a base for silver plating. The letters EPNS on cutlery stand for 'electroplated nickel silver'.

Nicol prism A device made from calcite crystals which produces plane polarized light.

nicotine A poisonous oily liquid found in tobacco leaves and in cigarettes. It is an alkaloid, $C_{10}H_{14}N_2$.

nicotinic acid See **niacin.**

nimbus A rain cloud. Nimbostratus is the typical dark-grey shapeless rain cloud from which steady rain usually falls. Cumulonimbus is the type of huge billowing cloud with a dark base associated with thunderstorms.

niobium (Nb) Once called columbium; a transition metal (rd 8·6, mp 2420°C) with excellent corrosion resistance. It is used in some stainless steels and super-conductors.

nitrates The salts or esters of nitric acid.

nitre See **saltpetre.**

nitric acid (HNO_3) One of the three strong mineral acids widely used in the chemical laboratory and in industry, particularly to make fertilizers and explosives. A colourless fuming liquid, it is the parent of the nitrates. Nitric acid is a powerful oxidizing agent, and will quickly attack organic tissue, staining skin brown. It is manufactured primarily by the catalytic oxidation of ammonia.

nitric oxide See **nitrogen oxides.**

nitride A compound in which nitrogen is combined with a metal.

nitriles See **cyanides.**

nitrocellulose A highly flammable substance made by reacting nitric acid with the cellulose in woodpulp or cotton linters. It is cellulose nitrate, and varies in properties according to the degree of nitration. Highly nitrated it becomes the explosive guncotton, which is used as a propellant. With less nitrogen, it is called collodion or pyroxylin, and is used as a solvent in lacquers.

nitrogen (N) The colourless, odourless and inactive gas that makes up 78% by volume of the air. It occurs also in minerals, such as Chile saltpetre (sodium nitrate). It is an essential element in the metabolism of living things, being found in proteins and nucleic acids. Nitrogen is obtained by the fractional distillation of liquid air. Most of it is made into ammonia by the Haber process and thence into fertilizers and nitric acid. The circulation of nitrogen in the biosphere is an important natural process (see **nitrogen cycle**).

nitrogen cycle A process by which nitrogen from the air is 'fixed', or taken from the atmosphere, and eventually returned there, by natural processes. Involved are the action of lightning, soil bacteria, plants and animals.

Nitrogen cycle

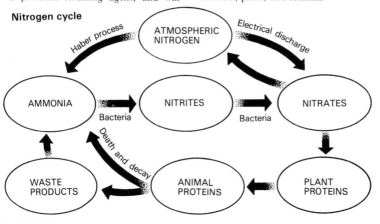

nitrogen oxides

nitrogen oxides Nitrogen forms several oxides. Nitric oxide (NO) is a toxic gas, which combines with more oxygen to form nitrogen dioxide (NO_2). Nitrogen dioxide is a reddish gas that is also toxic. It is found in smogs and car exhaust fumes. It coexists with its dimer, dinitrogen tetroxide (N_2O_4). The most useful oxide is nitrous oxide (N_2O), used as an anaesthetic and commonly called laughing gas.

nitroglycerin A colourless oily liquid, which is a powerful explosive, forming the essential ingredient of dynamite. It is the nitrate ester of glycerin, and is properly termed glyceryl trinitrate, $C_3H_5(ONO_2)_3$.

nitrous acid A weak acid (HNO_2), source of the nitrites.

nitrous oxide See **laughing gas.**

No The chemical symbol for nobelium.

nobelium (No) An artificial radioactive element (at no 102) obtained by the nuclear bombardment of curium.

Nobel prizes Annual awards in the sciences (physics, chemistry and medicine), literature and peace, established in 1901 at the bequest of inventor of dynamite Alfred Nobel, who had died five years earlier.

noble gases Also called rare or inert gases; gases in group O of the periodic table, distinguished by their chemical inertness. They are, from lightest to heaviest; helium, neon, argon, krypton, xenon and radon. They are stable because they have complete electron shells with no electrons available for combination. In general in chemical combination, an atom gains, loses or shares electrons so as to attain a noble-gas structure. This is a cornerstone of the electronic theory of valency.

noble metals Metals such as platinum, gold and silver, which do not corrode or tarnish readily. They are chemically very inactive.

nodes Of a wave form, points that are not displaced from the equilibrium position. In astronomy, two points where the orbit of the Moon or a planet intersects with the ecliptic. Eclipses of the Sun and the Moon can occur only when the Moon is at or near a node.

noise In acoustics, any unwanted sounds. In electronics, unwanted signals that cause interference with a selected signal.

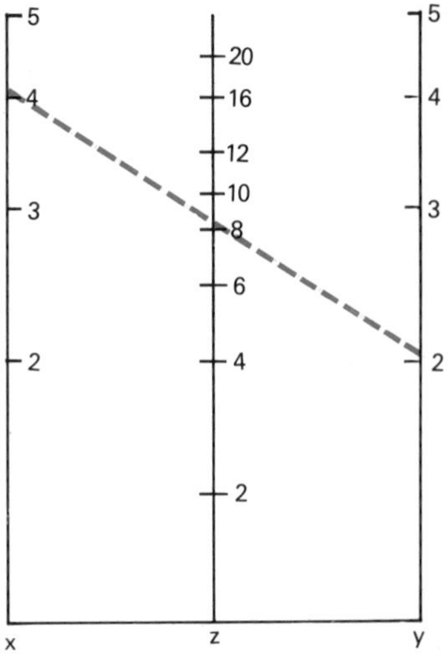

nomogram Or nomograph; a diagram used to carry out calculations involving more than two variables. The simplest type consists of three vertical scales representing values of the variables. The nomogram shown solves the equation $z = xy$.

non-ferrous metal Any metal other than iron and steel.

normal A line drawn at right-angles to a surface.

normal solution A solution containing the gram equivalent weight of a substance per litre. Normality was once widely used to express solution concentrations, but has been superseded by molarity, in which concentrations are expressed in moles per litre.

northern lights Or aurora borealis; see **aurora.**

north star See **Polaris.**

nova A star that suddenly brightens and then fades, thought to be caused by the star partly exploding and ejecting material into space. A nova may increase in brightness as much as a million times. See also **supernova.**

Np The chemical symbol for neptunium.

NTP An abbreviation for 'normal temperature and pressure', conditions of $0°C$ and 1 atmosphere pressure; now termed STP.

Nubecula Major and Minor See **Magellanic Clouds.**

nuclear bombs See **A-bomb; H-bomb.**

nuclear energy Or atomic energy; the energy released in nuclear fission or fusion when atoms split or combine to form new atoms. The energy comes from the conversion during these nuclear reactions of a small quantity of mass, according to Einstein's equation, $E = mc^2$. Nuclear energy is released uncontrolled in the immensely powerful nuclear bomb, and under control in nuclear reactors.

nuclear fission See **fission.**

nuclear fusion See **fusion, nuclear.**

nuclear magnetic resonance A recent analytical and diagnostic technique, which exploits the characteristic way substances absorb and emit radiation in the presence of a strong magnetic field.

nuclear reaction A reaction in which the nucleus of an atom undergoes change. Such a change may occur spontaneously in a nucleus because of its radioactivity. It may also occur as a result of bombardment with atomic particles, or from fusion with another nucleus.

nuclear reactor Once called atomic pile; a device designed to produce, sustain and control a nuclear fission reaction. The first successful reactor was built in a squash court at the University of Chicago in December 1942 under the direction of Enrico Fermi. Nuclear reactors are used experimentally for the manufacture of radioisotopes, and for producing power. Over 200 commercial power reactors producing electricity are in use worldwide. Other reactors power submarines. In a typical power reactor heat is produced in a core by the fission of uranium atoms. A gas or liquid coolant circulates through the core and extracts the heat. It transfers this heat to water in a heat exchanger, causing it to boil into steam. The steam then drives conventional steam turbogenerating machinery to produce electricity. Most reactors are known as thermal reactors because they require slow, or thermal neutrons to cause fission of the uranium. They incorporate a moderator (graphite or water) in the core to slow the neutrons down. Reactors are usually classified according to the type of coolant they use: magnox gas-cooled reactor; advanced gas-cooled reactor (AGR); pressurized-water reactor (PWR); boiling-water reactor (BWR); steam-generating heavy-water reactor (SGHWR). Another type of reactor now coming into limited use is the fast, or breeder reactor, which uses fast neutrons and thus requires no moderator. It is cooled by liquid sodium. Control over all types of reactor is exercised by means of control rods, made of neutron-absorbing material. When they are pushed further into the reactor core, they slow down the nuclear reaction.

Advanced Gas-cooled Reactor (AGR)

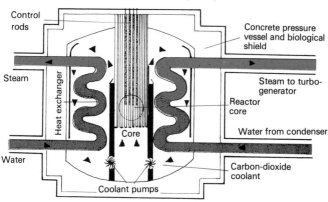

Control rods

Concrete pressure vessel and biological shield

Steam

Heat exchanger

Steam to turbo-generator

Reactor core

Core

Water from condenser

Water

Carbon-dioxide coolant

Coolant pumps

nucleic acids

nucleic acids See **DNA; RNA.**

nucleons The two stable elementary particles in the atomic nucleus—protons and neutrons.

nucleotides The substances that make up the units in the DNA and RNA long-chain molecules. They consist of a sugar group, a phosphate group and a base containing nitrogen. The long-chain molecules, or polymers, are made up of alternate sugar and phosphate groups, with the bases off to the side of the sugar groups.

nucleus, atomic The centre of an atom, which is positively charged. With the exception of hydrogen, all atoms are made up of positively charged protons and electrically neutral neutrons. The hydrogen nucleus consists of a single proton. The number of protons in the nucleus is called the atomic number; the number of protons and neutrons together, the mass number.

numerical control A means of controlling machine tools by means of a computer-generated tape carrying coded instructions in the form of numbers.

nylon A plastic and synthetic fibre developed by Wallace H. Carothers in the early 1930s. It is a polyamide, made by the polymerization of adipic acid and hexamethylene diamine. Nylon is tough, elastic, strong and inert.

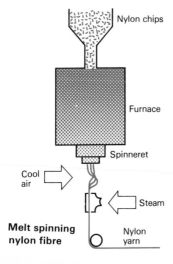

Nylon chips

Furnace

Spinneret

Cool air

Steam

Melt spinning nylon fibre

Nylon yarn

O The chemical symbol for oxygen.

objective The lens nearest the object under study in a microscope or telescope.

oblate Of a spheroid, flattened at the top and bottom. Because of their rotation, the planets are oblate, with their polar diameter smaller than their equatorial diameter. This is most pronounced with Jupiter and Saturn, which are largely fluid.

observatory A place built for astronomical observations, where optical, radio and solar telescopes may be used.

obsidian Volcanic glass, formed by very rapid cooling of lava. Usually jet-black, it is also found in red, brown and mottled varieties.

obtuse angle An angle between 90° and 180°.

occultation An eclipse of one celestial body by another, usually of a star or a planet by the Moon.

octane number A measure of the resistance of petrol to knocking, or pre-ignition. It is the percentage of isooctane in a mixture of isooctane and heptane that has the same knocking characteristics as the petrol in question. Most car engines run on petrol with an octane rating of about 97.

octet The set of eight electrons in the outer shell of noble gases, which gives them their chemical inertness.

ocular An alternative name for an eyepiece, as of a telescope.

oersted The unit of magnetic field strength in the cgs system. It is the strength of a magnetic field in a vacuum in which a unit magnetic pole would experience a force of 1 dyne. It was named after electromagnetism pioneer Hans Christian Oersted.

oestrogens The female sex hormones.

offset litho A common method of printing from a flat plate, see **lithography.**

ohm The SI unit of electrical resistance, named after Georg Simon Ohm. The resistance in a conductor is 1 ohm when a potential difference of 1 volt results in a flow of current of 1 amp.

Ohm's law The potential difference (V) between the ends of a conductor is directly proportional to the current (I) flowing through it. $V \propto I$, or $V = IR$, where R is the electrical resistance. Georg Simon Ohm established the law in 1827.

oil In general a substance that is insoluble in water but soluble in ether and feels greasy to the touch. The most familiar kind of oil is mineral oil, or petroleum (literally 'rock oil'). Animals and plants yield oils, which are liquid mixtures of glycerides of fatty acids. (Such substances are termed fats if they are solid.) They are called 'fixed' oils or 'volatile' oils according to the ease with which they can be evaporated. Typical fixed oils include olive, linseed, soya-bean and corn oils. Typical volatile, or essential oils, which have a distinctive fragrance, include clove, nutmeg and peppermint oils.

oil shales Or kerogenites; shale deposits rich in petroleum, which can be extracted (though expensively) by distillation. The biggest deposits are the Green River shales in Wyoming, Utah and Colorado in the United States.

olefines Or alkenes; a series of hydrocarbons containing one or more pairs of carbon atoms linked by a double bond. They are unsaturated. The simplest olefine is ethylene: $H_2C{=}CH_2$.

oleum See **sulphuric acid.**

Oligocene Epoch A period of geological time that spanned the period from about 38 million to 12 million years ago.

olivine A common group of silicate minerals found in igneous rocks such as basalt and gabbro. Olivine is often found as olive-green crystals, though it may appear in other colours as well.

onyx An attractive banded variety of the silica mineral agate, used as a gemstone and for carving.

opal A translucent silica mineral used as a gemstone. It is noted for its beautiful display of colours when viewed from different angles. It has a milky or pearly appearance, termed opalescence.

opaque Does not let light (or other wave motion) through; the opposite of translucent.

opencast mining Also called open-pit mining; the extraction of coal and mineral ores from the surface. The extraction of surface rock is usually termed quarrying.

open cluster A group of young stars held together by gravity and travelling together through space. Several open clusters can be seen with the naked eye, including the Pleiades and Hyades in the constellation Taurus.

open-hearth process Or Siemens-Martin process; until recently the main steelmaking process, developed in the 1860s by Pierre Martin and William Siemens. In the open-hearth process pig iron, steel scrap, iron ore and limestone are heated in a shallow hearth by the furnace flames. The air needed to burn the fuel is preheated by means of brick checkerwork regenerators.

Open-hearth process

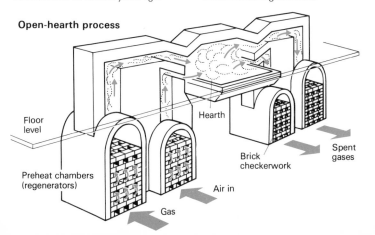

Floor level

Hearth

Preheat chambers (regenerators)

Brick checkerwork

Spent gases

Air in

Gas

ophthalmoscope An instrument for looking inside the eye at the retina and blood vessels.

opium A dangerous narcotic obtained from the unripe fruit of the opium poppy, *Papaver somniferum*. It is processed to yield painkilling drugs such as morphine and codeine.

opposition In astronomy, a situation in which two heavenly bodies appear in opposite directions in the sky. The outer planets are at opposition when they are on the opposite side of the Earth from the Sun, and they are best observed then.

optical activity The ability of some liquids and crystals to rotate the plane of polarization of a beam of light. The plane may be rotated either to the left (laevorotatory) or to the right (dextrorotatory).

Optical illusions

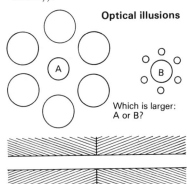

Which is larger:
A or B?

Are the two inner lines parallel?

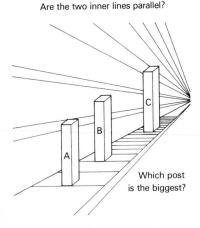

Which post
is the biggest?

optical illusion A scene or picture that deceives the eye, or rather falsifies the brain's interpretation of what the eye sees. It is an optical illusion, for example, that the Sun looks bigger when it is near the horizon than when it is high in the sky. The diagrams show some simple optical illusions.

optical isomerism Isomers that rotate the plane of polarized light in different directions. The mirror-image isomers (stereoisomers) of lactic acid rotate the plane respectively to the left and right, and are termed *l*- (for laevorotatory) and *d*- (for dextrorotatory) lactic acids. See **optical activity.**

optics The study of light.

orbit In astronomy and astronautics, the path through space of one body around another, such as the Earth around the Sun, or a satellite around Earth. See **Kepler's laws.**

orbital velocity The speed required by a natural or artificial satellite to remain in orbit. The orbital velocity for an orbit about 160 km (100 miles) above the Earth is about 28,000 km/h (17,500 mph).

Ordovician Period A geological time span in the Palaeozoic Era between about 500 million and 430 million years ago, during which time the seas teemed with life and primitive fish appeared.

ore A metallic mineral that occurs in deposits concentrated enough to be mined and processed economically into the metal.

organic chemistry One of the main divisions of chemistry, comprising the study of carbon compounds. Such compounds form the basis of living things. It was once thought that they could be derived only from living things, and they were thus termed 'organic'. Other materials were classed as inorganic—derived from non-living sources. See **urea.**

Orion nebula (M42) A bright nebula visible to the naked eye in the constellation Orion. It is about 1,000 light-years away.

Orlon Trade name for an acrylic synthetic fibre.

orrery A mechanical device for showing the relative motions of the heavenly bodies invented by George Graham (about 1710) and named after his patron, the 4th Earl of Orrery.

**ortho-xylene or
1,2-dimethylbenzene**

**meta-xylene or
1,3-dimethylbenzene**

**para-xylene or
1,4-dimethylbenzene**

ortho-, meta- and **para-** Prefixes used to denote the structure of benzene derivatives containing two similar radicals. Xylene is the dimethyl derivative of benzene; that is, two methyl groups ($-CH_3$) have replaced two hydrogen atoms in the benzene structure. They can occupy three different relative positions around the benzene ring, termed ortho-, meta- and para-. These three compounds are structural isomers. They can also be written using numbers to denote the positions of the radicals around the ring.

orthoclase A common feldspar mineral of various colours found in granite, pegmatite and other igneous rocks. It is a potassium aluminium silicate.

Os The chemical symbol for osmium.

oscillator An electronic device for producing alternating current, particularly high-frequency radio waves to act as carrier waves for radio transmission.

oscilloscope A cathode-ray tube device that produces a graphical display of variable electrical signals. Signals are fed to deflecting plates or coils, which manipulate a thin beam of electrons and cause it to trace a pattern on a fluorescent screen.

osmiridium A very hard alloy of osmium and iridium used, for example, to coat the tip of fountain-pen nibs.

osmium (Os) A rare hard metal (rd 22·5, mp 3,000°C) of the platinum group and densest of all the elements. It is sometimes used for hardening platinum alloys and in alloys such as osmiridium.

osmosis The diffusion of water or another solvent through a semipermeable membrane, leaving any dissolved substance (solute) behind. Osmosis is very important in living things, for example, allowing the passage of food and oxygen into the blood, and the removal of wastes from the blood. The flow of solvent through a semipermeable membrane is from a solution of low concentration to a solution of high concentration. Application of pressure can prevent osmosis taking place. This osmotic pressure can be remarkably high. A 30% sugar solution has an osmotic pressure of some 200 atmospheres.

OTEC An abbreviation for 'ocean thermal energy conversion', a scheme for producing power from the temperature difference between hot surface waters and the cold waters of the ocean deeps.

Otto cycle An alternative name for the four-stroke engine cycle, named after Nikolaus August Otto, who built the first successful four-stroke internal combustion engine in 1876.

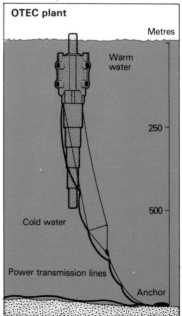

OTEC plant

Metres

Warm water

250

Cold water

500

Power transmission lines

Anchor

ounce

COMMON OXIDE MINERALS

Alumina	Aluminium oxide	Al_2O_3
Cassiterite	Tin oxide	SnO_2
Corundum	Aluminium oxide	Al_2O_3
Cuprite	Copper oxide	Cu_2O
Haematite	Iron oxide	Fe_2O_3
Ilmenite	Iron and titanium oxides	$FeTiO_3$
Litharge	Lead oxide	PbO
Magnetite	Iron oxide	Fe_3O_4
Pyrolusite	Manganese dioxide	MnO_2
Rutile	Titanium dioxide	TiO_2
Uranite	Uranium dioxide	UO_2

ounce A unit of weight in the avoirdupois system. 16 ounces (oz) = 1 pound (lb); 1 oz = 28·3 g. The fluid ounce is a unit of liquid capacity: 20 fluid ounces = 1 pint.

overburden The earth and rock over-laying a mineral deposit, which is removed during opencast mining.

overdrive An extra gear unit fitted to some cars, which gives higher gear ratios on (usually) third and fourth gears.

overtones In sound, faint notes of higher pitch, or frequency than the main note, or fundamental, produced when, for example, a string vibrates. The overtones are superimposed on the fundamental and give the sound its characteristic quality.

oxidation Originally, the combination of a substance with oxygen (as in combustion and corrosion) or the removal of hydrogen from a compound. In a broader sense, oxidation is a reaction in which an atom or group loses electrons. Oxidation is always accompanied by reduction, in which an atom or group gains electrons. The process is often called redox (reduction-oxidation).

oxidation state Or oxidation number; a number numerically equal to the valency of an element, but with a positive or negative sign. In a chemical compound the oxidation state of an element is the number of electrons the atoms have more (-) or less (+) than they do in their neutral state. When describing compounds containing elements that possess more than one oxidation state, it is common to indicate that state by means of Roman numerals. The compound cupric sulphate contains copper with an oxidation state of + 2, and is written copper(II) sulphate.

oxides Compounds of elements with oxygen. The oxides of most metals are ionic crystalline solids and are bases. The oxides of most non-metals are covalent, volatile compounds and are acidic. Some oxides, for example that of aluminium (Al_2O_3), are amphoteric, possessing acidic or basic properties according to circumstances. Some of the commonest minerals are oxides, eg alumina, magnetite, quartz, cuprite and cassiterite.

oxidizer The propellant in a rocket that provides the oxygen to burn the fuel. Common oxidizers are liquid oxygen and nitrogen tetroxide.

oxidizing agent A compound that brings about oxidation. Potassium permanganate, chlorine, concentrated sulphuric acid, concentrated nitric acid and hydrogen peroxide are common oxidizing agents, in addition to pure oxygen, of course.

oxyacetylene torch A device used for welding and cutting metal which burns a mixture of oxygen and acetylene at a temperature of over 3,000°C. In the cutting mode a jet of pure oxygen is directed onto the metal.

oxygen (O) The life-giving gas that makes up about one-fifth of the atmosphere. It is the most abundant element (bp —183°C) in the Earth's crust, where it occurs combined with other elements in oxide, carbonate, sulphate and other minerals. Oxygen in the atmosphere features in the respiration of practically all living things. In the upper atmosphere an allotrope of oxygen occurs—ozone. Oxygen is produced commercially by the fractional distillation of liquid air. Liquid oxygen is a major rocket propellant.

ozalid process A copying process often used to produce printing proofs. Paper coated with diazonium compounds is exposed to ultraviolet light through film containing the image and is developed with ammonia vapour.

ozone An allotrope of oxygen, consisting of three atoms per molecule (O_3). It is bluish in colour and is formed in small quantities when an electric discharge takes place in the air. There is a layer of ozone in the Earth's atmosphere between about 10–50 km (6–30 miles) high, which filters out dangerous radiation from the Sun's rays.

P

P The chemical symbol for phosphorus.

Pa The chemical symbol for protactinium.

pacemaker A device used in medicine to help a patient's heart beat regularly. It delivers minute electric shocks that cause the heart muscles to contract and thus bring about a pumping action. The latest pacemakers are powered by nuclear batteries and are implanted in the patient's chest.

paint A coating applied to protect and decorate a surface. A typical paint consists of a varnish, or vehicle, containing pigment; a solvent, or thinner, to give the paint the right consistency; and a drier to make the paint dry quickly. Modern paints are based on varnishes made from synthetic resins, such as alkyd, phenolic and polyurethane resins. Emulsion paints contain resins such as polyvinyl acetate dispersed in water. Non-drip paints are gels, which are normally jelly-like but flow under brush pressure.

palaeontology A branch of geology concerned with the study of fossils and how they relate to the evolution of life on Earth.

Palaeozoic Era The era of ancient life, being the span of geological time following the Precambrian Era. It extended from about 570 million years ago to about 225 million years ago. The Era is subdivided into six major periods—the Cambrian, Ordovician, Silurian, Devonian, Carboniferous and Permian.

palladium (Pd) One of the platinum-group elements (rd 12, mp 1,552°C), found in association with and resembling platinum. It is used as a hardener for gold, and is a valuable industrial catalyst. It has a great affinity for hydrogen, which it absorbs readily.

Pangaea The original Earth land mass, or supercontinent that split up eventually to form the present continents. See **continental drift**; **plate tectonics**.

panning A method of mining gold carried out by prospectors in the 'gold-rush' days. It involves swirling round gravel from a stream bed in a metal pan. This action washes out lighter material, leaving (hopefully) heavy specks of gold behind.

pantograph A drawing instrument for enlarging or reducing a figure. It consists of a number of hinged rules forming a parallelogram. A sprung arm called a pantograph is used on electric locomotives to make contact with overhead power lines.

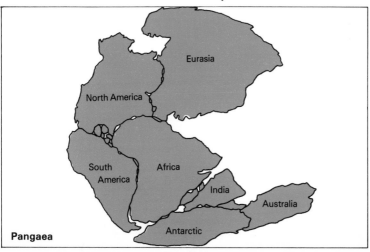

Pangaea

paper

Paper-making process

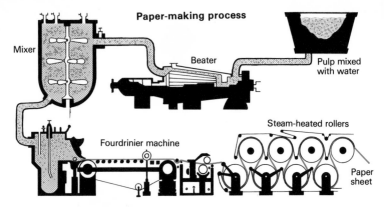

Mixer

Beater

Pulp mixed with water

Steam-heated rollers

Fourdrinier machine

Paper sheet

paper Material used for writing, printing and packaging, made usually from the cellulose in woodpulp. The pulp is mixed with water and beaten to make the fibres flexible. Size and clay are then added and the mixture is allowed to flow on to a moving wire-mesh belt. The water drains or is sucked away, and the damp web resulting is dried and pressed into sheets between heavy rollers. See **Fourdrinier machine**.

paper chromatography A simple form of chromatography using porous paper, such as filter paper. A drop of the mixture under test is placed on the paper, which is then dipped in a suitable solvent. As the solvent (mobile phase) moves through the paper (stationary phase), it gradually separates the components in the mixture.

para See **ortho, meta- and para**.

parabola One type of conic section. An object follows a parabolic curve when it is thrown into the air. Any point on a parabola is the same distance from a fixed line (the directrix) as it is from a fixed point (the focus), see diagram. When a parabola is rotated about its axis, it forms a paraboloid.

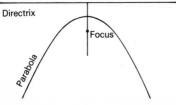

Directrix

Focus

Parabola

parachute An umbrella-like device used to slow down a body falling through the air. Though depicted by Leonardo da Vinci in the 15th century, it was not put to practical use for over 300 years. The modern parachute has a canopy made of nylon, which is attached by suspension lines to a harness.

paraffin A common name for kerosene.

paraffins Or alkanes; a series of hydrocarbons of the general formula C_nH_{2n+2}, beginning with methane (CH_4), ethane (C_2H_6), propane (C_3H_8) and butane (C_4H_{10}). The paraffins are saturated compounds, containing single bonds between the carbon atoms, and are relatively inert chemically. The paraffins mentioned above are gases, the next 11 members in the series are liquids, and the remainder are waxy solids. The paraffin wax used for making candles is a mixture of higher paraffins.

parallax The apparent change in position, or difference in direction, of an object when viewed from different points. When a nearby object is viewed first by one eye and then the other, it appears to change position against a distant background. From this apparent change in position, the distance to the object viewed can be calculated. Astronomers use a parallax method to measure the distances to some nearby stars, by observing them from opposite points in the Earth's orbit around the Sun. The parallax of a star (π) is defined as the angle that would be subtended by the radius of the Earth's orbit at the distance of the star.

parallelogram of forces A method of finding the resultant of two forces acting in different directions. If the magnitude and direction of the two forces are represented by vectors drawn from a point, then the resultant force is represented in magnitude and direction by the diagonal of the parallelogram having the two force vectors as two sides (see diagram). The same method can be used for determining the resultant of two velocities.

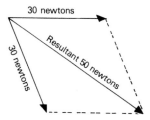

30 newtons

30 newtons

Resultant 50 newtons

paramagnetism A very weak form of magnetism displayed by most elements, especially the transition elements, in a strong magnetic field.

parity A term relating to the symmetry of an object and its mirror image in space.

parsec A unit used to express astronomical distances. It is the distance at which the radius of the Earth's orbit would subtend an angle of 1 second. It is equal to $3 \cdot 26$ light-years.

partial pressures, law of When two or more gases at constant temperature are mixed, each gas exerts the same pressure as if it alone occupied the whole of the containing vessel. The total pressure of the mixture is equal to the sum of the partial pressures. It was first stated by John Dalton in 1801.

particle accelerator See **accelerator, particle**.

pascal The SI unit of pressure, equal to 1 newton per square metre.

Pascal's law When pressure is applied anywhere to an enclosed body of fluid, it is transmitted equally and without loss in all directions. It is named after the French physicist Blaise Pascal.

pasteurization A method of temporarily sterilizing milk and other foods by brief heating. Milk is usually pasteurized by heating it to about 72°C for about 15 seconds. The process is named after the French chemist Louis Pasteur.

patent A legal right granted to an inventor of a device or process to prevent other people exploiting or copying it. Patents originated in England in the 1600s.

patina The attractive greenish-blue coating on copper and bronze that forms as a result of corrosion.

payload The cargo carried by a space rocket, for example, a satellite.

Pb The chemical symbol for lead, from the Latin 'plumbum'.

Pd The chemical symbol for palladium.

pearl A spherical lustrous gem that forms within certain shellfish, particularly the pearl oyster. It grows when layers of nacre, a kind of calcium carbonate secreted by certain cells, form around a grain of sand or other foreign body in the shell. See **mother of pearl**.

peat A spongy fuel and soil conditioner made up of partly decomposed vegetable matter. It represents an early stage in the formation of coal.

pectin A polysaccharide found in fruits and plant cell walls. It forms solutions that set like a jelly, and is responsible for making jams set.

pegmatite A coarse-grained igneous or metamorphic rock in which very large crystals occur, often of gems like beryl.

Peltier effect The heating or cooling that occurs at the junction of two dissimilar metals when electric current flows across it. The effect, discovered in 1834 by the French physicist Jean Peltier, is utilized in some refrigerators. Compare **Seebeck effect**.

Pelton wheel A water turbine, invented in 1889, consisting of a wheel with cup-shaped buckets around the edge.

pen The modern fountain pen was invented by Lewis E. Waterman in 1884, replacing the quill pen and steel pen. The first practical ballpoint pen was invented in the 1930s by the Hungarian Ladislao Biro. It uses a steel ball rolling in a socket to transfer a greasy, quick-drying ink to the paper.

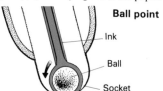

Ball point

Ink

Ball

Socket

Lithium 3	Beryllium 4	Boron 5	Carbon 6	Nitrogen 7	Oxygen 8	Fluorine 9	Neon 10
Sodium 11	Magnesium 12	Aluminium 13	Silicon 14	Phosphorus 15	Sulphur 16	Chlorine 17	Argon 18
					Selenium 34	Bromine 35	Krypton 36
					Tellurium 52	Iodine 53	Xenon 54

Periods

Groups

Part of the Periodic Table

pencil lead A mixture of graphite and clay, used instead of metallic lead for nearly 500 years.

pendulum In its simplest form a string or rod with a weight (bob) at the end, suspended so that it can swing freely back and forth. The pendulum was first investigated scientifically in 1584 by Galileo. The time of swing (T) of a pendulum is determined only by its length (l) and the local force of gravity (g), and is given by the expression, $T = 2\pi\sqrt{l/g}$. The constant time of swing of a given pendulum makes it useful as a regulator for clocks. See also **Foucault pendulum**.

penicillin The original antibiotic drug, discovered by Alexander Fleming in 1928 and manufactured since 1940. It is produced by moulds belonging to the genus *Penicillium*.

pentagon A five-sided figure.

pentane See **paraffins**.

penumbra See **shadow**.

pepsin An enzyme produced in the stomach, which breaks down proteins into peptones during digestion.

percentage A method of writing fractions using hundredths. 1 per cent (%) equals 1 hundredth (1/100); 50% equals 50/100, or ½.

perfect gas One that obeys the gas laws, whose molecules are assumed to have negligible size and to have no effect on one another, rebounding elastically during collisions.

perigee The point in its orbit when the Moon or a satellite is closest to Earth. Compare **apogee**.

perihelion The point in its orbit when the Earth or another planet is closest to the Sun. Compare **aphelion**.

perimeter The distance around a figure. The perimeter of a square is equal to four times the length of one side; the perimeter of a circle is equal to π times its diameter.

period In physics, the time interval in a regularly repeated motion after which the motion begins to repeat itself. In chemistry, a period is a horizontal sequence of elements in the periodic table.

periodic table An arrangement of the chemical elements into horizontal rows (periods) and vertical columns (groups) in order of their atomic number, so as to bring out relationships in their chemical properties. The periodic table was devised by Dmitri Mendeleyev in 1869. In general elements with similar chemical properties fall into the same group. This can be explained according to modern atomic theory, because elements in the same group have a similar outer electronic structure and valency, or combining power. A complete periodic table appears on pages 252/253.

periscope An optical device by which an observer can see over or around obstacles. In its simplest form it consists of a hollow tube containing two plane mirrors angled at 45°. The periscopes used by submarines contain reflecting prisms and lenses to provide magnification and wider-angle vision.

permanent gas A gas, such as nitrogen, which cannot be liquefied at ordinary temperatures by pressure alone. It is a gas above its critical temperature.

permanent magnet One that retains its magnetism, as opposed to an electromagnet, which is magnetic only when the energizing current is switched on.

permanganates Salts of the hypothetical permanganic acid. The commonest is potassium permanganate, $KMnO_4$, which is a powerful oxidizing agent.

permeability The capacity for transmitting fluid. Rock layers, for example, may be permeable and let water through or be impermeable and block its flow. Semipermeable membranes used in osmosis let through the solvent but not dissolved substances. In magnetism, permeability is a quantity used to describe how a material is affected by a magnetising field.

Permian Period A span of geological time between about 280 million and 225 million years ago, during which a diversity of reptiles flourished.

permutations and combinations Permutations are ordered arrangements of a set of objects (elements)—*xyz*, *xzy* and *yxz* are permutations of the set of symbols *x*, *y* and *z*. Combinations are permutations that include the same elements irrespective of the order in which they are arranged. So the permutations given above are all examples of the same combination.

peroxides Inorganic and organic compounds containing two oxygen atoms linked together. Sodium peroxide, Na_2O_2, is a bleaching agent and, like other peroxides, reacts with acids to yield hydrogen peroxide, H_2O_2.

perpetual motion machine One that would carry on working indefinitely without outside interference. Such a machine is not feasible because it violates the law of conservation of energy. The inventor of the endless-chain machine shown here thought that the extra weight of chain on the right would keep pulling the chain round.

Perseids A meteor shower, apparently originating in the constellation Perseus, that occurs on about 11 August.

persistence of vision The ability of our eyes to hold on to an image of what they see for a fraction of a second. This phenomenon makes possible the illusion of 'moving' pictures.

Perspex Also called Plexiglass and Lucite; trade names for a clear, tough acrylic plastic correctly termed polymethylmethacrylate (PMMA).

perturbations In astronomy, slight deviations in the orbital motion of heavenly bodies.

pesticides Substances used to kill pests such as insects, fungi and weeds. They include herbicides, which kill weeds; fungicides, which kill fungi and moulds; insecticides, which kill insects; and rodenticides, which kill rodents such as rats and mice. Many very effective pesticides, such as organic halogen compounds, have the drawback that they can build up in the environment and pose a potential threat to other forms of wildlife and also the human race.

petrified wood A fossilized form of wood whose fibres have over millions of years been replaced by minerals, particularly silica, so that the wood has become rock-like. In places large areas of primeval forest have been petrified, as in the Petrified Forest in Arizona.

petrochemicals Chemicals obtained from petroleum. They include a vast range of organic chemicals that form the main raw materials for the chemical industry. Petrochemicals are obtained by processing less valuable fractions from petroleum distillation and the gases and other by-products of further refinery operations.

petrol Also called gasoline; the lowest-boiling and most valuable fraction from petroleum distillation. It is a complex mixture of hydrocarbons that boils at temperatures up to about 200°C. Highly flammable, it is the main fuel used in car engines. It contains additives such as tetraethyl lead.

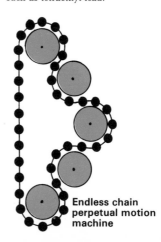

Endless chain perpetual motion machine

petrol engine

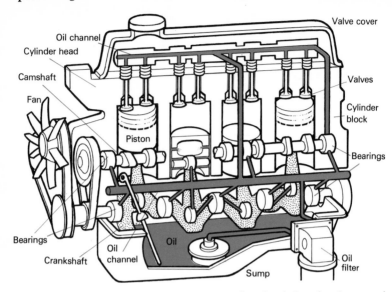

Valve cover
Oil channel
Cylinder head
Camshaft
Fan
Piston
Bearings
Crankshaft
Oil channel
Oil
Sump
Oil filter
Valves
Cylinder block
Bearings

petrol engine The main type of engine used to power cars, which uses petrol as fuel. It was developed in 1885 by the German engineers Gottlieb Daimler and Karl Benz. The engine is a reciprocating piston engine, which works on the four-stroke cycle (outlined on page 99). In most petrol engines the petrol is vaporized in a stream of air in a carburettor. The vapour is then burned in the engine cylinders, producing hot gases that expand and drive the pistons. The pistons (usually four or six) are connected to a crankshaft and drive that round. A flywheel at the end of the crankshaft passes on the motion to the car's transmission system. In some cars the petrol is injected into the cylinders instead of going via a carburettor. An ignition system uses an ignition coil and distributor to convert low-voltage electricity from a battery into a high-voltage pulse and to create a spark at the sparking plugs in each cylinder, which explodes the petrol vapour. The engine has a cooling system, usually incorporating a radiator, to keep the engine cool; and a lubrication system that provides oil to keep the moving parts running smoothly. See also **Wankel engine**.

petroleum Or crude oil; the world's most important fuel and source of organic chemicals, found trapped underground in the rock layers. Petroleum was formed over hundreds of millions of years as minute marine organisms died and decayed and were subjected to heat and pressure within the Earth's crust. It is a thick, greenish-black liquid, consisting of a complex mixture of hydrocarbons. The mixture can be separated into useful fractions by means of distillation, and the various fractions can then be further processed into a near-infinite variety of chemical raw materials, often termed petrochemicals (see **refinery**).

petrology The study of rocks.

pewter An alloy of tin and lead, together with a little copper and antimony, which has been used for centuries. Once used on a large scale for tableware, it is now used mainly for such things as tankards.

pH A measure of the acidity or alkalinity of a solution. The pH scale goes up to 14. A pH of 7 is the value for pure water, indicating a neutral solution. A pH greater than 7 indicates an alkaline, or basic solution; a pH less than 7 indicates an acidic solution. pH is defined as the $\log_{10}$ of the reciprocal of the hydrogen-ion concentration (H^+) in a solution; ie pH = $\log_{10} 1/(H^+)$.

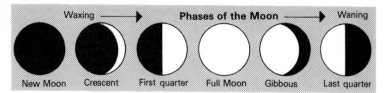

Waxing → **Phases of the Moon** → **Waning**

New Moon | Crescent | First quarter | Full Moon | Gibbous | Last quarter

pharmaceuticals Medicines and drugs.

pharmacology The study of drugs and their effects on living things. It encompasses such disciplines as chemotherapy—the treatment of disease by means of chemicals; and toxicology—the study of drug poisoning. Pharmacy is the practice of dispensing drugs.

phase In astronomy, the varying appearance of a heavenly body such as the Moon and Venus when illuminated from different directions by the Sun (see **phases of the Moon**). In chemistry, a phase is a physically distinct part of a mixture which can be separated out. An oil and water mixture has two distinct phases, with a definite boundary between them. But a solution of salt in water is only single phase because no boundary exists between the dissolved salt and water. Matter exists in one of three phases (or states)—solid, liquid or gaseous. In wave motion, phase is the portion of a cycle already completed. Phase differences between two waves are expressed in angles.

phases of the Moon The different shapes of the Moon seen from Earth during the month as more or less of its surface is illuminated by the Sun. The four main phases are new Moon, first quarter, full Moon and last quarter.

phenacetin A widely used painkilling drug, chemical formula $C_{10}H_{13}O_2N$.

phenobarbitone Or Luminal; a widely used sedative and hypnotic drug which is dangerous when taken in large doses. One of the barbiturates.

phenol Also called carbolic acid; first of a class of aromatic compounds (phenols) that contain one or more hydroxyl groups attached to a benzene ring. The formula for phenol is C_6H_5OH. It is weakly acidic and is widely used as a disinfectant. Large quantities are used to make synthetic resins and plastics.

phenol-formaldehyde resin See **Bakelite**.

phenolphthalein An indicator widely used in volumetric analysis of chemical formula $C_{20}H_{14}O_4$. It has laxative properties.

Petroleum

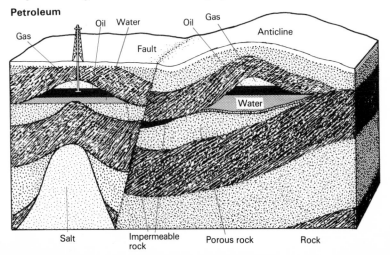

Gas, Oil, Water, Oil, Gas, Anticline, Fault, Water, Salt, Impermeable rock, Porous rock, Rock

phenylamine

phenylamine See **aniline**.

phenyl group The radical C_6H_5—, derived from benzene.

philosopher's stone A mythical substance sought by alchemists that would transform base metals into gold.

Phobos The largest of the two tiny moons of Mars. Probably a captured asteroid, it is a heavily cratered lump of rock of irregular shape: approximate dimensions $19 \times 21 \times 27$ km ($12 \times 13 \times 17$ miles). See also **Deimos**.

phon A unit of loudness in acoustics.

phonograph The original name given to the gramophone.

phosgene Or carbonyl chloride ($COCl_2$); a highly toxic poisonous gas, used against troops in World War 1.

phosphates Chemical compounds derived from phosphoric acid (H_3PO_4). Many phosphate minerals are known, the most important being forms of calcium phosphate, $Ca_3(PO_4)_2$. Phosphates are used as fertilizers, providing the phosphorus needed by growing plants. Animal bones are also a source of phosphates.

phosphine Or hydrogen phosphide (PH_3); a flammable poisonous gas with the odour of garlic.

phosphor A substance that emits visible light when exposed to ultraviolet radiation or beams of electrons. Television screens are coated on the inside with phosphors, as are fluorescent tubes. Zinc and cadmium sulphides are common phosphors.

phosphorescence Also called afterglow; the emission of light by a substance that has been exposed to radiation which continues after the source of radiation has been removed (compare **fluorescence**). A kind of phosphorescence is shown by certain marine creatures (see **bioluminescence**).

phosphoric acid An oxyacid of phosphorus. Most important is orthophosphoric acid, H_3PO_4, which is the parent of the phosphates. This acid is used in anti-rust preparations and for flavouring.

phosphorus (P) A non-metallic element best known in its waxy yellow form (rd $1 \cdot 82$, mp $44 \cdot 2°C$) which glows in the dark. It oxidizes rapidly in the air and may—catch fire. It is highly poisonous. When heated to a high temperature, yellow phosphorus changes to a dark red powder, the non-toxic and stable red phosphorus. A major use of phosphorus is in making matches, which have phosphorus compounds in the head and striking surface. The main source of phosphorus are the phosphate minerals.

photocell See **photoelectric cell**.

photochemistry Study of the effect of light on chemical reactions. Photosynthesis and photography involve photochemical reactions.

photocopier A machine for copying drawings, documents or other printed material. Copying may be done using light, heat, chemicals or electrostatic powders to create and develop an image. See **blueprint**; **ozalid process**; **xerography**.

photoelectric cell Also called photocell; an electronic device whose electrical properties change when light falls on it. There are three main types. In a photoemissive device electrons are emitted when light falls on it. In a photoconductive device the conductivity alters according to the amount of light striking it. This type is used in many light meters and TV camera tubes. In a photovoltaic device the incoming light sets up a voltage. The solar cells used on spacecraft are of this type.

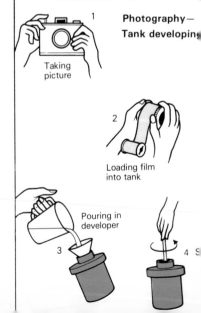

Photography—
Tank developing

1 Taking picture

2 Loading film into tank

3 Pouring in developer

4 S

photoengraving A method of producing printing plates for letterpress and gravure printing by means of photography and etching. For relief, or letterpress printing, the copy to be reproduced is photographed and a negative made. This is placed over the printing plate, which is covered with a light-sensitive coating, and exposed to light. The coating hardens where the light shines through the transparent parts of the negative. Washing removes the unaffected coating. Treatment with acid cuts into the unprotected metal, leaving the printing surface in relief. For gravure printing the printing surface is etched into the printing plate by a similar method.

photogrammetry The use of photographs in map-making. Generally, overlapping aerial photographs are used.

photography The art and craft of making pictures using light-sensitive film in a camera. Photography was pioneered in 1826 by the Frenchman Joseph Nicéphore Niépce, who produced an image in a coating of bitumen on pewter. The modern negative positive process was introduced in 1839 by the Englishman William Henry Fox Talbot. In the camera a lens focuses an image on a film, covered with a light-sensitive emulsion containing grains of silver salts. The grains alter chemically when light strikes them and invisibly record the image. Treatment with a developer converts the invisible, latent image into a visible one, creating a negative. When light is shone through the negative on to light-sensitive paper and that is developed, a positive picture results which is a replica of the original scene viewed. Colour photography is carried out using a multilayered film composed of three emulsions and a yellow filter. Each emulsion is sensitive to one of the primary colours. During development, the image in each emulsion is dyed and the colours of the dyes merge in the finished transparency or colour print to recreate the colouring of the original scene photographed. See also **Polaroid camera.**

photometer An instrument for measuring brightness, or the intensity of light. Many photometers are similar to photoelectric cells.

photon A basic unit, or quantum of electromagnetic radiation, which can sometimes be regarded as an elementary particle.

photosphere The visible surface of the Sun whose temperature is about 5,700°C. When examined closely, the surface is granular in appearance, the 'grains' being cells of hot gas.

photosynthesis The process by which green plants make their food. They manufacture carbohydrates from the carbon dioxide in the air and from water absorbed from the soil, using the energy in sunlight. The green pigment chlorophyll is responsible for absorbing solar energy and converting it into chemical energy that the plants can use.

physical chemistry The branch of chemistry that deals with such things as the study of the properties and structure of compounds and the various states of matter; the nature and mechanisms of chemical reactions; the relationship between chemistry and electricity.

physical sciences The sciences concerned with the study of the nature of the universe from individual atoms to groups of galaxies. The main physical sciences are astronomy, chemistry, geology and physics.

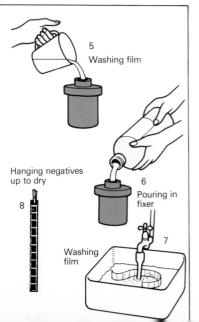

5
Washing film

Hanging negatives up to dry

8

6
Pouring in fixer

7

Washing film

physics

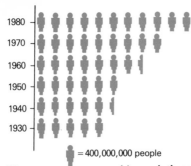

1980
1970
1960
1950
1940
1930

👤 = 400,000,000 people

Pictogram: approx world populations

physics One of the major sciences, which is concerned with the study of matter and energy. Major branches of physics include mechanics, heat, light, sound, electricity and magnetism, and atomic and nuclear power.

pi (π) A symbol that appears in many mathematical formulae. It is the ratio of a circle's circumference (C) to its diameter (D): $\pi = C/D$. It is an irrational number, which cannot be stated exactly, and approximates to 22/7 or $3 \cdot 14159$.

picric acid Or trinitrophenol, $HOC_6H_2(NO_2)_3$; a yellow crystalline solid used as a dye, antiseptic and explosive.

pictogram A pictorial method of representing statistics, using symbols to represent quantities.

piechart A pictorial method of representing statistics, in which quantities are shown as sectors of a circle.

Piechart: Energy sources of the world

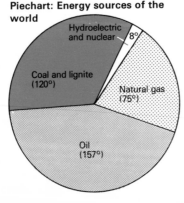

Hydroelectric and nuclear 8°
Coal and lignite (120°)
Natural gas (75°)
Oil (157°)

piezoelectric crystals Crystals such as quartz and Rochelle salt in which electricity is produced when they are compressed or stretched. Conversely, applying electricity to the crystals will cause them to deform. Gramophone pick-ups often contain piezoelectric crystals, which convert mechanical vibrations from the record into varying electrical signals. Digital watches contain piezoelectric crystals that vibrate at a precise frequency when electricity is applied to them.

pig iron The impure iron produced in a blast furnace, which contains too much carbon (about 4%) to be generally useful. It may be slightly refined into cast iron or converted into steel.

pigment A fine white, black or coloured powder used to colour materials such as paints and plastics. It may be inorganic (eg red lead) or organic (eg azo compounds).

pile In building construction, a timber, steel or reinforced-concrete column sunk into the ground to act as foundations to a structure. Some are driven into the ground by huge mechanical hammers called pile-drivers; others are formed in position ('cast in situ').

pile, atomic See **nuclear reactor**.

pion See **meson**.

Pioneer probes A series of American interplanetary probes, two of which (10 and 11) took the first close-up pictures of Jupiter in 1973/4. See **space probes**.

pipeline A tube used to transport liquids, gases and even solids long distances. The biggest pipeline networks transport oil and natural gas. One of the longest is the 1,285-km (800-mile) Trans Alaskan Pipeline, opened in 1977.

pipette A device used in the chemical laboratory to deliver a precise volume of liquid. It consists usually of a glass tube, bulbous in the middle.

Pisces The Fishes; an uninteresting constellation of the zodiac, lying between Aries and Aquarius. The vernal equinox now lies in Pisces instead of Aries because of precession.

piston A cylindrical device that slides back and forth in a cylinder. In steam and internal combustion engines, expanding steam or hot gases move the piston, and a connecting rod passes on this motion to a crankshaft, which transmits power to wheels or other

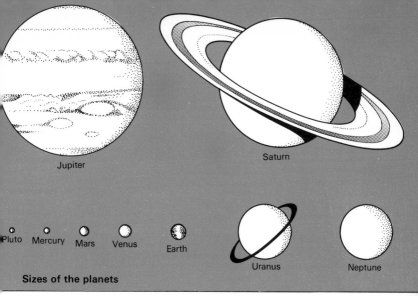

Jupiter

Saturn

Pluto Mercury Mars Venus Earth Uranus Neptune

Sizes of the planets

machinery. In a pump or compressor the piston moves air or liquid. The piston is made tight-fitting by means of rings.

pitch The dark tarry residue obtained after distilling coal tar or wood tar. It is used to caulk ships' timbers, as road tar and to make roofing materials.

pitch, musical A term relating to the frequency of vibration of a sound. A high-pitched sound has a high frequency; a low-pitched sound, a low frequency.

pitch of a screw See **screw**.

pitchblende Also called uraninite; the main ore of uranium, containing mainly uranium oxide. A dark, pitch-like mineral, it also contains other radioactive elements, including radium and polonium.

pitot tube An instrument that measures the speed of fluid flow, invented by the Frenchman Henri Pitot in the 1730s. It is used to measure the speed of ships and aircraft. It works by means of pressure variations in different directions in the fluid stream.

placer deposit A deposit of heavy minerals found in stream beds, for example, where flowing water has washed away lighter materials. Early gold prospectors worked placer deposits by panning. Gold, diamonds and cassiterite (tin ore) are still obtained from placers.

plagioclase A series of feldspar minerals made up of sodium and calcium aluminium silicates.

plane A flat surface; one with only two dimensions—length and breadth.

plane of polarization See **polarized light**.

planet A heavenly body that orbits the Sun. The Earth is one of nine planets. In order, going out from the Sun, the planets are Mercury, Venus, Earth, Mars, Jupiter, Saturn, Uranus, Neptune and Pluto. They are the main bodies in the solar system. Other stars are thought to have planets, including Barnard's star.

planetarium A device for showing the movements of the heavenly bodies, or the building in which this takes place. The main instrument is a complex projector with numerous axes of rotation which is able to project many images at once. German engineer Karl Zeiss pioneered the planetarium in 1913. The orrery is a simple planetarium.

planetary gear See **epicyclic gear**.

planetoid See **asteroid**.

plasma In physics, a fourth state of matter found in the searing hot interior of the Sun and stars where matter exists as clouds of electrons and charged ions. Scientists are experimenting with plasma in their quest to control nuclear fusion.

175

plaster of Paris A quick-setting plaster used to set bones in hospital. Made by heating gypsum, it is calcium sulphate hemihydrate, $CaSO_4.\frac{1}{2}H_2O$. When mixed with water, it changes back to gypsum ($CaSO_4.2H_2O$) and sets hard.

plasticity The ability of materials to flow under stress.

plasticizer A substance added to plastics and other materials to make them more pliable and easier to mould. Camphor is added to celluloid as a plasticizer.

plastics Substances that have long-chain molecules and are easy to shape by the application of heat. Rubber, shellac and bitumen are natural plastic materials, but most plastics are man-made. Some are derived from plant cellulose, including celluloid and cellulose acetate, but most are synthetic—made from chemicals. Polythene, PVC, polystyrene, nylon and the polyesters are among common synthetic plastics. These plastics soften when heated, and are termed thermoplastics. Other plastics, including Bakelite, are termed thermosetting plastics because they set hard when formed and do not soften again when heated. Many thermosetting materials are called synthetic resins because of their resin-like appearance before moulding, the commonest plastics shaping process. Plastics are also shaped by casting, extrusion and laminating.

plate glass High-quality flat glass made by grinding and polishing both surfaces accurately flat. It has now been almost entirely superseded by float glass.

plate tectonics In geology, a theory that the Earth's crust is made up of a number of thin segments, or plates, which are moving relative to one another. Movement of the plates gives rise to the observed phenomenon of continental drift. Where the plates meet, earthquakes and volcanoes may occur.

platinum (Pt) The most abundant of a closely related group of rare metals (rd 21·5, mp 1,769°C). It is a precious metal that resists corrosion and is invaluable as a catalyst. It is widely used in expensive jewellery. It occurs native and is mined mainly from placer deposits. The platinum group also includes the metals palladium, osmium, iridium, ruthenium and rhodium.

Pleiades Or Seven Sisters; the most conspicuous open star cluster in the heavens, in the constellation Taurus. It is made up of several hundred young stars, though only six or seven are visible to the naked eye. It lies about 400 light-years away.

Pleistocene Epoch The period of geological time from about 2,500,000 to 10,000 years ago. It was a time of ice ages, when glaciers advanced over vast areas of the northern hemisphere.

Plexiglass See **Perspex.**

Plimsoll line A load-line marking on ships' hulls that indicates how much cargo a ship can safely carry in certain waters at certain times. It is named after the English reformer Samuel Plimsoll and dates from 1875.

Pliocene Epoch The span of geological time between about 7,000,000 and 2,500,000 years ago, notable for the emergence of man-like apes and ape-like men in Africa.

Plough, The The best-known northern star group, known in North America as the Big Dipper. It is part of the constellation Ursa Major, the Great Bear. It is useful as a pointer to Polaris, the Pole Star.

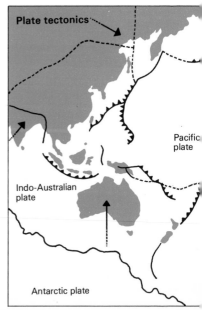

Plate tectonics

Pacific plate

Indo-Australian plate

Antarctic plate

Pluto The smallest and outermost of the nine planets, an icy ball only about 2,500 km (1,500 miles) across. It has a very eccentric orbit, which at times takes it 5,900 million km (3,700 million miles) from the Sun but at others takes it inside the orbit of Neptune. It orbits the Sun every 248 years. Pluto was the last planet to be discovered, by American astronomer Clyde Tombaugh in 1930. A satellite of Pluto, Charon, was detected in 1978.

plutonium (Pu) An artificial radio-active element (at no 94), which is one of the most toxic substances known. Resembling uranium, it is prepared by irradiating uranium with neutrons. Its most important isotope, Pu-239, is a nuclear fuel, readily undergoing fission.

plywood A building material made from thin sheets (plies or veneers) of wood glued together. In plywood the grain of each sheet is at right-angles to the grain of the sheets on either side. An odd number of plies is used so as to get a balanced structure that has equal strength in all directions.

Pm The chemical symbol for promethium.

PMMA See **Perspex.**

pneumatic tools Those operated by compressed air. Best-known is the pneumatic road drill. This has a piston that is forced by compressed air repeatedly against the drill bit.

Po The chemical symbol for polonium.

poise A unit of viscosity in cgs units.

polarimeter An instrument for measuring the angle of rotation of the plane of polarized light through transparent materials. See **optical activity; polarized light.**

Polaris (α Ursae Minoris) Also called Pole Star and North Star; a 2nd magnitude star in the constellation Ursa Minor (Little Bear). It is very close to the north celestial pole and scarcely appears to move at all in the night sky. It is a triple star, including a Cepheid.

polarization Of electric cells; a reduction in voltage after a cell has been operating for some time, caused usually by the formation of gas bubbles on the electrodes. Polarization also refers to the slight separation of positive and negative electric charges in atoms or molecules that may occur in an insulator in an electric field. Some molecules, including water, are permanently polarized.

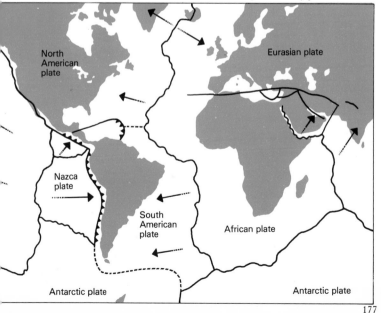

polarized light

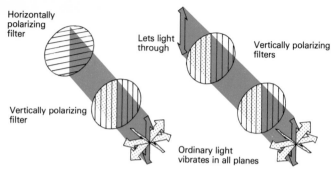

Horizontally polarizing filter

Lets light through

Vertically polarizing filters

Vertically polarizing filter

Ordinary light vibrates in all planes

polarized light Light that vibrates in one plane only, contrasting with ordinary light, which vibrates in many planes. Polarizing filters pass light vibrating in only one plane.

Polaroid camera The original 'instant-picture' camera invented by the American Edwin H. Land in 1947. Photographs are developed and printed immediately after they have been taken. Each piece of film contains its own chemicals that start to work when the film is removed or ejected from the camera.

pole Of a magnet; the place where the magnetism appears to be concentrated.

Pole Star See **Polaris.**

pollution The poisoning of the environment—land, water and air—by undesirable or harmful substances. Pesticides, factory effluents, oil spillage, raw sewage, smoke and fumes from chimneys, nuclear radiation, noise and even light can be sources of pollution.

Pollux (β Geminorum) A first magnitude ($1 \cdot 2$) reddish star in the constellation Gemini (the Twins), brighter than its twin, Castor.

polonium (Po) A very rare radioactive element (rd $9 \cdot 5$, mp $254°C$), discovered in pitchblende by Marie Curie in 1898.

polyamides A class of polymers, the best known of which is nylon.

polyesters A class of polymers, the best known of which is the synthetic fibre Terylene. Other polyesters can be cured, or made to cross-link, to form rigid plastics. They are reinforced with glass fibre in 'fibreglass' construction, more accurately termed GRP (glass-reinforced plastic).

polyethene See **polyethylene.**

polyethylene Or polythene; one of the most widely used of all plastics, made by polymerizing ethylene gas (C_2H_4). It was first made by ICI in 1935.

polygon A plane figure with three or more sides, the simplest of which is the triangle. A regular polygon has sides of equal length and internal angles equal.

polygraph See **lie detector.**

polyhedron A three-dimensional solid figure, whose faces are polygons.

Four regular polyhedrons

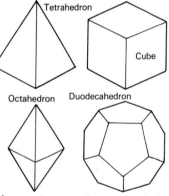

Tetrahedron

Cube

Octahedron

Duodecahedron

polymer A natural or synthetic substance composed of very large molecules (macromolecules) built up of long chains of repeated small units (monomers). Cellulose is a natural polymer, made up of repeated glucose units ($C_6H_{12}O_6$). Polystyrene is a synthetic polymer made up of repeated styrene units ($C_6H_5CHCH_2$). See also **copolymer; homopolymer; polymerization.**

polymerization A process in which small molecules (monomers) combine to form very large molecules (polymers). Synthetic plastics are made by polymerization reactions. In addition polymerization, monomers combine together and no other products result. In condensation polymerization, a by-product such as water is also formed.

polymethylmethacrylate Or PMMA, see **Perspex**.

polymorphism In chemistry, a phenomenon in which some minerals can appear in more than one crystal form. Calcium carbonate may form hexagonal crystals, when it is known as calcite. When it forms orthorhombic crystals, it is known as aragonite. Calcite and aragonite are called polymorphs of calcium carbonate. Compare **allotropy**.

polypropylene A transparent plastic similar to, but tougher than polyethylene. It is made by polymerizing propylene, CH_2CHCH_3.

polystyrene A widely used plastic made by polymerizing styrene, $C_6H_5CH=CH_2$. In the form of clear plastic it is used for containers; as foam it is used for heat insulation and packaging.

polytetrafluoroethylene See **PTFE**.

polythene See **polyethylene**.

polyurethane A synthetic resin used in paints and to make plastic foams and elastic materials. It is made by polymerizing diisocyanates with substances such as glycol.

polyvinyl acetate See **PVA**.

polyvinyl chloride See **PVC**.

porcelain The finest type of pottery, which is translucent and glass-like, has a clear ring when struck, and is fired at a high temperature (up to 1,500°C). The secret of making porcelain has been known in China since the T'ang dynasty (618–907, but was not discovered in Europe (by J. F. Böttger at Meissen) until about 1707. The essential ingredient of true, or hard-paste porcelain is kaolin—pure china clay. Imitation, or soft-paste porcelain contains an alternative ingredient such as bone ash (see **bone china**). Industrially porcelain is widely used for electricity insulators.

portland cement The usual kind of cement, made from limestone, clay and shale. It was named by Joseph Aspdin in 1824, who likened the set cement (concrete) to the stone obtained from the Isle of Portland, Dorset.

positive ion See **cation**.

positron An elementary particle with the same mass as an electron but with a positive electric charge. It is the anti-particle of the electron.

potash A common name given to potassium salts, eg caustic potash (potassium hydroxide) and sulphate of potash (potassium sulphate).

potassium (K) A highly reactive alkali metal (rd 0.86, mp 63.2°C), which decomposes water. It was the first metal prepared by electrolysis, by Humphry Davy in 1807. It occurs abundantly in the Earth's crust as the chloride (KCl) in sea water and brines, and in the minerals sylvite and (with magnesium) in carnallite. It is a very important trace element in plant metabolism, which is why potassium salts are applied as fertilizers.

potassium ferricyanide Also called potassium hexacyanoferrate (III), $K_3Fe(CN)_6$; a complex compound used in chemical analysis to identify ferrous (Fe^{2+}) and ferric (Fe^{3+}) ions. With ferrous ions, it forms a blue precipitate (Turnbull's blue). With ferric ions, it forms a brown solution.

Ethylene monomer

Part of polyethylene polymer

Carbon atoms

Hydrogen atoms

potassium ferrocyanide Also called potassium hexacyanoferrate (II), $K_4Fe(CN)_6$; a complex compound used, like the ferricyanide (see above), in chemical analysis to identify ferrous and ferric ions. With ferrous ions, it forms a white precipitate; with ferric ions, a blue precipitate (Prussian blue).

potassium nitrate Formula KNO_3; occurs naturally as the mineral nitre, or saltpetre. It is used as a fertilizer and, as an ingredient of gunpowder, in fireworks.

potassium permanganate Formula $KMnO_4$; a powerful oxidizing agent used in the laboratory. It forms dark purple crystals.

potential difference The difference in electric potential of two points, measured in volts. If the two are connected by a conductor, electric current will flow between them. Potential difference between two points is defined as the amount of work done when a unit positive electric charge moves from one point to the other.

potential, electric A measure of the energy stored in an electric field, measured in volts.

potential energy Stored energy; energy contained in a body by virtue of its position. A compressed or stretched spring has potential energy; so has a ball on a table.

potentiometer An instrument for the accurate measurement of potential difference.

pound A unit of weight in the avoirdupois system. 1 pound (lb) = 7,000 grains = 16 ounces (oz) = 454 grams (g); 1 kilogram (kg) = 2·2 lb.

poundal A unit of force in the foot-pound-second system of units. It is the force that will impart to a mass of 1 pound an acceleration of 1 foot per second per second. 1 poundal = 0·138 newtons.

powder metallurgy A method of shaping metals in powdered form. This technique is needed for metals with a high melting point, such as tungsten. They are pressed in a mould and then sintered.

power In mathematics, the number of times a quantity is multiplied by itself. y to the power 3 (y^3) means $y \times y \times y$. In physics, power is the rate of doing work, expressed in horsepower or watts.

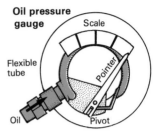

Oil pressure gauge

Pr The chemical symbol for praseodymium.

praseodymium (Pr) One of the rare-earth elements (rd 6·8, mp 935°C), obtained from monazite.

Pre-Cambrian Era The great span of geological time from the birth of the Earth some 4,600 million years ago to the beginning of the Cambrian Period, about 570 million years ago, during which time primitive life forms developed.

precession Of a spinning body, such as a top or gyroscope; a tendency for the axis of rotation itself to rotate. The axis describes a circular cone. The Earth, spinning in space, precesses in a similar way, causing an apparent shift in the constellations around the celestial sphere and a change in the equinoxes. This is termed the precession of the equinoxes.

precipitation In chemistry, the formation of an insoluble solid (precipitate) in a solution as a result of a chemical change. In meteorology, precipitation refers to the many forms in which solid and liquid water fall from the clouds to the ground—drizzle, rain, snow, sleet, hail and so on.

press See **hydraulic press**.

pressure Force per unit area. The SI unit of pressure is the pascal. In engineering, pressures are often expressed in pounds per square inch (psi). Atmospheric pressure is often measured in millibars or millimetres of mercury.

pressure cooker A saucepan with an airtight lid in which food can be cooked quickly. Under pressure, the water inside boils at a much higher temperature than usual, and the food cooks more quickly as a result. The pressure can be regulated by means of a weighted safety valve in the lid.

pressure gauge An instrument for measuring fluid pressure. The commonest instrument is the Bourdon gauge, which contains a curved metal tube. Changes in pressure cause the tube to coil or uncoil, and this movement moves a pointer over a scale.

pressurized-water reactor (PWR) See **nuclear reactor.**

primary cell The simplest type of electric cell, in which electricity is produced when two electrodes, immersed in an electrolyte, and connected by a conductor, undergo chemical change. See also **battery**; **Daniell cell**; **Leclanché cell**; **standard cell.**

primary colours Of light, the colours red, green and blue. Light of any colour can be produced by combining light of the primary colours in certain proportions. When the three are combined in equal proportions, white light is produced. Of pigments (paints), the primary colours are red, yellow and blue. When these three are combined in equal proportions, black is produced.

prime number A number that has no factors, being divisible only by itself and 1.

Principia The name usually given to what is arguably the single most important book on science ever published—Isaac Newton's book *Philosophiae Naturalis Principia Mathematica* (1687). It includes discussion of Newton's laws of motion and theory of gravitation.

printed circuit An electrical circuit consisting of a thin film of copper on an insulating board. It is produced by a modified photoengraving technique.

printing Reproducing words and pictures, usually on paper. The Chinese introduced wood-block printing in the AD 700s. But the story of modern printing began in about 1440, when Johannes Gutenberg in Germany pioneered the use of reusable metal type and invented a printing press that used it. Gutenberg's printing method, in which the printing surface is raised, is called letterpress. It has now been largely superseded by litho (see **lithography**) in which the printing surface is flat. A third major method of printing is gravure, in which the printing surface is recessed.

Triangular prism

Rectangular prism

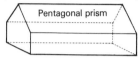

Pentagonal prism

prism A solid whose ends are congruent polygons and whose sides are parallelograms. Glass prisms may be used in optical instruments to reflect light (see **binoculars**) or to split up light into a spectrum (see **spectroscope**).

probability In mathematics, an indication of the likelihood that an event will take place. A certainty has a probability of 1; an impossibility, a probability of 0. The probability that a coin will come down heads (or tails) is $\frac{1}{2}$.

Procyon (α Canis Minoris) The brightest star (mag $0 \cdot 35$) in the constellation Canis Minor (Little Dog), and eighth brightest in the entire heavens. It is a double star and lies some 11 light-years away.

producer gas A low-grade fuel gas widely used in industry and made usually by the partial combustion of coal in air and steam. Carbon monoxide ($25-30\%$) and hydrogen ($10-15\%$) are the chief flammable gases.

Printing methods
Letterpress printing plate

Litho printing plate

Gravure printing plate

program(me)

Map projections

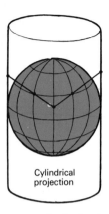

Cylindrical projection

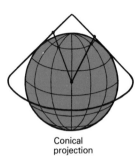

Conical projection

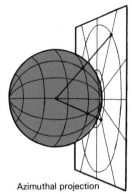

Azimuthal projection

program(me) In electronics, a set of instructions fed to a computer, written in a simplified computer language.

projection In map-making, a method of representing the surface features of the Earth (which is of course curved) on a flat surface. There are three basic types—azimuthal, cylindrical (which includes the Mercator) and conic projections.

projector A device that projects, or throws an enlarged image of (usually) a piece of film onto a screen. The common slide projector contains a powerful lamp, a condenser lens to ensure even illumination of the slide, and a projection lens to form the image on the screen. The ciné projector for showing 'movies' contains a mechanism to stop the film intermittently so that each frame is projected as a still picture. A rotary shutter covers the film while it is being moved between frames.

promethium (Pm) One of the rare-earth metals (mp $1,000°C$), which is radioactive.

prominence A huge eruption on the Sun which shoots gases hundreds of thousands of kilometres high.

propane A major constituent of natural gas and the third paraffin (alkane) hydrocarbon, C_3H_8. It is readily liquefied (bp $-42·1°C$) and is sold in liquid form as bottled gas.

propanone See **acetone**.

propellant In rocketry, the substances burned to propel a rocket. A solid propellant consists typically of a fuel (eg powdered aluminium) and oxidizer (eg ammonium perchlorate) in a synthetic rubber matrix. Commonest among liquid propellants are the combinations liquid hydrogen (fuel) and liquid oxygen (oxidizer); and hydrazine (fuel) and nitrogen tetroxide (oxidizer). The latter are hypergolic. In aerosol cans, a gaseous propellant such as Freon is used to create the spray.

propeller A screw-like device used to propel a ship or aeroplane. A propeller has several blades, which usually have an aerofoil cross-section. When it rotates, it develops forward thrust by accelerating fluid backwards.

Ciné projector

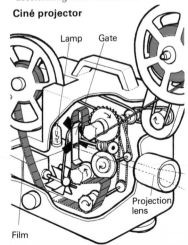

Lamp   Gate

Projection lens

Film

182

propene See propylene.

proper motion The change in a star's position on the celestial sphere, as a result of its motion through space. Only about 200 of the nearest stars have a proper motion greater than 1 second of arc per year. Compare radial motion.

propylene Or propene; second member of the olefine, or alkene series of unsaturated hydrocarbons, $CH_2=CHCH_3$. It is a gas, which is readily polymerized into polypropylene.

prosthetics A branch of medicine concerned with the use of artificial aids, organs and other body parts. It covers the use of such things as artificial limbs, hip joints and heart pacemakers. See also bionics.

protactinium (Pa) A very rare radioactive metal (rd 15·4, mp 1,200°C) always found in uranium ores; one of the actinides.

proteins One of the major groups of food substances found in living things. They are involved particularly in the building up of tissues. Their very large molecules contain nitrogen and are made up of different combinations of mainly about 20 amino acids.

proton A positively charged elementary particle found in the nucleus of every atom. The hydrogen nucleus consists of a single proton. The mass of a proton, about $1·67 \times 10^{-30}$g, is 1,836 times that of an electron.

prototype The first working full-scale model of, say, an aeroplane, used to evaluate the performance of the design. As a result of tests on the prototype, the design may be modified before it goes into production.

Proxima Centauri The nearest star to us, a red dwarf star that lies some 4·28 light-years away in the southern constellation Centaurus. It is part of the Alpha Centauri multiple star system.

Prussian blue See potassium ferrocyanide.

prussic acid Hydrocyanic acid, the acid that gives rise to the cyanides. A volatile liquid (bp 25·7°C) with the smell of bitter almonds, it is one of the most poisonous substances known.

psychrometer A device for measuring the relative humidity and dew-point temperature of the atmosphere. It usually consists of wet and dry bulb thermometers.

Pulley systems

Pt The chemical symbol for platinum.

PTFE The plastic polytetrafluoroethylene. Known by such trade names as Teflon and Fluon, it is made by polymerizing tetrafluoroethylene, C_2F_4. It is heat resistant and has low friction, which fits it for its major use as the coating on 'non-stick' pans.

Pu The chemical symbol for plutonium.

puddingstone An alternative name for the sedimentary rock known as conglomerate, which can look rather like a plum pudding.

puddling process A method of making wrought iron by reheating pig iron with iron ore in a reverberatory furnace and stirring it frequently.

pulley A simple machine, consisting of a grooved wheel with a rope or belt passing over it. A number of pulleys can be mounted together in a block. The block-and-tackle (tackle means ropes or chains) is widely used in engineering to help lift loads. It has a mechanical advantage.

pulsar A star that emits rapid pulses of radio waves. Some also emit X-rays and light as well. Pulsars are thought to be the remnants of stars that have blasted themselves apart as supernovae. They may be rapidly rotating neutron stars of small dimensions.

pulse jet The type of jet engine used in the German 'revenge weapon', V-1. In this engine combustion takes place intermittently—the fuel is injected at intervals into the hot combustion chamber.

pumice

pumice A kind of porous, frothy natural glass formed when lava from a volcano cools very quickly. Pumice is widely used as an abrasive.

pump A machine for raising, moving or compressing fluids. A simple pump is the lift pump, once used widely for raising water from wells. It operates by means of two one-way valves, one in the inlet pipe and one in the pump piston. Rotary pumps consist of a gear wheel or rotor with vanes around the edge. Diffusion pumps are used to achieve high vacuum (see **vacuum pump**).

putty A traditional sealant for glazing windows, consisting of powdered chalk and boiled linseed oil.

PVA The synthetic resin polyvinyl acetate, prepared from vinyl acetate, $CH_2=CHOOCCH_3$. It is widely used to make water-based paints and in lacquers and adhesives.

PVC The versatile plastic polyvinylchloride, made by polymerizing vinyl chloride, $CH_2=CHCl$. A plasticizer is added to it to make it flexible. Typical uses of PVC are gramophone records, guttering, piping and rainwear.

pyknometer A device for measuring the density and coefficient of expansion of a liquid.

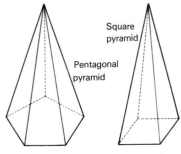

Square pyramid

Pentagonal pyramid

pyramid A solid whose base is a polygon and whose sides are triangles that meet at a common point (vertex). The Great Pyramids of Egypt have a square base, a shape termed a square pyramid. The Great Pyramid of Cheops, built in the 2500s BC, is some 137 m (450 ft) high and covers an area of more than 5 hectares (13 acres).

Pyrex Trade name for a heatproof borosilicate glass, widely used for laboratory-ware and kitchen-ware.

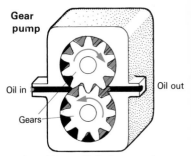

Gear pump

Oil in

Oil out

Gears

pyrite Also called iron pyrites; a common mineral form of iron sulphide, FeS_2. It is a major source of sulphur. It is also called fool's gold because it is relatively heavy (rd about 6·5) and shiny yellow.

pyridine An organic compound (C_5H_5N) with a highly unpleasant smell. Its structure contains a ring of five carbon atoms linked by a nitrogen atom. It is added to industrial ethanol to make it unfit to drink.

pyroelectricity A phenomenon displayed by certain crystals (eg tourmaline), which become electrically charged when they are heated.

pyrolusite A common ore of manganese, being manganese dioxide (MnO_2). It is a soft dark-grey mineral.

pyrolysis A chemical change brought about by heating.

pyrometer An instrument for measuring high temperatures. Some pyrometers use the light or other radiation a hot body gives out when it is heated. Some exploit the change in electrical resistance with temperature, while others use the principle of the thermocouple.

pyroxylin See **nitrocellulose**.

Pythagoras's theorem One of the best-known mathematical theorems, which states that, for a right-angled triangle, the square on the hypotenuse is equal to the sum of the squares on the other two sides. In the diagram: $c^2 = a^2 + b^2$.

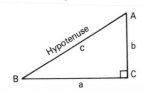

Q

quadrant An ancient astronomical instrument used for measuring the altitude of the heavenly bodies. It is so called because it is a quarter segment of a circle.

quadratic equation In mathematics, an algebraic equation which includes variables raised to the second power (squared). A typical quadratic equation is $ax + bx^2 + c = 0$, where a, b and c are constants and x is the variable.

quadrilateral A plane figure with four straight sides. The square, rectangle and parallelogram are different types of quadrilaterals.

qualitative analysis The branch of analytical chemistry concerned with the identification of the elements, ions or groups present in a sample.

quantitative analysis The branch of analytical chemistry concerned with finding the quantity of particular constituents in a sample. Two traditional methods are used—gravimetric (weighing) and volumetric (see **volumetric analysis**). More recent methods involve such techniques as mass spectrometry and gas chromatography.

quantum In physics, a basic unit, or 'packet' of energy or other physical property. The quantum theory was put forward by the German physicist Max Planck in 1900 to explain the precise way radiation is emitted by a hot body. He suggested that radiation is emitted, not in a continuous stream, but in the form of discrete quanta. The theory was developed further and provided the basis of the present discipline of quantum mechanics, which deals with the behaviour of all atomic particles and other small-scale effects.

quarks Hypothetical particles which are thought to be the true basic particles of matter. Elementary particles such as protons, neutrons and mesons are thought to be made up of different combinations of four different quarks. These have different properties and are termed 'up', 'down', 'charmed' and 'strange' quarks.

quarry A place where stone and rock are mined at the surface.

quart A unit of liquid capacity in the Imperial system, equal to one-quarter of a gallon, or 2 pints, or $1 \cdot 14$ litres.

quartz One of the commonest minerals, being a crystalline form of silica – silicon dioxide, SiO_2. It is often found as hexagonal crystals. Some varieties of quartz are gemstones, including amethyst and cairngorm. The purest variety is rock crystal. Pure quartz crystal is now widely used in electronics in piezoelectric devices and as a regulator in watches and clocks.

quartzite A very hard metamorphic rock, formed when sandstone is subjected to heat and pressure.

quartz watch A watch regulated by the vibrations of a wafer-thin quartz crystal. When a tiny current is applied to the crystal, it vibrates exactly 32,768 times a second. A microcircuit converts this 'time base' in the quartz digital watch to a digital display, such as LCD or LED. In the quartz analog watch the time base drives a gear train to move the hands.

quasars Or quasi-stellar sources; cosmic sources of very high energy that appear star-like in photographs but have the radiation emission typical of galaxies. Their spectra exhibit very large red shifts, indicating that they are very far away. The farthest lie at a distance of over 15,000 million light-years.

Quaternary Period The most recent span of geological time; which began about 2,500,000 years ago.

quenching A type of heat treatment in which hot metal is suddenly cooled by being plunged into cold water or oil. It makes the metal hard.

quicklime Calcium oxide, CaO, made by roasting chalk in a limekiln. It reacts vigorously with water to form slaked lime, calcium hydroxide, $Ca(OH)_2$.

quicksilver A common name for mercury.

quinine An alkaloid drug $(C_{20}H_{24}O_2N_2.3H_2O)$ obtained from the bark of the South American cinchona tree; used for treating malaria.

quinones A group of aromatic compounds containing oxygen linked to two of the carbon atoms in the benzene ring. The simplest one is p-benzo-quinone, $C_6H_4O_2$, a substance used to make dyes and developers.

R

Ra The chemical symbol for radium.

rack-and-pinion A common way of transmitting mechanical motion, used for example in the steering system of many cars.

rack railway A type of railway that uses the principle of the rack-and-pinion to enable locomotives to climb steep gradients. The locomotive drives a toothed cog or pinion, which engages with a toothed rack on the track.

rad In mathematics, the abbreviation for radian. In physics, a unit used to measure the amount of radiation absorbed by a body. It is the amount of absorbed radiation that releases 100 ergs of energy per gram. The word is an abbreviation for 'radiation absorbed dose'. See also **rem**.

radar Abbreviation for radio detection and ranging; a major technique used in navigation, developed first in Britain in the late 1930s. Radar works by sending out radio pulses and then detecting the echoes when the pulses are reflected by objects in their path. The echoes are displayed as 'blips' of light on a calibrated cathode-ray screen. From the position of the blips on the screen, the actual position of the objects causing the echoes can be determined.

radial motion The component of a star's motion through space that is in line of sight, that is, towards or away from an observer. It can be detected and measured by observing the shift in the spectral lines of the star's light. See **blue shift**; **red shift**.

radian A unit of measuring angle. 2π radians are equivalent to $360°$; 1 rad $= 360/2\pi = 57\cdot3°$.

radiant In a meteor shower, the point from which the meteors appear to originate.

radiation The emission of rays from an object. The Sun and the stars give out various forms of electromagnetic radiation, such as light, heat, and radio waves. Radioactive elements give out particle radiation, such as streams of electrons or alpha-particles. In heat physics, radiation is one of the three main mechanisms of heat transfer.

radiation belts See **Van Allen radiation belts**.

radiation sickness The illness brought about by excessive exposure to ionizing radiation from nuclear sources and also X-ray machines. Symptoms include loss of appetite, nausea, vomiting and diarrhoea. In large doses the red blood cells may be damaged, leading to leukaemia, and the body becomes defenceless against disease.

radical In chemistry, an atom or group of atoms that retains its identity during chemical reactions, eg the ethyl radical, C_2H_5—. See also **free radical**.

Plan position indicator screen

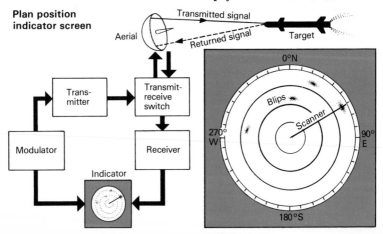

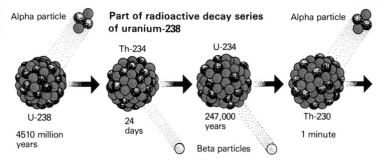

Alpha particle

Part of radioactive decay series of uranium-238

Alpha particle

Th-234

U-234

U-238
4510 million years

24 days

247,000 years

Th-230
1 minute

Beta particles

radio The use of radio waves to transmit signals. In radio broadcasting, sounds are converted by a microphone into variable electrical (audio) signals. These are superimposed on a radio carrier wave, a process known as modulation, and the carrier wave is transmitted. In a radio receiver, a suitably tuned aerial picks up the carrier wave, and detector circuits remove the audio signal from it. These signals are then fed to a loudspeaker, which reproduces the sounds that entered the microphone originally.

radioactivity A process in which unstable atoms spontaneously emit radiation or streams of particles from their nucleus. In so doing the atoms change into other atoms, with different particles in the nucleus. The radiation given out during radioactive decay may be alpha-particles, beta-particles or gamma-rays. Henri Becquerel in France pioneered the study of radioactivity when he discovered in 1896 that uranium gives out radiation. During radioactive breakdown, or decay, the atoms change. The atoms of one element change into those of another. The new element may also be radioactive, and decay in turn. This process continues until a stable element results. This successive decay process is known as a radioactive series. A well-known radioactive series is uranium-238 to lead-206, part of which is illustrated above. The whole series involves 14 stages, alpha- or beta-particles being given out at each stage.

radio astronomy A branch of astronomy that studies the radio waves emitted by heavenly bodies. It was pioneered in 1931 by Bell Telephone engineer Karl Jansky, who discovered that the heavens are a source of radio interference. See **radio telescope**.

radiography The use of X-rays or gamma-rays to take photographs. Radiography with X-rays is done in hospitals to look inside a patient's body. Radiography with gamma-rays is used in metallurgy to check for flaws inside metal parts.

radioisotope An isotope that is radioactive. Most radioisotopes used in science and industry today are made artificially by irradiation in nuclear reactors.

radiometer A device for measuring the intensity of radiation. The simplest type is Crooke's radiometer, which is often sold as a toy. It has a number of vanes which can rotate on a pivot inside an evacuated glass bulb. They are black on one side and white on the other. The black sides absorb more radiant energy than the white and get hotter. Air molecules rebound with greater energy from the warm side than they do from the white side, and this makes the vanes rotate.

radiometric dating Using the known rate of decay of certain natural radioisotopes to estimate the age of rocks and archaeological remains. Uranium, radioactive potassium and radioactive carbon are widely used. See **carbon dating**;

radiosonde A balloon-carried instrument package used by meteorologists to take 'soundings' of the upper atmosphere. It includes instruments to measure temperature, pressure and humidity, and a radio transmitter to radio the measurements back to the ground.

radio telescope A telescope that collects radio waves from the heavens; the basic tool of radio astronomy. Most radio telescopes are of the dish type. They have a parabolic metal bowl, which collects the radio waves and focuses them on an antenna. Because the incoming radiation is so faint, huge dishes must be used. The world's largest radio telescope is the 305-metre (1,000-ft) dish at Arecibo in Puerto Rico, which is fixed. The largest steerable dish, 100 metres (328 ft) across, is at Effelsberg near Bonn in West Germany.

radiotherapy The treatment of illness by means of radiation. It is used, for example, to treat cancerous tumours.

radio waves Electromagnetic waves of wavelengths from a few centimetres to many thousands of metres, or of a frequency from about 3,000 megahertz (millions of cycles per second) to about 3 kilohertz (thousands of cycles per second). Television transmissions use the highest radio frequencies; long-wave radio the lowest.

radium (Ra) A rare radioactive metal (rd 5, mp 700°C), one of the rare earths. Found in pitchblende, it was first isolated by Marie Curie in 1898.

radius See **circle**.

radon (Rn) Also called emanation; a radioactive noble gas, formed when radium decays.

railways A major land transportation system that developed from the tramways used since the 16th century in mines. Richard Trevithick pioneered steam locomotion at Penydarran in South Wales in 1804, but it was George Stephenson who founded the modern railway system by building the Stockton and Darlington Railway in 1825 and designing the prototype modern locomotive, 'Rocket', four years later. The gauge (distance between the rails) Stephenson used, 1·435 metres (4 ft 8½ in) remains the standard gauge throughout the world. Most trains today are hauled by electric and diesel locomotives. Among the fastest are the French Trains à Grande Vitesse (TGVs) which touch 260 km/h (160 mph).

rain Water droplets that fall from the clouds; the commonest form of precipitation. The size of raindrops varies from about 0·5–4 mm across.

rainbow A colourful arc spanning the sky when the Sun shines during showery weather. An observer can see a rainbow only when he has his back to the Sun and it is raining in front of him. The rainbow, which shows the colours of the spectrum, is produced as a result of successive refractions and reflections inside the raindrops.

ram jet A simple kind of jet engine that has no moving parts, being basically a hollow tube. Fuel is sprayed into air entering the front of the engine and burned to produce hot gases, which escape as a propulsive jet. The ram jet works efficiently only when it is travelling at speeds over about twice the speed of sound.

Ranger A series of American space probes sent to explore the Moon. The first successful one was Ranger 7 (1964), which photographed the surface before crashlanding.

rare-earths Or lanthanides; a series of closely related metallic elements. In order of their atomic number they are lanthanum, cerium, praseodymium, neodymium, promethium, samarium, europium, gadolinium, terbium, dysprosium, holmium, erbium, thulium, ytterbium and lutetium. The transition metals scandium and yttrium are also usually considered rare-earths.

rare gases See **noble gases**.

ratchet A mechanism consisting of a toothed wheel and a pivoted pawl, which allows movement of the wheel in one direction only.

ratio In mathematics, a comparison of two quantities of the same kind. The ratio between the lengths 6 m and 3 m is 6:3, also written 6/3. In practice this would be simplified to 2:1, or 2/1.

rayon The most widely used of all man-made fibres, prepared from the cellulose in woodpulp or cotton linters. There are now two main types—viscose and acetate rayon (see **acetate**). Viscose rayon is made by treating cotton or woodpulp in turn with caustic soda, carbon disulphide and caustic soda again. The resultant syrupy solution of pure cellulose is then forced through the holes of a spinneret, into an acid bath. The fine streams of solution are converted by the acid into continuous filaments of pure cellulose—rayon.

Rb The chemical symbol for rubidium.

The world's biggest radio telescope at Arecibo, Puerto Rico, which has a dish 305 metres across.

RDX

RDX Also called cyclonite, hexogen or T4; a powerful explosive used in blasting caps and plastic explosives, prepared from formaldehyde and ammonia.

Re The chemical symbol for rhenium.

reactance In electricity, a measure of the opposition of a circuit to alternating current. Together with the resistance, it makes up the impedance of the circuit.

reaction principle The name often given to Newton's third law of motion: to every action there is an equal and opposite reaction. This is evident when a gun is fired. As the bullet shoots forwards, the gun is forced backwards by reaction. Jet and rocket engines work by reaction. They are thrust forwards by reaction to a stream of gases shooting backwards.

reactor, nuclear See **nuclear reactor.**

reagent A substance that can bring about a chemical reaction.

realgar A major ore of arsenic; arsenic disulphide, As_2S_2.

real gas As opposed to a perfect gas, which obeys the gas laws. A real gas deviates from the gas laws to a lesser or greater degree, because its molecules tend to attract one another and occupy some space. Real gases obey the gas laws best at low pressure and high temperature.

reciprocal Of a number; 1 divided by the number. The reciprocal of 25 is 1/25.

reciprocating engine An engine, such as a petrol and diesel engine, whose pistons travel back and forth in a cylinder.

recording See **sound recording.**

record player See **gramophone.**

rectangle A four-sided plane figure with right-angles. A square is a rectangle with equal sides.

rectification The refining of a liquid by distillation. Ethanol is purified by rectification, resulting in what is termed rectified spirit.

rectifier A device that converts alternating electric current into direct electric current (AC into DC). It may take the form of a valve (electron tube) or crystal.

recycling Re-using materials in an industrial process. In the chemical industry, materials are often recycled as part of the chemical processing (eg ammonia in the ammonia-soda process).

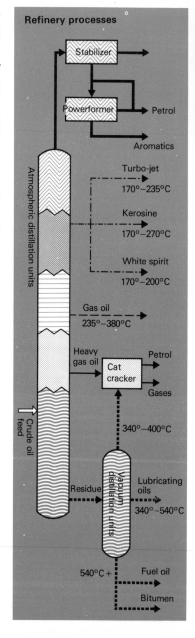

Refinery processes

red giant A class of large, low-density stars, which radiate mainly red light. They represent a late stage in stellar evolution. Arcturus, Pollux and Aldebaran are typical red giants.

red lead The lead oxide Pb_3O_4 (trilead tetroxide), an orange-red compound widely used in anti-rust primer paints for metals.

redox See **oxidation; reduction.**

Red Planet See **Mars.**

red shift The shift of lines in the spectrum of starlight towards the red end of the spectrum. It is interpreted as a Doppler effect, caused by the star receding from us. Compare **blue shift.**

Red Spot A persistent red oval marking in Jupiter's southern hemisphere, now considered to be a violent storm centre. In the early 1980s it measured about 28,000 km (17,000 miles) long and about 14,000 km (9,000 miles) wide.

reducing agent A substance that removes oxygen from, or adds hydrogen to, another substance. In a broader sense, it is a substance that donates electrons. Good reducing agents include hydrogen, carbon, zinc dust, sulphur dioxide and formaldehyde.

reduction A process in which oxygen is removed from or hydrogen added to a substance; or more generally one in which a substance gains electrons. It is always accompanied by oxidation. The whole process is termed redox (reduction—oxidation). See also **reducing agent.**

re-entry In astronautics, the time during which a spacecraft returning to Earth from orbit re-enters the atmosphere. Due to friction, the air surrounding the craft becomes hot and ionized and prevents the passage of radio waves. This causes a communications blackout of about 20 minutes. See also **heat shield.**

refinery A chemical plant where petroleum (crude oil) is refined, or purified. The major refinery process is distillation, in which the crude is separated into a number of useful fractions, including petrol and kerosene. Subsequent refinery processes, such as cracking, reforming and polymerization, convert less useful fractions into higher-grade petrols and valuable chemical raw materials, or petrochemicals. Refineries are highly automated.

refining Purifying, or converting into a more useful form. Crude oil needs refining in a refinery before it is useful. Pig iron from the blast furnace needs refining in furnaces to make steel.

reflection The rebound of a wave of energy, such as light and sound, after striking a surface. Mirrors work because of the reflection of light from a silvery surface. For a light ray striking a regular plane mirror, the angle of incidence (between the incident ray and the normal) equals the angle of reflection (between the reflected ray and normal). See also **echo.**

reflector A reflecting telescope, which uses a curved mirror to gather and focus light from the stars. Most amateur instruments are Newtonian reflectors. They have a concave main mirror, which reflects light to a plane mirror near the top of the telescope tube, which in turn directs an image into an eyepiece. Large instruments, such as the Hale 5-m (200-in) telescope at Palomar Observatory and the 3·9-m (154-in) Anglo-Australian telescope (AAT) in New South Wales, have alternative means of focusing, such as the Cassegrain and Coudé. (The AAT is illustrated on page 225.)

reflex camera A camera that uses a mirror to form an image in the viewfinder. In the single-lens reflex (SLR) camera, the mirror swings out of the way when the shutter is pressed, and allows light to fall on the film. The twin-lens reflex (TLR) camera has a separate lens for viewing.

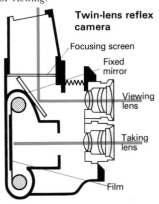

Twin-lens reflex camera

Focusing screen

Fixed mirror

Viewing lens

Taking lens

Film

reflux condenser

reflux condenser One in which vapour is condensed and the condensate is returned to the distillation vessel.

refraction The change in the direction in which waves travel when they pass from one medium to another. Light waves are refracted when they pass from air into water. This is what makes straight sticks floating in the water look bent, and pools look shallower than they really are. Every medium bends rays to a different extent, expressed as a refractive index (μ).

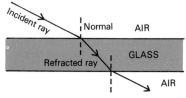

refractive index An indication of the way in which a wave is refracted when passing from one medium to another. If the angle of incidence of the wave is i, and the angle of refraction is r, the refractive index (μ) is given by the following expression, which is a statement of Snell's law: $\mu = \sin i / \sin r$.

refractor A refracting telescope, which uses lenses to gather light and form an image. In a refractor a convex objective lens forms an image which is then enlarged by an eyepiece. The image is inverted (upside-down). Refractors suffer from several defects, including spherical and chromatic aberration (see **aberration**). Large lenses are difficult to manufacture and mount, and even the world's largest refractor, at Yerkes Observatory near Chicago, has a lens diameter of only 102 cm (40 in).

refractory A substance that is resistant to high temperatures. Widely used refractories, used to line furnaces for example, include silica and dolomite. Alumina is an even better refractory, as are cermets.

refrigerant A substance, used as the working fluid in refrigerators, which will vaporize readily and condense readily when compressed. The commonest refrigerants are ammonia and the freons.

refrigerator A device for keeping food and other substances cool. It works on a cycle of vaporization and compression of a volatile fluid (refrigerant). The liquid refrigerant is made to expand through a valve and evaporate, extracting heat from the contents of the refrigerator. The vapour formed is then compressed by a compressor and passed through a condenser, whereupon it loses heat and liquefies. It is then recycled.

regelation The reforming of ice after it has been melted by the application of pressure (eg from ice skates).

regenerator In metallurgy, a heat exchanger that uses the heat of hot gases leaving a furnace to preheat gases going into the furnace. It usually takes the form of brick checkerwork, through which the exhaust gases and incoming gases are blown in turn.

regulator The part of a timepiece which regulates the mechanism that indicates the time. The pendulum, balance wheel and quartz crystal are common regulators. The atomic clock is regulated by vibrating atoms.

Regulus (α Leonis) The brightest star (mag $1 \cdot 35$) in the constellation Leo (the Lion) and 21st brightest in the heavens.

reheat See **afterburner.**

relative atomic mass See **atomic mass.**

relative density Also called specific gravity; the density of a substance compared with that of water—or more accurately the density of water at 4°C, when it has maximum density.

relative humidity The ratio of the pressure of water vapour in the air at any time to the pressure of water vapour that would saturate the air (the saturated vapour pressure).

relative molecular mass See **molecular weight.**

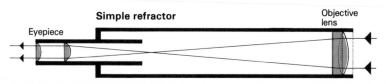

Simple refractor

Essential features of a refrigerator

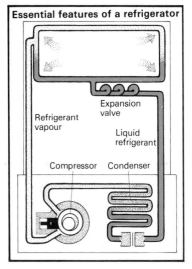

Refrigerant vapour

Expansion valve

Liquid refrigerant

Compressor   Condenser

relativistic mass The mass of a moving body, taking into account the velocity at which it is travelling. When a body travels at a velocity close to that of light, it acquires a significant extra mass because of its velocity, in accord with relativity theory. This increased mass is called the relativistic mass. Compare **rest mass.**

relativity A revolutionary theory developed by Albert Einstein in 1905 (the special theory) and 1915 (the general theory) to describe more accurately the nature of the universe. It involves the interrelationships between space, time, motion, mass and gravitation. Two consequences of the theory are that the velocity of light is a fundamental constant and is the highest velocity obtainable; and that mass and energy are equivalent (see **Einstein's equation**).

relay An electromagnetic device in one circuit that can be used to operate a switch in another circuit, usually remotely. The Morse buzzer is a simple example of a relay.

rem A measure of the intensity of ionizing radiation, similar to the rad. But it takes into account the effect of a particular type of radiation on the human body. Rem stands for 'roentgen equivalent man'.

remote control The control of a mechanism or process from a distance.

rennin An enzyme that curdles milk, found in the stomach of cattle and other cud-chewers.

reservoir An artificial pool or lake for storing water. It is usually made by damming a river.

resins Natural gum-like substances exuded by, or otherwise obtained from trees or insects. Rosin is a resin obtained from pine trees; shellac, a resin made from lac insects. Synthetic resins are resin-like polymers and plastics materials (see **synthetic resins**). Both natural and synthetic resins are used in paints, printing inks and varnishes.

resistance The property of an electrical circuit to oppose the flow of electricity. The resistance to flow results in the generation of heat. If a current i flows in a circuit under a potential difference of V volts, the resistance (R) is given by the formula: $R = V/i$. It is measured in ohms. The specific resistance, or resistivity, is the resistance offered by a unit cube of conductor.

resistor An electronic component with a specified resistance.

resolving power The ability of an optical device, such as a telescope, to form separate images of objects located close together.

resonance In chemistry, a term used to describe the phenomenon when the structure of a molecule can be considered to alternate between a number of possibilities. The classic example of resonance may be seen in benzene's ring structure (see **Kekulé structure**). In physics, a vibrating or oscillating system is said to be in resonance when the application of a force or vibration of the right frequency causes a large response. An opera singer singing a prolonged note can cause a glass to shatter, if that note is of the natural frequency of vibration of the glass. The glass begins to resonate and may eventually shake itself to pieces.

resorcinol Or meta-dihydroxybenzene, $C_6H_4(OH)_2$; an aromatic organic compound widely used for making ointments, resins, plastics and dyes.

respiration The process in which living things absorb oxygen (to 'burn' their food and release energy) and give out carbon dioxide (as a waste product).

restitution, coefficient of The ratio of the relative velocity of two colliding bodies after impact to their relative velocity before impact. It is a measure of their elasticity.

rest mass The mass of a body that is at rest with respect to an observer. Compare **relativistic mass.**

retardation Also called deceleration; a rate of decrease in velocity; a negative acceleration.

retort In the laboratory, a bulbous glass vessel with a long tapering neck used for distillation; in industry, any vessel in which a chemical process takes place.

retrobraking In astronautics, using the thrust from a rocket in the direction of travel to achieve a braking effect. Retrobraking is carried out at the end of a space mission to slow down a spacecraft prior to re-entry.

retrograde motion In astronomy, motion in the opposite direction from usual. Some of the planets display apparent retrograde motion in the heavens, but this is because of the relative movement of the Earth. The four outermost moons of Jupiter display actual retrograde motion for they orbit clockwise (as seen from the north) around the planet. Usually moons orbit around their planet, and planets orbit around the Sun in a counterclockwise direction (seen from the north).

retting Exposing flax (or other fibre-bearing plants) to the action of bacteria and moisture so as to loosen the fibres in the stems. See also **linen.**

reverberatory furnace One in which the charge is out of contact with the fuel, and is heated by a flame blown over it and from heat radiated down by the furnace roof.

reversible reaction A chemical reaction that can take place in either direction, depending on the conditions. It is in equilibrium when the rate of the forward reaction (reactants into products) equals the rate of the reverse reaction (products into reactants). The Haber process for the manufacture of ammonia (NH_3) from nitrogen (N) and hydrogen (H) utilizes the reversible reaction:

$$N_2 + 3H_2 \rightleftharpoons 2NH_3$$

The external conditions may affect the equilibrium point of a reversible reaction (see **Le Chatelier's principle**).

Reynold's number A dimensionless number of great importance in fluid mechanics. For $Re < 2000$ flow is usually steady, or streamline; for $Re > 2000$ flow is usually turbulent.

Rf The chemical symbol for rutherfordium.

Rh The chemical symbol for rhodium.

rhenium (Re) A very rare transition metal (rd 21, mp 3180°C).

rheology The science concerned with the way materials deform and flow under stress.

rheostat A variable resistance, used to control the flow of electric current. A simple type consists of a coil of resistance wire with a sliding contact.

rhodium (Rh) A hard white metal (rd 12·4, mp 1960°C) of the platinum group, used for plating jewellery and in thermocouples.

rhombus A quadrilateral with parallel sides of equal length. A rhomboid is a solid figure, each of whose six faces is a rhombus.

rhyolite A fine-grained igneous rock of similar composition to granite.

riboflavin Vitamin B_2, $C_{17}H_{20}N_4O_6$.

ribonucleic acid See **RNA.**

Richter scale A 0–10 scale on which the strength of earthquakes is measured, devised by the American seismologist C. F. Richter. It relates to the amplitude of the ground waves generated by the earthquake. The largest earthquakes register over 8 on the scale.

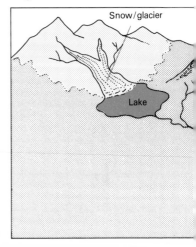
Snow/glacier
Lake

rift valley A valley formed when the land sinks between roughly parallel faults in the Earth's crust. The best-known example is the Afro-Arabian rift system, which runs from Mozambique to Israel and contains the Red Sea and many of Africa's lakes.

Rigel (β Orionis) The brightest star (mag $0 \cdot 11$) in the constellation Orion and seventh brightest in the heavens. It is a blue-white supergiant and lies about 815 light-years away.

Rigil Kent See **Alpha Centauri.**

right ascension One of the two main stellar coordinates that fix the position of a star in the heavens. It is the equivalent to terrestrial longitude, and is the angular distance along the celestial equator east of the First Point of Aries. It is usually measured in hours and minutes. See also **declination**.

rille A long, narrow valley on the Moon's surface.

river A body of water flowing in a natural channel, typically from inland hills and mountains into other rivers, lakes or the sea. A river and its branches, or tributaries, make up a river system. The region a river system drains is called a river basin. The rivers Amazon and Nile vie to be the longest in the world. The Amazon certainly drains the largest river basin (some 7,050,000 sq km, 2,720,000 sq miles) but is probably some 225 km (140 miles) shorter than the Nile (6670 km, 4145 miles).

riveting A method of joining two pieces of metal, typically steel plates, by means of rivets—threadless metal bolts. A rivet is inserted through matching holes in overlapping plates and the narrow end is hammered into a head to hold it fast.

Rn The chemical symbol for radon.

RNA The abbreviation for ribonucleic acid. It is an essential constituent of the living cell, where it carries 'instructions' from the DNA. It is made up of long chains of nucleotides.

robot A device or mechanism that can replace human beings in certain situations. The word comes from the Czech word 'robota', meaning slave. It was first used by the Czech playwright Karel Capek in his play *Rossum's Universal Robots* (1921). The introduction of robots has brought about increased automation in many industries.

Rochelle salt The compound sodium potassium tartrate tetrahydrate, $KNaC_4H_4O_6.4H_2O$. The crystalline form has useful piezoelectric properties.

rocks The materials that make up the Earth's crust. They are composed of one or more minerals, but unlike minerals rocks do not usually have a uniform composition. They are usually, but not always, hard. Clay and gravel are classed as rocks, but they are not solid in the way, say, granite is. For the three main types of rock, see **igneous rocks, sedimentary rocks** and **metamorphic rocks.** The study of rocks is petrology.

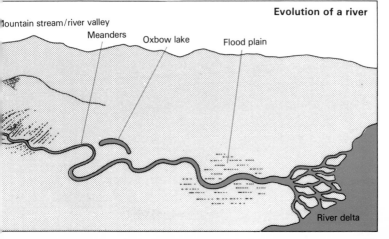

Evolution of a river

Mountain stream/river valley
Meanders
Oxbow lake
Flood plain
River delta

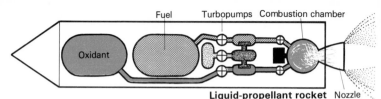

Liquid-propellant rocket Nozzle

rock crystal A very pure transparent form of quartz.

rocket A missile or launch vehicle propelled by reaction, which carries both fuel and the oxygen to burn the fuel. Because of this, it can work in airless space. The fuel and oxidizer (oxygen-provider) are called the propellants. The Chinese are thought to have used rockets first some 750 years ago, employing gunpowder as a propellant. Modern fireworks rockets still have a similar propellant. Most space launch vehicles, however, have liquid propellants. The largest successful rocket was the American Saturn V, 111 m (365 ft) long with a take-off thrust of 3·4 million kg (7·5 million lb), built for the Apollo Moon-landing missions. See also **Ariane; step rocket.**

rock salt See **halite.**

roentgen A unit used to measure the intensity of radiation in terms of its ionizing effect on air. It is named after the German physicist who discovered X-rays, Wilhelm Conrad Roentgen. See also **rad; rem.**

rolling One of the main methods of shaping metals—by 'mangling' them with heavy rollers. In steel rolling mills, red-hot ingots are passed between many sets (stands) of rollers in succession, the thickness of the metal being reduced a little each time. The emergent sheet metal may finally be cold rolled.

rosin A natural resin obtained by distilling the gum that exudes from pine trees. It is the residue that remains after terpentine vapour is driven off.

rotor The part that rotates in a generator, electric motor or turbine. Compare **stator.**

Ru The chemical symbol for ruthenium.

rubber An elastic material obtained from the juice or latex of the rubber tree, *Hevea brasiliensis*. The latex, obtained by tapping the rubber tree, is usually treated with weak acid (such as formic

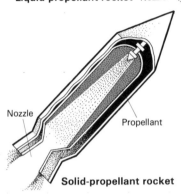

Solid-propellant rocket

acid), which coagulates it into crude rubber. Rubber is a natural polymer of isoprene, $CH_2CHC(CH_3)CH_2$. See also **synthetic rubber; vulcanization.**

rubidium (Rb) One of the reactive alkali metals (rd 1·5, mp 38·8°C). Soft and silvery white, it burns spontaneously in the air and decomposes water vigorously.

ruby A rare and valuable red gem. It is a variety of corundum, coloured red by traces of chromium oxide. Synthetic ruby is used to make one type of laser and jewel watch bearings.

rust The reddish-brown product formed when iron and steel corrode in the presence of air and moisture. It consists mainly of hydrated iron oxide, $Fe_2O_3.xH_2O$.

ruthenium (Re) A rare, hard metal (rd 12·4, mp 2300°C) of the platinum group, used to harden platinum and palladium.

rutherfordium (Rf) An artificial radioactive element (at no 104), also known as kurchatovium.

rutile The common ore of titanium, titanium dioxide (TiO_2), which is usually brown to black in colour. When refined, it becomes pure white and is widely used as a pigment in paints.

S

S The chemical symbol for sulphur.

Sa The chemical symbol for samarium.

saccharides See **disaccharides; monosaccharides; polysaccharides.**

saccharin A white powder used for sweetening food and drink, over 500 times sweeter than sugar. It is ortho-sulphobenzoic acid imide, $C_6H_4SO_2NHCO$.

sacrificial protection A way of protecting one metal (particularly steel) from corrosion by the use of another metal (such as magnesium or zinc), which is designed to corrode away instead.

safety glass Glass made so that it does not shatter into sharp fragments when broken; fitted, for example, to cars. Toughened glass, made by rapidly cooling hot glass with an air blast, forms into rounded pieces when shattered. Laminated glass, made up of two sheets of glass with a film of clear plastic in between, simply cracks under impact, the glass adhering to the plastic.

safety lamp See **Davy lamp.**

safety valve A pressure-operated valve fitted to boilers and other vessels, which opens when the pressure inside exceeds a certain level.

Sagittarius The Archer; a zodiacal constellation in the southern hemisphere, lying between Capricornus and Scorpio. The centre of our galaxy lies in Sagittarius, which is notable for its dense star clouds.

St Elmo's fire An ethereal glow seen at the top of spires or ships' masts during stormy weather. It is caused by the discharge of atmospheric electricity.

sal ammoniac The compound ammonium chloride, NH_4Cl, used commonly as the electrolyte in dry cells. It sublimes readily.

salt The common name for sodium chloride, NaCl, found abundantly dissolved in the oceans and in massive form as rock salt, or halite. Its uses are legion, from flavouring and preserving food to manufacturing soap, chlorine and caustic soda. In chemistry, a salt is the product of an acid and a base.

salts Compounds produced, together with water, when acids react with bases. Salts are ionic compounds that generally dissolve in water and conduct electricity when molten and when in solution —they are good electrolytes. Among the commonest types of salts, named after the acid from which they are derived, are the sulphates, chlorides, nitrates, phosphates and carbonates.

saltpetre Also called nitre; the compound potassium nitrate, KNO_3, which is an excellent oxidizing agent.

Salyut A successful series of Russian manned space stations, beginning with Salyut 1 in 1971. Salyut 6, launched in 1977, was outstandingly successful, being visited by many cosmonaut crews in Soyuz spacecraft and also by unmanned Progress supply craft. In 1982 cosmonauts Anatoly Berezovoy and Valentin Lebedev in Salyut 7 established a space endurance record of 211 days.

samarium (Sm) One of the rare-earth metals (rd 7·5, mp 1050°C) of little importance.

San Andreas fault A fracture in the Earth's crust in California that is a region of potential earthquakes. It occurs where two crustal plates—the Eastern Pacific and North American— are moving relative to one another (see **plate tectonics**). The 1906 San Francisco earthquake was caused by movement along the fault.

sand Fine particles of rocks and minerals formed as a result of prolonged weathering of solid rocks. The most common mineral in sand is silica, or quartz.

sandpaper See **abrasive.**

sandstone A common sedimentary rock made up of particles of sand cemented together with clay, calcite or other minerals.

saponification The hydrolysis of an ester by an alkali to produce an alcohol and a salt. Saponification of a fat by caustic soda produces glycerol and soap.

sapphire A blue-tinted variety of corundum (aluminium oxide, Al_2O_3) valued as a gem. Other coloured tints are also found, the red ones being called rubies.

saros In astronomy, a period of 18 years 11 days, after which time eclipses of the Sun and Moon occur in the same order and at the same time.

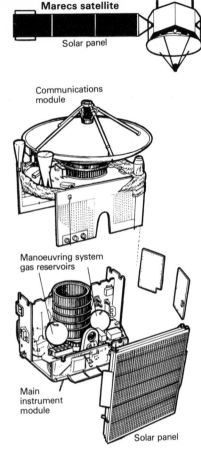

Marecs satellite

Solar panel   Solar panel

Communications
module

Manoeuvring system
gas reservoirs

Main
instrument
module

Solar panel

satellite A small body, or moon, that orbits around a planet. The Earth has one satellite, the Moon. Since 1957 it has had many artificial satellites, man-made objects launched into space by rockets. They carry a variety of sensing instruments and a radio to relay instrument readings back to Earth; or they may act as remote telecommunications relay stations. See **communications satellite; Landsat; Sputnik; weather satellite.**

satellite galaxy A small galaxy that is associated with a larger one. Our own Milky Way galaxy has two satellite galaxies, the Magellanic Clouds.

saturated compounds Organic compounds that have no double or multiple bonds between the carbon atoms. Paraffins (alkanes) are saturated compounds, their carbon atoms being linked by single bonds. Compare **unsaturated compounds.**

saturated solution A solution in which dissolved solute is in dynamic equilibrium with undissolved solute.

saturated vapour pressure The pressure exerted by a vapour in equilibrium with its liquid. The term most commonly refers to water vapour in the air.

Saturn The sixth planet in the solar system going out from the Sun, noted for its magnificent ring system. A rapidly rotating ball of hydrogen and helium, it is the second largest planet after Jupiter with a diameter of 120,600 km (74,950 miles). The main ring system measures about 272,000 km (169,000 miles) across; the Voyager space probes found the three rings visible from Earth to be made up of thousands of separate ringlets composed of lumps of ice-covered rocks of various sizes. The planet turns on its axis every 10 hr 39 min, and revolves around the Sun in 29·5 years. It is the least dense of the planets (rd 0·7). At favourable oppositions, it reaches a magnitude of − 0·3. It has at least 21 satellites, the biggest of which is the orange Titan, notable because it has an atmosphere.

Saturn V rocket See **rocket.**

Sb The chemical symbol for antimony; from the Latin, stibium.

Sc The chemical symbol for scandium.

scalar A quantity that can be fully described by its magnitude, expressed in appropriate units. Weight, density, volume, speed (but not velocity) are scalar quantities. Compare **vector.**

scales A device for weighing objects. Many scales work on the principle of balance. The weighing pans rest on or are suspended from each end of an arm pivoted in the centre. Many laboratory balances are of this type. Spring scales work by means of the extension or compression of a spring under load.

screw

scandium (Sc) One of the rarer of the transition metals (rd 3·0, mp 1400°C), usually considered a rare-earth. Its existence was predicted by Dmitri Mendeleyev (who called it ekaboron).

scanning In electronics, see **television.**

scarp See **escarpment.**

scattering In physics, the change in direction of motion of particles or radiation because of collisions with other particles. Sunlight, for example, is scattered by molecules in the atmosphere. Blue light is scattered more than red light, which is why the sky appears blue.

scheelite A major tungsten ore, being calcium tungstate ($CaWO_4$).

schist A coarse crystalline metamorphic rock that displays a great tendency to split into layers (a property called schistosity). Schists contain flaky minerals such as mica, talc and graphite, and often gemstones such as garnets.

schlieren photography A photographic technique that reveals flow patterns in fluids by exploiting changes in refractive index.

Schmidt telescope A wide-angle telescope that has a spherical main mirror and is capped by a correcting plate, which compensates for defects such as spherical aberration and astigmatism.

science The broad field of human knowledge, acquired by systematic observation and experiment, and explained by means of rules, laws, principles, theories and hypotheses. The three broad categories of science are the physical sciences (eg physics, chemistry, geology, astronomy); the biological sciences (eg botany, zoology, biochemistry); and the social sciences (eg sociology, anthropology, economics). In addition mathematics may be considered, if not a science in itself, as an essential tool of science. This dictionary is concerned mainly with the physical sciences and applied science, or technology.

scintillation A flash of light given out when radiation strikes a phosphor. A scintillation counter is a device that detects ionizing particles or radiation by counting the flashes emitted when the radiation strikes a phosphor.

scopalamine Also called hyoscine; an alkaloid drug obtained from deadly nightshade and other plants that causes a 'twilight' sleep. It is used widely as premedication before anaesthesis. Its chemical formula is $C_{17}H_{21}NO_4$.

Scorpio The Scorpion; a zodiacal constellation, whose brightest star is Antares. It lies between Libra and Sagittarius.

SCP Abbreviation for single-celled protein, an artificial foodstuff made by processing such simple organisms as algae, bacteria, yeasts and moulds.

scraper An earth-moving vehicle that scrapes up surface material, used mainly during road construction work. It consists essentially of an open bowl with a cutting blade at the lower front edge.

scree Or talus; the rock debris that collects at the foot of cliffs, produced by the weathering of the solid rock above it.

screw Any cylinder or cone carrying a spiral ridge, or thread; commonly used as a fastener. The distance between the threads is called the pitch. The screw is in effect a very simple machine, being an inclined plane spirally wound around a cylinder. The mechanical advantage of the screw is utilized in the screw jack.

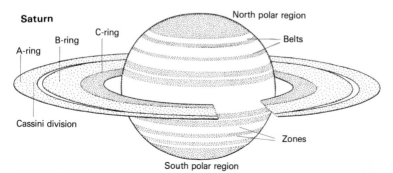

Saturn

North polar region · A-ring · B-ring · C-ring · Belts · Cassini division · Zones · South polar region

199

Se

Se The chemical symbol for selenium.

seawater The water of the oceans contains approximately 3·5% dissolved salts, including about 2·8% sodium chloride, 0·4% magnesium chloride and 0·2% magnesium sulphate. Calcium sulphate, potassium chloride, magnesium bromide and calcium carbonate are also present in appreciable quantities. The water in the Dead Sea contains up to 25% of dissolved salts.

seas, lunar See **maria.**

seasons Natural divisions of the year, marked by regular changes in the weather and the lengths of the days and nights. The rhythm of the seasons is most evident in temperate regions, where the four seasons—spring, summer, autumn and winter—are most distinct. The seasons are caused by the inclination (23½°) of the Earth's axis to the plane of its orbit around the Sun. This allows a certain point on the Earth to receive more solar radiation at some times than at others. It is summer in the northern hemisphere (winter in the southern) when the north pole is tilted towards (south pole away from) the Sun; and winter in the northern hemisphere (summer in the southern) when the north pole is tilted away (south pole towards) the Sun. At times in between, when the Earth's axis is tilted neither towards nor away from the Sun, the equinoxes occur. Then day and night are of equal length throughout the world. See also **equinoxes; solstices.**

seasoning Of timber; allowing it to dry before use, to prevent warping and shrinkage later.

second The basic unit of time, now defined as 9,192,631,770 cycles of the radiation emitted by caesium atoms during the transition between two levels of the ground state.

secondary cell See **accumulator.**

sedative A drug that reduces nervous tension and has a calming effect on the body. In larger doses it may induce sleep. The barbiturates, chloral hydrate and tranquillizers such as Valium have a sedative effect.

sedimentary rocks Rocks formed when accumulated layers of sediment are cemented and compacted under pressure in the Earth's crust. The 'sediments' may be fragments of pre-existing rocks of all types that have been broken up by weathering. Rocks of this type, termed clastic rocks, include the coarse-grained breccia, the medium-grained sandstone and the fine-grained shale. Or the sediments may consist of minerals precipitated from ancient seas; or deposited when those seas evaporated. The latter deposits are termed evaporites. Rocks so formed, termed non-clastic rocks, include most limestones, dolomite and gypsum. Other non-clastic sedimentary rocks have an organic origin. They are formed of accumulations of the shells and skeletons of animals and even plants. Chalk is such a rock, made up of the shells of

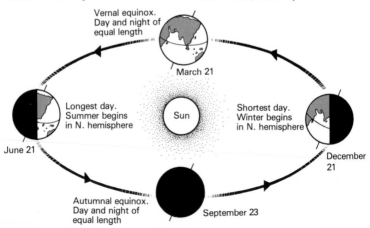

Vernal equinox. Day and night of equal length

March 21

Longest day. Summer begins in N. hemisphere

June 21

Sun

Shortest day. Winter begins in N. hemisphere

December 21

Autumnal equinox. Day and night of equal length

September 23

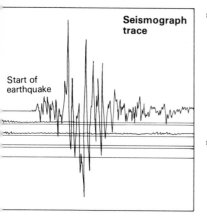

Seismograph trace

Start of earthquake

microscopic organisms. Some lime-stones are made up of fossil fish bones, shells, coral and crinoids (sea lilies). Coal is a special sedimentary 'rock', formed from the decayed remains of ancient forests. Fossils abound in sedi-mentary rocks.

sedimentation In geology, a process whereby solid material is deposited, for example, after being transported by such agents as wind, water and glaciers. The deposited sediments may over mil-lions of years be compacted into rocks.

Seebeck effect When the ends of two wires made from dissimilar metals are joined together, and the two junctions are held at different temperatures, an electric current flows through the wires. This thermoelectric phenomenon, discovered in 1821 by the German physicist T. J. Seebeck, is utilized in the thermocouple.

seeding The addition of fine particles to a solution to bring about crystallization, or to a cloud to trigger off precipitation.

seeing In astronomy, the quality of the viewing conditions when observing, usually recorded on a scale devised by E. M. Antoniadi of I (perfect) to V (very poor).

seismic survey A technique used in oil and mineral prospecting to explore the rock layers beneath the surface. It involves making an explosion, or other-wise creating shock waves in the ground, and detecting these waves (by geo-phones) after they have been reflected from the various rock layers.

seismograph An instrument for detec-ting ground vibrations, as experienced during an earthquake. Seismographs usually utilize the inertia of a heavy weight. This is suspended in a frame by a delicate spring. When the ground moves, so does the frame, but the weight tends to remain stationary because of its inertia. The relative movement of frame and weight is magnified and fed to a recording pen, which makes a trace on a moving sheet (a seismogram).

seismology The science of earthquakes, concerned particularly with the propa-gation and transmission of shock waves (seismic waves) through the Earth's crust. Earthquakes give rise to four main types of seismic waves—two body waves (P and S) in the crust and two surface waves. The P (primary) waves are compression, or longitudinal waves; the slower S (secondary) waves are shear, or transverse waves. See also **wave.**

selenium (Se) A useful metalloid (rd 4·8, mp 217°C) related to sulphur. It can be obtained as red crystals or powder and as a grey metallic solid. Metallic selenium is used in many photoelectric devices because it can con-vert light into electricity. It is also useful as a rectifier, for converting alternating into direct current.

selenite See **gypsum.**

semiconductor A material whose electrical properties lie between those of a conductor and those of an insulator. The group of elements known as the metalloids are semiconductors. They include silicon, germanium, tellurium and selenium. The former is best-known in the guise of the silicon chip. Pure silicon does not conduct electricity, but when certain impurities (such as phos-phorus) are added to it (by a process called doping) an excess of electrons is produced, which allows the silicon to conduct electricity. This type of silicon is known as the *n*-type. When certain other impurities (such as boron) are added to it, a lack of electrons is produced, creating what are termed 'holes'. These holes can pass from atom to atom, and a flow of holes also constitutes an electric current. The silicon so formed is termed *p*-type. Electronic devices such as the transistor are made by sandwiching crystals of *p*- and *n*-type silicon together (see **transistor**).

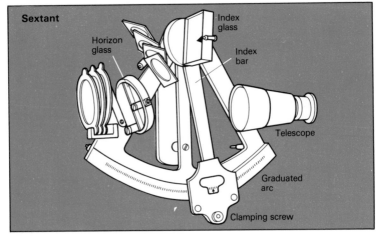

Sextant

Horizon glass

Index glass

Index bar

Telescope

Graduated arc

Clamping screw

semipermeable membrane One that allows some substances through but not others. Such membranes feature in osmosis, allowing the passage of a common solvent between solutions of different concentrations.

serpentine A common green or yellow mineral of attractive appearance, named after its likeness to a snake's skin. It is magnesium silicate, $Mg_3Si_2O_5(OH)_4$, and appears in three main forms—flaky antigorite, fine-grained lizardite and fibrous chrysotile, which is the commonest variety of asbestos.

servomechanism An automatic control device that relies on feedback to correct the performance of a mechanism.

Seven Sisters See **Pleiades.**

sextant A navigation instrument used for measuring the altitude of heavenly bodies. An observer looks through the telescope at the horizon through the clear half of the horizon glass (see diagram). He moves the index bar, with mirror (index glass) attached, until an image of the Sun (say) in the horizon glass appears to be on the horizon. He then takes the reading, noting the exact time.

shadow The darkness behind an object that is blocking the light. If the light source is large, a partial shadow (penumbra) may be formed around the region of complete shadow (umbra). This effect is experienced during eclipses of the Sun.

shale A common fine-grained sedimentary rock formed from deposits of silt. Shale is often rich in fossils and may be impregnated with hydrocarbon oils. See also **oil shale; slate.**

shear The deformation of a body in which parallel planes of the body move in relation to one another.

shellac A natural resin obtained from the lac insect, used to make varnishes. Filled with black pigment, it was used to make early gramophone records.

SHM See **simple harmonic motion.**

shock absorber Or damper; a device used, for example, in car suspensions to damp out (limit) vibrations of springs. A common type consists of a piston inside an oil-filled chamber.

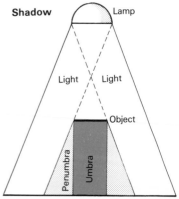

Shadow

Lamp

Light   Light

Object

Penumbra   Umbra

shock wave A strong pressure wave created by an explosion, lightning, or an aircraft travelling at supersonic speed.

shooting star See **meteor.**

short circuit What happens when the terminals of a battery or generator, or the live and neutral wires of the mains are connected by a conductor of little resistance.

shortsightedness Or myopia; a defect of the eyes that results in blurred vision in which objects are brought into focus in front of the retina. It can be corrected by the use of a diverging lens.

Shortsightedness

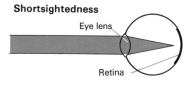

Eye lens

Retina

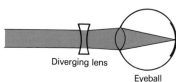

Diverging lens

Eyeball

shuttle In astronautics, see **space shuttle;** in weaving, see **loom.**

Si The chemical symbol for silicon.

sial A term used to describe the material that makes up the rocks in the Earth's upper crust—the rocks of the continents—derived from the two main constituents, silicon and aluminium. Compare **sima.**

sidereal time Time based on the rotation of the Earth on its axis with respect to the stars. The sidereal day is 23 hr 56 min 4 sec long, about 4 minutes less than the solar day.

siderite Or chalybite; a major ore of iron, being iron carbonate, $FeCO_3$.

siemens The SI unit of electrical conductance, once termed the reciprocal ohm, or mho.

Siemens-Martin process An alternative name for the open-hearth process.

silage See **silo.**

silanes A series of silicon hydrides, analogous to the paraffin hydrocarbons (alkanes).

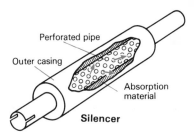

Perforated pipe

Outer casing

Absorption material

Silencer

silencer Also called muffler; a device used to reduce the noise of the exhaust from an engine. Two types of car silencers are used. One uses baffles to slow down and silence the exhaust; the other ('straight-through') uses perforated pipes and expansion chambers.

silica The commonest mineral in the Earth's crust, being silicon dioxide, SiO_2. It occurs in the form of sand, quartz, flint, chert, chalcedony, agate and onyx.

silica gel An amorphous (non-crystalline) form of silica used as a drying agent because it is very porous and has a remarkable capacity for absorbing water.

silicates The most abundant group of minerals in the Earth's crust, which is composed particularly of sodium, potassium, calcium, magnesium and aluminium silicates. Clays and feldspars are typical silicate minerals. Silicates are derived from the weak silicic acids, which are often considered hydrates of silica. Metasilicic acid, for example, may be written H_2SiO_3 or $SiO_2.H_2O$. See also **water glass.**

silicon (Si) After oxygen, the most common element (rd $2 \cdot 3$, mp $1410°C$) in the Earth's crust, of which it comprises nearly 28%. It occurs combined with oxygen in silica (silicon dioxide, SiO_2) and with oxygen and various metals in silicates. It is a grey crystalline solid with a metallic lustre, closely related to carbon. It is a metalloid and, like many metalloids, is a semiconductor. It is now widely used in solid-state electronics for making components such as transistors and silicon chips.

silicon carbide Also called carborundum; a very hard material used as an abrasive. Of chemical formula SiC, it is made by heating together pure sand and coke in an electric furnace.

silicon chip Also called the microchip; a thin sliver of silicon a few millimetres square that contains thousands of microscopic electronic components etched into its surface, forming complete electronic circuits (see **integrated circuits**). Silicon chips are used in digital electronic calculators and watches and desktop computers (microcomputers). Chips containing circuits that carry out most of the functions of a computer are termed microprocessors.

silicones A series of synthetic resins, elastomers and fluids consisting of polymers based on chains of alternate silicon and oxygen atoms. They are noted for their resistance to heat and cold and their chemical stability.

silk A strong, lustrous textile fibre, obtained by unravelling the cocoons of the silkworm, the larva of the moth *Bombyx mori,* which feeds on the leaves of mulberry trees. It is the only natural fibre obtained as long filaments (threads), which are gathered together (a process called throwing) into a yarn thick enough for weaving.

silk-screen printing A method of printing with a fine silk screen, using the principle of the stencil. The silk is processed so that non-printing areas are coated with glue and prevent the printing ink going through.

sill A table-like intrusion of igneous rock running parallel to the strata of the surrounding rocks.

silo In agriculture, an airtight cylindrical tower in which silage is made by partial fermentation of fresh chopped plant material, such as grass and lucerne. In military technology, a silo is an underground installation for a ballistic missile.

silt Fine-grained soil or sediment that may be transported by flowing water or the wind.

Silurian Period The span of geological time that lasted from about 430 million to about 395 million years ago. During this time plants first appeared on the land, while the seas teemed with shelled creatures.

silver (Ag) A white precious metal (rd $10 \cdot 5$, mp $960 \cdot 8°C$), which is the finest conductor of heat and electricty. It is found native in ores such as argentite (silver sulphide, Ag_2S) and is obtained as a by-product in smelting other metals,

such as lead and nickel. It is highly malleable and ductile and resistant to corrosion, though it does tend to tarnish in air that contains sulphur. The metal itself is widely used for jewellery. Silver bromide (AgBr) and iodide (AgI) are used in photography, because they are light-sensitive. Silver nitrate ($AgNO_3$) is a widely used laboratory reagent and is used industrially to prepare other silver salts.

sima The rocks of the lower part of the Earth's crust—the ocean floors—composed predominantly of silica and magnesium minerals. Compare **sial.**

simple harmonic motion A type of motion in which a body moves in a path symmetric about a central point. At any point the acceleration of the body towards the centre is always proportional to its displacement from the centre.

sine (sin) See **trigonometry.**

sintering Consolidating powdery materials into bigger lumps by heating strongly. In powder metallurgy, a metal is shaped by sintering metal powder in moulds under pressure at temperatures below the metal's melting point.

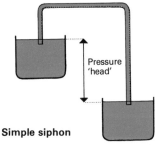

Simple siphon

siphon A device, consisting usually of a bent tube, used to transfer liquid out of one. vessel and into another at a lower level. The water closet flushes by siphon action. In the simple siphon illustrated here, the liquid flows into the lower vessel because of the pressure 'head' of liquid.

siren A device that produces a piercing sound, used as a warning signal. It consists of a perforated disc or cylinder, onto which jets of air or steam are blown. When the disc or cylinder is rapidly rotated, the jets are interrupted, generating sound vibrations of definite pitch.

Magnified picture of a silicon chip, held between finger and thumb. The actual size of the chip is about 7 mm square.

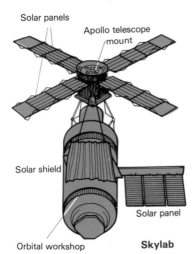

Solar panels

Apollo telescope mount

Solar shield

Solar panel

Orbital workshop

Skylab

Sirius (α Canis Majoris) The brightest star (mag − 1·45) in the heavens, in the constellation Canis Major (the Great Dog), from which its alternative name, Dog Star, is derived. It is a binary star whose other component, called the companion of Sirius, was the first white dwarf discovered. Sirius lies some 8·8 light-years away.

sirocco A warm, humid wind that blows from the south and south-east in southern Europe, bringing rain and fog. It originates in the Sahara Desert and gathers moisture across the Mediterranean.

sisal The name of a tropical plant of the Agave family and the valuable coarse fibres obtained from its leaves, which are made into ropes and sacking.

SI units The international system of units of measurement (Système International d'Unités) now preferred in science, based on the fundamental units metre (for length), kilogram (for weight), second (for time), ampere (for electric current), kelvin (for temperature), and candela (for luminous intensity).

sky The appearance of the heavens above us, bright in the daytime and dark at night. The colour of the daytime clear sky—blue—is caused by the scattering of light by molecules in the air. Blue light is scattered the most and the sky thus

appears blue. At sunset, however, the sunlight has to travel a long distance through the lowest part of the atmosphere, which can be heavily dust-laden. These larger particles tend to scatter red light more than the blue, and a 'red sky at night' results.

Skylab An early (1973/4) and most successful experimental American space station, fashioned from hardware left over from the Apollo Moon-landing project. With the Apollo ferry craft attached, the Skylab cluster measured over 36 m (120 ft) long and weighed some 90 tonnes. It was visited by three three-man teams for periods of 28, 59 and 84 days respectively. It returned to Earth in 1979, breaking up over Western Australia.

skyscraper A very high building, constructed usually of a steel skeleton and concrete. The first true skyscraper was completed in 1885 in Chicago. That city currently again boasts the tallest skyscraper—the 443 m (1454 ft) high Sears Tower, which has 110 storeys.

slag Impure non-metallic material formed during the smelting of iron, copper and other metallic ores. See **blast furnace**.

slaked lime The common name for calcium hydroxide, $Ca(OH)_2$, formed by slaking (adding water to) calcium oxide.

slate A black, grey or coloured fine-grained metamorphic rock produced by the action of heat and pressure on clays and shale. The minerals (typically mica, chlorite and quartz) form thin parallel layers. It can readily be split into sheets, used for example for roofing.

slide rule A mathematical device used for making rapid calculations, especially multiplication and division. It generally takes the form of a ruler with a sliding middle section. The scales are logarithmic.

slip rings See **electric generator**.

slurry A watery suspension of solid particles.

Sm The chemical symbol for samarium.

Small Magellanic Cloud See **Magellanic Clouds**.

smelting The extraction of a metal from its ore by means of heat. In a typical smelting process an oxide ore is reduced to metal by means of carbon, which also acts as fuel. See also **blast furnace**.

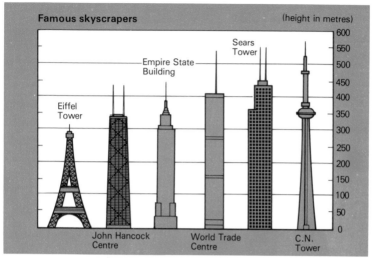

Famous skyscrapers (height in metres)

Eiffel Tower — Empire State Building — Sears Tower — John Hancock Centre — World Trade Centre — C.N. Tower

smog A thick smoky fog that may form in industrial regions or in traffic-choked cities in certain weather conditions.

smoke A suspension of fine solid particles in a gas, typically of particles of carbon in air.

Sn The chemical symbol for tin, from the Latin, stannum.

Snell's law See **refraction**.

snow Ice crystals that form in the atmosphere when the humidity is high and the temperature is below freezing. The crystals appear in a variety of beautiful hexagonal shapes.

soap A cleansing agent made by boiling fats or oils with an alkali, such as caustic soda. The soap-making process is one of saponification. Ordinary soap is made up of a mixture of the sodium salts of stearic, palmitic and oleic acids. See also **detergent**.

soapstone See **talc**.

soda A term used to describe several sodium salts. Washing soda is sodium carbonate decahydrate, $Na_2CO_3.10H_2O$; soda ash is anhydrous sodium carbonate, Na_2CO_3; bicarbonate of soda, or baking soda is sodium bicarbonate, or sodium hydrogen carbonate, $NaHCO_3$. Caustic soda is sodium hydroxide, NaOH. Soda water, on the other hand, is unrelated, being water charged with carbon dioxide.

sodium (Na) A soft, highly reactive alkali metal (rd 0·97, mp 97·8°C), which occurs widely in nature in the form of compounds, particularly as the chloride, NaCl. It reacts violently with water, liberating hydrogen and forming caustic soda, or sodium hydroxide, NaOH. Important uses of the metal itself is in the form of a vapour in sodium-vapour lamps and as a coolant in some nuclear reactors.

sodium compounds Among the most important of the widely used compounds of sodium are the following: sodium azide, NaN_3, used in explosives; sodium bicarbonate, or sodium hydrogen carbonate, or baking soda, $NaHCO_3$; sodium carbonate, or washing soda, $Na_2CO_3.10H_2O$; sodium chloride, NaCl, see **salt**; sodium chlorate, $NaClO_3$, a powerful oxidizing agent; sodium fluoride, NaF, used for water fluoridation; sodium hydroxide, NaOH, see **caustic soda**; sodium hypochlorite, NaOCl, a bleaching and oxidizing agent; sodium nitrate, $NaNO_3$, used to make fertilizers and explosives, see **Chile saltpetre**; sodium peroxide, Na_2O_2, an oxidizing agent; sodium silicate, Na_2SiO_3, see **water glass**; sodium sulphate, $Na_2SO_4.10H_2O$, see **Glauber's salt**; sodium tetraborate, $Na_2B_4O_7.10H_2O$, see **borax**.

207

software The programs and systems in a computer. Contrast **hardware**.

soil mechanics A branch of engineering science concerned with the nature and properties of soils, especially in relation to their suitability as foundations in construction work.

solar cell A device, usually made from thin slivers of silicon, that produces electricity directly from sunlight. Most satellites are powered by solar cells.

solar energy Energy derived from the Sun. The amount of energy reaching the ground on Earth is on average only about 1 kW per sq metre. It goes into heating the ground and the waters of the oceans and driving the global wind systems. Many schemes to harness solar energy have been proposed, and some houses already have solar water-heaters.

solar system The Sun and all the bodies that circle around it in space and bound to it by gravity. They include the planets and their moons, the asteroids, the comets and the smaller lumps of rocks that burn up in the Earth's atmosphere as meteors. From one extreme of the orbit of Pluto (the most distant known planet) to the other, the solar system spans a distance of some 12,000 million km (7,500 million miles).

solar wind A stream of charged particles, mainly protons and electrons, emitted by the Sun which intensifies at times of solar flares.

solder Soft solder is an alloy of tin and lead that melts and solidifies at a low temperature (about 200°C). It is used mainly to join copper wires in electrical circuits. Hard solder is a kind of brass (copper and zinc) that melts at much higher temperatures. It is used in brazing to join steel and other metals.

solenoid A cylindrical wire coil often used as a kind of electromagnet to operate switches and relays.

solid One of the three main states of matter, a solid is characterized by a fixed size and shape. Most solids are crystalline, and their crystals are made up of repetitions of a basic space lattice formed by regular arrays of atoms or ions. Crystalline solids have a definite melting point, at which temperature they change into another state of matter—liquid. Non-crystalline, or amorphous solids (such as glass) have no definite melting point.

solstices The times in the year when the Earth's axis is tilted most towards or away from the Sun; or from an observer's point of view, the times when the Sun is highest and lowest in the sky at noon. In the northern hemisphere the Sun is highest in the noon sky on about June 21 each year (summer solstice); and lowest on about December 21 (winter solstice). On these dates occur, respectively, the longest and shortest periods of daylight.

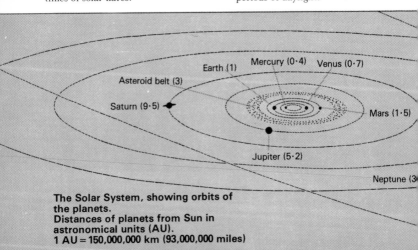

**The Solar System, showing orbits of the planets.
Distances of planets from Sun in astronomical units (AU).
1 AU = 150,000,000 km (93,000,000 miles)**

Earth (1) Mercury (0·4) Venus (0·7)
Asteroid belt (3)
Saturn (9·5)
Mars (1·5)
Jupiter (5·2)
Neptune (3

solubility Of a substance (the solute); the extent to which it will dissolve in a liquid (the solvent) at a given temperature. It is usually given as the number of grams of solute that will dissolve in 100 g of solvent to give a saturated solution. The solubility of most solid substances increases as the temperature rises. The solubility of a gas decreases with temperature rise.

solute See **solution**.

solution A homogeneous mixture on a molecular scale of two different substances. The substance being dissolved is called the solute, and the substance in which the solute is dissolved is called the solvent. The most familiar type of solution is solid-in-liquid, eg salt in water. Gases dissolve in liquids to form solutions, as do liquids in other liquids. Many alloys are solid solutions of metals soluble in one another when molten.

Solvay process See **ammonia-soda process**.

solvent See **solution**.

sonar The use of sound waves for detection, navigation and communication under water. Sonar (sound navigation and ranging) is the sound analogue of radar. Sonar equipment transmits ultrasonic (very high frequency) sound waves and detects the echoes when the waves are reflected from objects. Sonar techniques are also used in metallurgy for detecting flaws in metals and in obstetrics for looking at unborn babies in the womb. (Bats and dolphins use sonar for navigation.)

sonic boom A noise like a thunderclap caused by the shock wave produced when an aircraft, for example, exceeds the speed of sound.

sonic speed The speed of sound, about 1,220 km/h (760 mph) at sea level, but progressively less at higher altitudes. Speeds below sonic speed are termed subsonic; speeds up to about five times sonic speed are termed supersonic; higher speeds are termed hypersonic. See also **Mach number**.

sound A physical vibration that our ears can detect, usually transmitted through the air. The study of sound is called acoustics. Sound waves are propagated through the air as a series of compressions (as the air molecules vibrate in the forward direction) and rarefactions, or regions of low pressure (as the molecules vibrate back again). The wave moves outwards from the source of the sound. The air molecules do not move bodily with the wave, but merely vibrate as the wave passes. Sound travels in air at sea level at a speed of about 1,220 km/h (760 mph) and in water faster four times faster.

sound barrier A term relating to the difficulty of exceeding the speed of sound in an aircraft. When an aircraft reaches the speed of sound (Mach 1), it is subject to severe buffeting and shock, rather as if it is passing through a physical barrier of some sort. By careful design planes can now slip through the 'barrier' with ease. C. E. Yeager first 'broke the sound barrier' in 1947 in a rocket plane.

sound recording Sound may be recorded in several ways for playback on a number of different devices. It may be recorded in the form of wavy grooves on a plastic disc (record) and played back on a gramophone. It may be recorded as magnetic 'patterns' on tape and played back on a tape recorder. It may be recorded optically on a film sound track. The latest gramophone recording techniques include digital recording. In this process sounds are given precise numerical, or digital values, which improves the quality of reproduction. Laser methods are also coming into use. See **gramophone**; **sound track**; **tape recorder**.

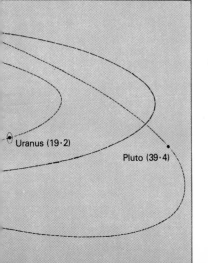

Uranus (19·2)

Pluto (39·4)

sound track A narrow band at the edge of a film that carries the record of sounds to accompany the pictures. Most sound tracks are optical, consisting of a black-and-white strip of variable area or density. To record the sound track, signals from a microphone are made to modulate, or vary, the output of a beam of light. The variable light signals are thus recorded photographically on the film. The sound is recreated by shining light through the sound track. The variable light signals produced fall on a photocell, which generates corresponding electrical signals to feed to a loudspeaker.

southern lights See **aurora**.

Soyuz spacecraft The standard Russian manned spacecraft used to ferry cosmonauts into space, for example, to space stations like Salyut. The first Soyuz flight took place in 1967.

spacecraft See **satellite; space probe; space station**.

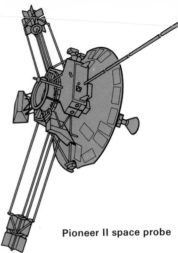

Pioneer II space probe

Spacelab A reusable space laboratory built by the European Space Agency, designed to fit into the payload bay of the space shuttle. It usually consists of a pressurized experiment compartment and an unpressurized instrument platform carried on a pallet. See **ESA**.

space probe A spacecraft that journeys into deep space to explore the planets. See **Mariner; Pioneer; Viking; Voyager.**

space shuttle The main American space transportation system, which uses reusable components. The major part of the system is a delta-winged orbiter, which is about the same size as a medium-range airliner. This carries the astronaut crew and the cargo, which is installed in a huge payload bay measuring about $18 \cdot 3 \times 4 \cdot 6$ m (60×15 ft). The orbiter rides into space on an external tank, which carries fuel for its engines. To the sides of the external tank are strapped solid rocket boosters. The boosters fire with the main engines at launch, and then separate and parachute back to Earth to be used again. The external tank is jettisoned just before the orbiter goes into orbit. The orbiter returns from orbit as a glider and lands on an ordinary runway. The first shuttle flight from the Kennedy Space Center, took place on April 12 1981. In November the orbiter, *Columbia*, became the first spacecraft ever to return to space.

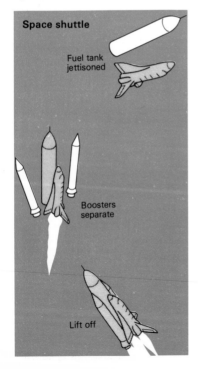

Space shuttle

Fuel tank jettisoned

Boosters separate

Lift off

Space shuttle *Columbia* soars from the launch pad at the Kennedy Space Center, Florida, on a pillar of flame.

211

space station A large manned spacecraft designed for a long stay in orbit. Salyut and Skylab were early experimental space stations.

spacesuit A multilayered garment worn by astronauts as protection against the hostile environment of space. Water is circulated through an inner garment to regulate temperature. Oxygen is supplied to a pressure suit for the astronaut to breathe. An outer garment acts as protection against tiny meteoroid particles and solar radiation. The suit may be self-contained or be supplied via an umbilical tube from a spacecraft.

space-time The concept that the universe can be described in terms of four dimensions—the three spatial dimensions, length, breadth and height (or coordinates x, y and z) and time.

spark The sudden flash of light that occurs during an electric discharge when the insulation between two conductors breaks down. In a petrol engine the fuel mixture is ignited by the spark from a sparking plug.

spark chamber A device for detecting particle radiation. It consists of a chamber filled with a noble gas (eg neon) in which there are thin parallel plates charged with high voltage. When a particle enters the chamber, it ionizes the gas and triggers off an electric discharge as a spark.

specific gravity See **relative density**.

specific heat The amount of heat required to raise the temperature of the unit mass of a substance by 1 degree. It is expressed in such units as joules/kg/kelvin; calories/gram/°C.

specific impulse In rocketry, the ratio of thrust to the rate of consumption of propellants, usually expressed as pounds thrust per pound of propellant used per second.

spectacles Or glasses; lenses worn by people with defective vision to enable them to see normally. Convex or concave lenses may be used. See **long-sightedness**; **shortsightedness**.

spectral lines Lines observed in the spectrum of light from certain sources. They are caused by electrons moving from one energy level to another. They may be bright lines, separated by dark spaces. Each element has a characteristic bright-line, or emission spectrum, by which it can be identified. Dark lines are produced when light with a continuous spectrum passes through a gas. Certain wavelengths are absorbed, producing a so-called dark-line, or absorption spectrum.

spectroscope An instrument used to produce and observe a spectrum. It is called a spectrograph if it is used to photograph the spectrum, and a spectrometer if it is used for the measurement of individual wavelengths. The simplest type of spectrometer incorporates a prism to produce the spectrum. Light from the source under investigation is passed through a narrow slit and collimator and emerges as a thin parallel beam. This is passed through the prism, which disperses the wavelengths in the light into a spectrum, which is then viewed by a telescope. More accurate instruments use a diffraction grating to produce a spectrum. Spectroscopy has become a vital investigative tool in all branches of science.

spectrum The spread of colours produced when light is passed through a prism or diffraction grating. White light from the Sun produces a near continuous spectrum in which the colours (from short wavelengths to long) are violet, indigo, blue, green, yellow, orange and red. They are the colours of the rainbow, which is a spectrum produced by the dispersion of sunlight by raindrops. Electromagnetic radiations other than light can also be split up into individual wavelengths to form a spectrum. We use the term electromagnetic spectrum to refer to the wide range of radiations that take the form of electric and magnetic vibrations and differ in wavelength (see diagram).

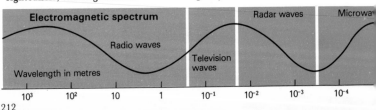

Electromagnetic spectrum

Radio waves

Television waves

Radar waves

Microwa

Wavelength in metres

| 10^3 | 10^2 | 10 | 1 | 10^{-1} | 10^{-2} | 10^{-3} | 10^{-4} |

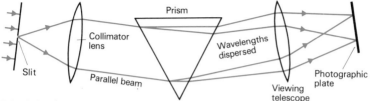

Principle of spectroscope

speed The rate at which a body is moving, given by the ratio of the distance travelled to the time taken. It is measured in such units as kilometres per hour (km/h); miles per hour (mph); and metres per second (m/s). Speed is a scalar quantity, compare **velocity**.

speedometer An instrument used to measure the speed of a vehicle. It also usually incorporates an odometer, which records the distance travelled. The usual type of speedometer is operated by a drive cable from the gearbox.

sphalerite See **blende**.

sphere A solid figure produced by rotating a circle around one of its diameters as an axis. For a sphere radius r the surface area is $4\pi r^2$, and the volume is $4\pi r^3/3$.

spherical aberration See **aberration**.

spheroid A solid figure produced by rotating an ellipse around one of its axes.

sphygmomanometer An instrument for measuring blood pressure. It consists of a mercury manometer, one leg of which is set against a scale. The other leg is connected to an inflatable armband, or cuff, which is wound around the patient's arm.

Spica (α Virginis) The brightest star (mag $0 \cdot 96$) in the constellation Virgo, the Virgin, and the 15th brightest in the sky. It is a blue-white binary and lies at a distance of about 260 light-years.

spiegeleisen An alloy of pig iron and up to 30% manganese, added during steelmaking to deoxidize and improve the working properties of the steel.

spin A basic property of elementary atomic particles. Part of the energy they possess is due to their angular momentum as they spin around their own axis. This spin energy is given by a spin quantum number of $+ \frac{1}{2}$ or $- \frac{1}{2}$.

spinels A group of oxide minerals, often of gemstone quality, found in igneous and metamorphic rocks. They are mixed oxides of two metals of the general formula MON_2O_3. The commonest variety is spinel itself, magnesium aluminium oxide, $MgO.Al_2O_3$, whose red forms are often confused with rubies.

spinning The process of drawing out and twisting fibres into a strong thread suitable for weaving. The first machine for spinning was the spinning wheel (1500s), which was a mechanical means of drawing out and twisting the fibres and winding the thread so formed on a bobbin. In the 18th century the invention of three spinning machines made textile making an industry and launched the Industrial Revolution. They were James Hargreaves' spinning jenny (1767); Richard Arkwright's spinning frame (1769); and Samuel Crompton's spinning mule (1779). The mule is still used today. It has a movable carriage holding a rotating bobbin. The thread is drawn out as the carriage moves forwards and then wound on the bobbin as the carriage moves back. The other main modern machine is the ring-spinning frame. It has a rotating bobbin that draws out the thread, while a 'traveller' revolves around the bobbin on a ring and imparts the twist.

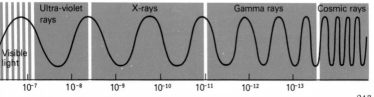

spirit level Or bubble level; a simple device for finding the horizontal level. It consists of a liquid in a glass tube in which a bubble is trapped. When the tube is horizontal, the bubble moves to the centre of the tube.

spontaneous combustion The burning of a substance without external application of heat. It results from an exothermic reaction taking place inside the substance. It often occurs in hay, for example, as a result of internal oxidation or fermentation.

spring The season of the year between winter and summer, extending from (in the northern hemisphere) the spring, or vernal equinox on about March 21 to the summer solstice on about June 21. In the southern hemisphere spring extends from about September 23 to December 21. See also **seasons**.

spring, mechanical A device that returns to its original shape when deformed—it exhibits elasticity. Springs are widely used in machinery—for example, as dampers (shock absorbers) in vehicles. In clocks and watches they are used to store energy.

Sputnik 1 The first artificial satellite, launched by the Russians on October 4 1957. It was an aluminium sphere 58 cm (23 in) in diameter and weighing 84 kg (184 lb). On November 3 the Russians launched the 500-kg (1,120-lb) Sputnik 2, carrying a dog called Laika.

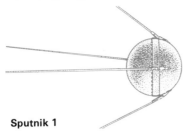

Sputnik 1

square root Of a number; is the number that when multiplied by itself gives the original number. The square root of 4 when multiplied by itself gives 4. We write the square root of 4 as $\sqrt{4}$.

Sr The chemical symbol for strontium.

stable In chemistry, a stable compound is one that is not easily decomposed. In physics, a stable system is one that is not easily disturbed.

stabilizers In aircraft, the surfaces that confer stability on the design. The tail fin is known as the vertical stabilizer; the tailplane as the horizontal stabilizer. In ships, stabilizers are large fins that project from the hull underwater. They are moved hydraulically to counteract the rolling tendency of a ship in rough water.

stainless steel A useful alloy steel containing chromium and usually nickel as well. It is resistant to heat, corrosion and chemical attack. A common stainless steel contains 18% chromium and 8% nickel.

stalactites and stalagmites Icicle-like growths of calcium carbonate that form in limestone caves when lime-laden water drips from the roof (stalactites) or drips onto the floor (stalagmites). Specks of calcium carbonate come out of solution when water in the drips evaporates.

stalling In aircraft flight, a condition in which the wings suddenly lose lift. This happens, for example, when the wings are angled too steeply into the airstream for a particular speed, and the airflow round the wing becomes turbulent instead of streamlined.

standard cell A primary cell that maintains a constant voltage (emf) over a long period of time; used to calibrate electrical instruments. See **Weston cell**.

standard gauge See **railways**.

standard temperature and pressure See **STP**.

stannic A term used to describe compounds of quadrivalent tin or tin(IV); eg stannic chloride, $SnCl_4$, used to make mordants.

stannous A term used to describe compounds of divalent tin, or tin(II); eg stannous sulphide, SnS_2, used as a gold pigment.

starch A carbohydrate food obtained from plants. Starches from potatoes and cereals form a large part of the human diet. Starch consists of polysaccharides, which during digestion are broken down into simple sugars such as dextrose and glucose.

stars Huge balls of glowing gases, which give out all forms of electromagnetic radiation—heat, light, X-rays, radio waves, etc. The energy comes from thermonuclear (fusion) reactions taking place in the centre of the star, the main reaction being the conversion of hydrogen into helium. The Sun is a typical medium-size star, classed as a yellow dwarf. Supergiant stars are many hundreds of times bigger than the Sun, while white dwarfs are several hundred times smaller. Stars vary in mass from about one-tenth to about 10 times the mass of the Sun. Their surface temperatures vary from about 20,000°C for blue-white stars like Spica to less than 3,000°C for red dwarfs such as Proxima Centauri. Proxima Centauri is the closest star to the Sun and is about 4·28 light-years away. Most information about the stars is obtained by studying the lines in their spectrum. In the heavens two out of three stars have at least one companion star. On a broader scale, stars cluster together in space into galaxies. See also **binary; constellations; galaxies; giant stars; globular cluster; Hertzsprung-Russell diagram; nova; open cluster; pulsar; variable star; white dwarf**.

static The build-up of electric charge; in radios, static in the atmosphere causes interference.

static electricity Electricity at rest, as opposed to current, that is, flowing electricity. Static electricity is concerned with the effects of electric charges and the electrostatic fields they produce.

statics The branch of mechanics that deals with the forces acting in structures, where there is no movement involved. Contrast **dynamics**.

THE BRIGHTEST STARS		
Name	*Constellation*	*Mag*
Sirius	Canis Major	−1·45
Canopus	Carina	−0·73
Rigil Kent	Centaurus	−0·20
Arcturus	Boötes	−0·06
Vega	Lyra	0·04
Capella	Auriga	0·08
Rigel	Orion	0·11
Procyon	Canis Minor	0·35
Achernar	Eridanus	0·48
Hadar	Centaurus	0·60
Altair	Aquila	0·77
Betelgeuse	Orion	0·80
Aldebaran	Taurus	0·85
Acrux	Crux	0·90
Spica	Virgo	0·96
Antares	Scorpio	1·00
Pollux	Gemini	1·15
Fomalhaut	Piscis Austrinus	1·16
Deneb	Cygnus	1·25

stationary orbit Also called geostationary orbit and 24-hour orbit; a circular orbit 35,900 km (22,300 miles) above the equator in which the orbital period (time to complete one orbit) is exactly 24 hours. If a satellite is launched into such an orbit, it appears to an observer on Earth that it remains stationary in the sky.

statistics A branch of mathematics concerned with the collection, manipulation, analysis and display of numerical data.

stator The stationary part of a rotating machine such as an electric generator or turbine.

steady-state theory In cosmology, a theory developed by Fred Hoyle, Hermann Bondi and Thomas Gold that the universe has always existed in the form in which it is now and will continue to exist for ever. Matter is continually being created so that the average density of the universe remains constant as it expands. Contrast **big-bang theory**.

steam Water in the vapour state at or above its boiling point (100°C). It is actually colourless; the white cloud of 'steam' seen when a kettle boils actually consists of condensed droplets of water. Steam occupies 1,670 times the space of the water from which it comes.

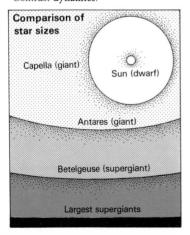

Comparison of star sizes

Capella (giant)
Sun (dwarf)
Antares (giant)
Betelgeuse (supergiant)
Largest supergiants

steam engine A machine that exploits the power of expanding steam. It is an external combustion engine. Steam, the working fluid, is produced outside the engine by heating water in a boiler. The engine consists essentially of a piston in a cylinder. Steam is introduced to each side of the piston in turn ('double-action'), driving it back and forth. The motion is carried by a connecting rod to, say, the wheels of a locomotive. The first practical steam machine, used to pump water from mines, was Thomas Savery's steam pump (1698), which operated by means of valves. But the forerunner of the reciprocating steam engine that was to power the Industrial Revolution was Thomas Newcomen's beam engine of 1712. James Watt brought out a greatly improved design in 1769 and by the 1790s had produced a powerful and reliable power source, capable of rotary motion and controlled by a governor. Steam engines were also successfully introduced to power ships and locomotives. See **Flywheel**.

steam turbine A turbine driven by expanding steam. Steam turbines are the main power source in electricity generating stations and large ships. The British inventor Charles Parsons developed the steam turbine in 1884. The steam turbine consists of a multibladed rotor able to rotate inside a casing. Fixed to the casing in between the rotor blades are stator blades, which direct the steam onto the rotor blades at the correct angle. The steam travels through the turbine, spinning the rotor blades, and then passes into a condenser, where it is cooled and condensed back into water. This ensures a big pressure drop.

Steam locomotive

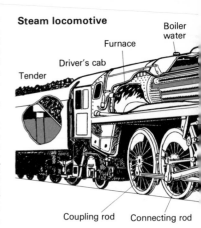

stearic acid A long-chain fatty acid, $CH_3(CH_2)_{16}COOH$, found in many fats and oils. It is used in making cosmetics and pharmaceutical preparations. Common soaps contain the sodium and potassium salts of stearic acid.

steel The most valuable of all alloys, being iron containing a small quantity of carbon (up to about $1 \cdot 5\%$) and also other metals as well. There are two main types—carbon steels, whose properties depend largely on the amount of carbon they contain; and alloy steels, whose properties depend mainly on the alloying elements. The commonest carbon steel is mild steel (up to $0 \cdot 25\%$ carbon). Stainless steel, containing chromium and nickel, is one of the commonest alloy steels. Steel is made by refining pig iron in a variety of furnaces; see **basic-oxygen process**; **electric furnace**; **open-hearth process**. The Bessemer process, which pioneered cheap steelmaking, has now been superseded.

step rocket Or multistage rocket; a rocket launch vehicle consisting of a number of rockets (stages) joined together either end to end, side by side (in tandem), or both. Step rockets are required for launching spacecraft in order to get a favourable power-to-weight ratio. Each stage fires in turn, boosts the ones above it to high speed, and then separates. The stages above then have less weight to carry.

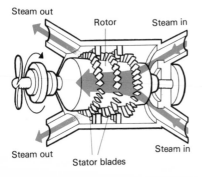

Steam out
Rotor
Steam in
Steam out
Stator blades
Steam in

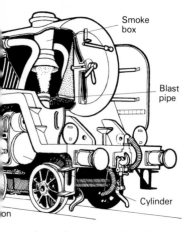

Smoke box

Blast pipe

Cylinder

stereoisomerism A condition in which compounds of similar molecular structure have their atoms arranged differently in space. There are two types—*cis-trans* (or geometric) isomerism and optical isomerism.

stereophonic sound A sound recording and reproduction system that utilizes two separate channels for 'right' and 'left' sounds. Most record players now offer stereo sound. The original sounds are recorded by 'right' and 'left' microphones in front of, say, an orchestra. The signals are played back through 'right' and 'left' speakers in front of a listener, who hears a much more realistic sound than he would from a single speaker. Quadraphonic sound from four speakers is also now available.

stereoscopic vision Seeing objects in three-dimensions (3D). We see stereoscopically because we have binocular (two-eye) vision. Each eye sees an object from a slightly different angle, creating a sense of depth. A stereoscope, an instrument invented by the English scientist Charles Wheatstone in 1832, creates a 3D effect using a pair of two-dimensional pictures. Experiments with 3D colour television were made, using red and green filter spectacles, in 1982.

stereotype A duplicate printing plate made by casting metal in a mould made from the made-up type.

sterling silver An alloy used in most silver jewellery, consisting of 925 parts of silver to 75 parts of copper.

steroids A group of organic alcohols of biological and medical importance. They include a variety of hormones, including the sex hormones (eg androgens and estrogens) and compounds called sterols, the best-known of which is cholesterol, $C_{25}H_{45}OH$.

stethoscope An instrument used by doctors to listen to noises produced inside the body, particularly by the heart and lungs; invented by the French doctor René Laënnec in 1816. It consists of a chest-piece, or contact piece, which is placed on the patient's body, and tubes that lead from the chest-piece into the doctor's ears.

stibnite The main ore of antimony—antimony sulphide, Sb_2S_3.

still An apparatus—glass in the laboratory, often copper in a distillery—used for the distillation of liquids.

Stirling engine A piston engine that uses hot air as the working fluid, invented by Scottish clergyman Robert Stirling in 1876 and produced until the 1920s. Interest in this engine has recently been revived because it works efficiently and creates no exhaust fumes.

stoichiometry Also spelt stoicheiometry; a branch of chemistry dealing with the composition of substances and the proportions in which the constituent elements combine together.

STOL The abbreviation for 'short take-off and landing'; applied to aircraft.

stoneware A type of pottery that is fired at a temperature of 1,200°C or more, which is higher than the firing temperature for earthenware. It is hard and non-porous.

stop Or f-stop; see **aperture**.

stop bath In photographic processing, a bath, usually of water, to wash away excess developer before fixing.

STP The abbreviation for 'standard temperature and pressure', which are 0°C and 760 mm mercury. The properties of gases are generally quoted at STP.

strain The relative change in dimensions that occurs when an object is deformed. It is the ratio of the change in a dimension (length, area, volume) to the original dimension. See also **stress**.

strangeness The name given to a certain property of elementary particles, including some quarks.

stratification Layering, as of rocks.

stratosphere The layer of atmosphere above the troposphere (the lowest layer). It is a region of near constant temperature, about $-55°C$ in mid-latitudes, and is also called the isosphere. It extends up to an altitude of about 50 km (30 miles) from a base about 8 km (5 miles) high over the poles and 16 km (10 miles) high over the equator. It is bounded at the bottom by the tropopause and at the top by the stratopause. It includes the ozone layer.

stratum Plural strata; a layer.

streamline flow Steady flow in a fluid that can be represented by imaginary lines that trace the path of the fluid particles. It occurs at low velocities, but breaks up into turbulent flow at high velocities. Streamlined shapes are those that offer the least drag when they move through a fluid.

stress Force per unit area. When stress is applied to a body, strain is produced, the two being related by Hooke's law.

stroboscope An instrument that uses a rapidly flashing light to make rotating machinery (eg a wheel) appear to stand still. This happens when the frequency of flashing equals the rate of rotation.

strontium (Sr) One of the alkaline-earth metals (rd $2·6$, mp $77·0°C$), closely related to calcium. Its compounds, which colour a flame a vivid crimson, are used in fireworks. The radioactive isotope strontium-90 is present in the fallout from a nuclear explosion. Since it is chemically similar to calcium, the strontium is absorbed by growing vegetation.

strychnine A powerful alkaloid poison with a bitter taste; obtained from the seeds of species of far-eastern vines and trees of the *Strychnos* genus. Its chemical formula is $C_{21}H_{22}N_2O_2$.

styrene Also called phenylethylene, $C_6H_5CHCH_2$; a flammable aromatic liquid made from ethylbenzene ($C_6H_5C_2H_5$). It polymerizes readily into the widely used plastic polystyrene, and is also made into synthetic rubber.

sub-atomic particles See **elementary particles**.

sublimation The change of a substance from the solid to the vapour state without an intervening liquid state. Dry ice (solid carbon dioxide) sublimes, as do iodine, camphor and ammonium chloride.

submarine A ship capable of operating underwater. The first prototype modern submarine did not appear until 1898. Powered by electric motors underwater and a petrol engine on the surface, it was designed by J. P. Holland. Later submarines had a diesel engine for use on the surface and to charge the batteries. In 1954 the Americans launched the first nuclear-powered submarine, *Nautilus*. Nuclear submarines have a boiler to raise steam and steam turbines for propulsion. Submarines dive and surface by letting water into and expelling it (with compressed air) out of ballast tanks.

submersible The name usually given to small submarine craft used, for example, to make underwater surveys and to assist divers. They may be equipped with mechanical arms and diving chambers.

subsonic See **sonic speed**.

sucrose Common table sugar, extracted from sugar cane and sugar beet. It is a disaccharide, $C_{12}H_{22}O_{11}$.

sugars Sweet-tasting carbohydrates, the commonest of which is sucrose. See **monosaccharides; disaccharides**.

sulpha drugs Or sulphonamides; powerful chemical drugs used to combat bacterial infection. They are so called because they contain the sulphonamide group, SO_2NH_2, or derivatives of it. Sulpha drugs are used to treat such diseases as pneumonia, dysentery, meningitis and blood poisoning.

sulphates Salts derived from sulphuric acid, H_2SO_4. Since the acid is dibasic, there are two series of salts—normal salts, containing the ion, SO_4^{2-}; and bisulphates, or hydrogen sulphates, containing the ion, HSO_4^-. Most sulphates are stable crystalline compounds that are soluble in water.

sulphides Compounds of an element with sulphur, which may also be regarded as salts of hydrogen sulphide, H_2S. Sulphides are found widely in the Earth's crust.

sulphites Salts of the weak sulphurous acid, H_2SO_3.

sulphonamides See **sulpha drugs**.

sulphonic acids Strong organic acids of general formula RSO_2OH, where R is a hydrocarbon radical. Typical is benzene sulphonic acid, $C_6H_5SO_2OH$. The acids, which give rise to salts and esters called sulphonates, are used to make dyes, detergents and drugs.

Lock-out submersible

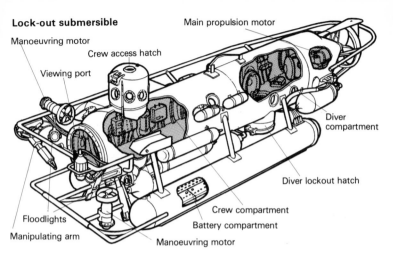

Main propulsion motor · Manoeuvring motor · Crew access hatch · Viewing port · Diver compartment · Diver lockout hatch · Floodlights · Crew compartment · Battery compartment · Manipulating arm · Manoeuvring motor

sulphur (S) A solid non-metallic element (rd 2·1, mp 119°C). It may be found native in volcanic regions and in subterranean deposits, from which it is extracted by the Frasch process. It can exist in various allotropic forms. At normal temperature it forms yellow rhombic crystals. Above about 96°C, a monoclinic form is stable. Sulphur vapour condenses into a fine yellow powder known as flowers of sulphur. Sulphur combines with most other elements. It is used to make gunpowder, vulcanize rubber and to make sulphuric acid, dyes and a host of other products.

sulphur dioxide A gas with a choking smell made by burning sulphur. It occurs in the air around volcanoes and in industrial regions. It is a strong bleaching and sterilizing agent, and is the important intermediate in the production of sulphuric acid. Its chemical formula is SO_2.

sulphuretted hydrogen See **hydrogen sulphide**.

sulphuric acid The most important of all the mineral acids and the single most important chemical in industry; formula H_2SO_4. It is made industrially by the contact process. When concentrated it is an oily liquid, which is highly corrosive and reacts with water vigorously. In water solution it is a strong acid that gives rise to the series of salts called the sulphates. Concentrated acid containing dissolved sulphur trioxide gas is called fuming sulphuric acid, or oleum. A common use for dilute sulphuric acid is in lead-acid car batteries. Vast amounts are used for 'pickling' steel and for making superphosphate fertilizer.

sulphur trioxide See **contact process**; **sulphuric acid**.

summer The season of the year between spring and autumn. In the northern hemisphere it lasts from the summer solstice, on about June 21, to the autumnal equinox, on about September 23. In the southern hemisphere, it begins on about December 21 and ends on about March 21.

summer triangle A prominent triangle of stars visible in the summer night sky. They are Vega, Deneb and Altair.

Sun The star around which the Earth and the other planets revolve. It is a gaseous body (about 90% hydrogen and the rest mainly helium), some 1,392,000 km (865,000 miles) across and it lies at an average distance from Earth of 149 million km (93 million miles). It is classed as a yellow dwarf star of spectral type G2. The temperature of its visible surface, the photosphere, is about 6,000°C. In its core, where nuclear fusion reactions take place, the temperature is about 15,000,000°C. Its mass is 333,000 times that of Earth. See also **chromosphere**; **corona**; **eclipse**; **prominence**; **sunspots**.

219

sundial An early device for telling the time by means of shadows. It consists of a pointer, or gnomon, which casts the shadow, set in the middle of a flat dial. The gnomon must slant upwards so that it makes an angle with the dial face equal to the latitude of the location.

sunspots Dark patches visible on the Sun's disc, which move across the surface as the Sun rotates. They are slightly cooler than their surroundings, and are associated with such phenomena as aurora and magnetic storms here on Earth. The number of sunspots visible varies widely according to a regular cycle, and reaches a maximum about every 11 years.

supercharger See **turbocharger**.

superconductivity The complete loss of electrical resistance in a conductor at temperatures within a few degrees of absolute zero. Once an electric current is introduced into a superconductor, it continues flowing indefinitely, as long as the temperature remains at the very low value. Lead, tin, niobium and vanadium are among metals that become superconductors.

supercooling A condition in which a liquid remains liquid even though it has been cooled below its freezing point.

superfluidity A phenomenon exhibited by liquid helium at temperatures below $2 \cdot 18K$ (about $2°$ above absolute zero), in which it flows without friction and can climb out of its containing vessel. It also displays very high heat conductivity.

supergiant In astronomy, the largest type of star, which has a very high luminosity, low density and a diameter many hundreds of times greater than that of the Sun. The prominent stars Rigel, Deneb, Canopus, Polaris, Antares and Betelgeuse are all supergiants.

superheated steam Steam raised to a temperature greater than 100°C by heating water under pressure.

superheating A condition in which a liquid is heated above its boiling point without boiling; or a solid is heated above its melting point without melting.

supernova A star that suddenly becomes millions of times brighter. The process is thought to be triggered off by the gravitational collapse of a star much larger than the Sun. A violent explosion results, making the star shine nearly as brightly as a whole galaxy. Most of the mass of the star is ejected in the process, leaving behind a small dense body—probably a neutron star. Three supernovae have been detected in our galaxy in the past 1,000 years—in Taurus in AD 1054 (we now see the remnants as the Crab Nebula); in Cassiopeia in 1572; and in Ophiuchus in 1604.

superphosphate One of the most widely used fertilizers, consisting mainly of calcium hydrogen phosphate, $CaH_4(PO_4)_2$, and gypsum. It is made by treating calcium phosphate with sulphuric acid.

supersonic See **sonic speed**.

surface-active agent Or surfactant; a substance such as a detergent that alters the wetting properties of a liquid.

surface tension A force that exists at the surface of a liquid that makes the surface act as if it had a skin. It arises because there is a net downward attraction at the surface from the molecules in the body of the liquid. Because of surface tension, pond skaters can walk on water, for example.

surveying Accurate measurement of the Earth's surface. Surveyors use instruments such as levels and theodolites to measure heights and angles between certain points. Using geometry in general and triangulation in particular, they can convert their measurements into distances. Geodetic surveys take into consideration the curvature of the Earth. Aerial surveys rely on photogrammetry, which uses photographs taken from the air.

Surveyor spacecraft A series of American space probes sent to soft-land on the Moon between 1966/68. They sent back close-up pictures of the surface and dug into the lunar surface.

suspension A type of colloid consisting of very fine particles in a liquid medium. When the particles are of another liquid, the suspension is called an emulsion.

suspension bridge A bridge in which the deck, or roadway, is hung from overhead cables that go up and over towers on either side of the gap to be bridged. Suspension bridges can have larger spans than any other type of bridge. Currently the world's largest is Britain's Humber Bridge (illustrated opposite), which has a main span of 1,410 m (4,626 ft).

The Humber Bridge in north-east England, opened in 1981, is the world's longest-span suspension bridge.

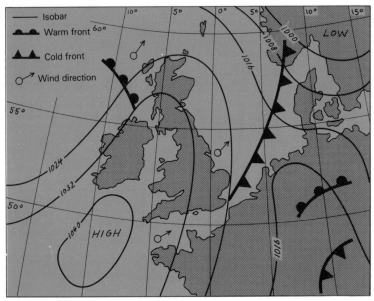

Isobar
Warm front
Cold front
Wind direction

swing bridge A common type of movable bridge, which pivots on a pier in the middle of the gap to be spanned.

switch A device for opening and closing an electric circuit. Domestic light switches generally consist of a spring-loaded tumbler that makes and breaks the circuit quickly to prevent arcing. Switches at power stations, which have to cope with very heavy voltage and currents may be immersed in oil to prevent arcing, or they may use a blast of gas to extinguish the arc.

sylvite A common mineral form of potassium chloride, KCl.

symbols See **chemical symbols**

symmetry A term applied to (among other things) geometric figures and crystal shapes. A figure is symmetrical about a certain axis, plane or point, if it appears unchanged when it is, say, rotated about that axis, plane or point. A circle, for example, is symmetrical about any diameter. Crystals are classified by their elements of symmetry.

sympathetic vibration A vibration that sets up a vibration in another object. This happens when the original vibration is the same as the natural frequency of the object. See **resonance**.

synchrocyclotron An improved form of cyclotron in which the accelerating voltage is changed in frequency to match the frequency of the orbiting particles.

synchromesh A term applied to the gears in a modern gearbox. The gears have conical surfaces on the sides which mate before the gears mesh and ensure that the gears are running at the same speed when the teeth interlock.

synchronous Occurring at the same time as; or more generally being in step with. A synchronous electric motor has its current alternating in step with the frequency of the power supply.

synchrotron A particle accelerator in which particles are accelerated in a doughnut-shaped chamber and kept within the chamber as they accelerate by the application of an increasing magnetic field.

syncline A U-shaped fold in the strata of sedimentary rocks. Compare **anticline**.

synodic period In astronomy, the time it takes for a heavenly body to return to the same position in the heavens relative to the Sun. The synodic months is the time it takes for the Moon to go through its phases, 29½ days.

synoptic chart In meteorology, a chart that provides a synopsis, or summary of the weather at a particular time. It uses symbols to represent weather features such as wind speed and direction and cloud cover. Lines called isobars connect regions of equal pressure. Fronts show the boundaries between masses of air at different temperatures.

synthesis In chemistry, the formation of a compound from simpler compounds or elements. The term 'synthetic' is often used now to contrast products with a chemical origin with those of natural origin, hence synthetic resins, synthetic fibres and synthetic rubber (see below). 'Synthetic' is not synonymous with 'man-made'. Man-made products include not only synthetics but also products made by processing natural products. Thus rayon, made by processing plant cellulose, is a man-made, but not synthetic product.

synthetic fibres Types of plastics that can be produced in the form of fibres and made into textiles. Nylon, polyester and acrylic fibres are examples. To make fibres the plastic is melted (or dissolved) and forced through a spinneret—a disc perforated with holes. The emergent streams then harden (or form as the solvent evaporates) into continuous filaments suitable for making yarn.

synthetic resins Resin-like materials made from chemicals. They are used in place of natural resins in paints and varnishes. They are also used to make plastic products. The original, and still widely used synthetic resin, is Bakelite, made from phenol and formaldehyde. Urea and melamine also form useful resins with formaldehyde. These are all thermosetting materials, which cross-link when they are heated to form rigid products that cannot be remelted afterwards.

synthetic rubber Elastic plastic material resembling rubber. One of the first successful synthetic rubbers, neoprene, was produced in the United States by the same team that developed nylon, headed by W. H. Carothers. It is still widely used, as is styrene-butadiene rubber (SBR).

syzygy In astronomy, a time when three bodies are in a straight line in space, such as the Sun, Earth and Moon at new or full Moon.

T

Ta The chemical symbol for tantalum.

tachometer An instrument that measures the speed of rotation of a shaft, eg the crankshaft in a petrol engine, in revolutions per minute (rpm).

talc One of the softest of all minerals, used in powdered form in talcum powder. It is hydrated magnesium silicate ($3MgO.4SiO_2.H_2O$). In solid form it is called soapstone or steatite.

tangent In geometry, a straight line that touches a curve at only one point.

tannin Or tannic acid; a substance obtained mainly from galls on oak trees. Tannins are used to tan hides, as mordants in dyeing and in making inks.

tanning Treating animal hides to make leather. In vegetable tanning the hides are treated with tannin, while in chrome tanning they are treated with solutions of chromium salts.

tantalum (Ta) A metallic element (rd 16·6, mp 3,000°C) used in high-temperature alloys for rocket engines. Highly ductile and chemical resistant, it is obtained from the minerals tantalite and columbite.

tape recording Recording sound by means of magnetic tape. Valdemar Poulsen in Denmark discovered the basic principles in the 1890s, recording on magnetized wire. Modern recordings are done on plastic tape coated with magnetic iron oxide or chromium oxide particles. In recording, sounds are converted by a microphone into variable electrical signals. These are then fed through the coils of an electromagnet and cause its magnetism to fluctuate in sympathy. This fluctuating magnetism is recorded as an invisible pattern on magnetic tape moving past the electromagnet. In playing back, the magnetic tape is moved past the electromagnet and induces in it variable electrical signals. These mimic the original signals produced by the microphone and are fed through a loudspeaker to reproduce the original sounds. In video tape recording, signals from a television camera are recorded on magnetic tape, and are played back through a television receiver to reproduce pictures.

tar A thick, dark oily liquid obtained when coal is distilled. Coal tar is a valuable source of organic chemicals, such as phenol, benzene and toluene.

tarmac see **macadam**.

tartaric acid An organic chemical, $COOH(CHOH)_2COOH$, found in fruits and often obtained commercially as a by-product of wine fermentation. It is used in baking powder and health salts and as a mordant in dyeing.

Taurus The Bull; a prominent and interesting constellation of the zodiac between Gemini and Aries. It includes the giant red star Aldebaran, the Pleiades and Hyades star clusters, and the Crab Nebula.

tautomerism In chemistry, the existence of a compound as a mixture of two isomers. The proportions of the isomers in the compound remains constant. Vinyl alcohol (CH_2CHOH) and acetaldehyde (CH_3CHO) are tautomers.

Tb The chemical symbol for terbium.

Tc The chemical symbol for technetium.

Te The chemical symbol for tellurium.

tear gas Or lachrymator; a vapour that affects the eyes, nose and throat, producing tears and causing coughing. Most are synthetic organic halogen compounds such as α-chloroacetophenone.

technetium (Tc) The first artificial element (rd $11\cdot4$, mp $2,100°C$) produced in 1937 by bombarding molybdenum with particles accelerated in a cyclotron. It is radioactive and now obtained as a by-product from nuclear reactors.

technology The practical application of science.

Teflon See **PTFE**.

tektites Small glass-like objects, often dark green or black, thought to have been produced by meteorite impact. Another theory holds that they originated on the Moon.

telecommunications Long distance ('tele') communications by wire or radio.

telegraph A means of transmitting information in the form of electrical impulses over long distances by wire. William F. Cooke and Charles Wheatstone in Britain and Samuel F. B. Morse in the United States developed practical telegraphs in 1837. Morse's telegraph and code for transmitting messages

became widely adopted in the 1840s (see **Morse code**). In the simple telegraph, pressing a key sends an impulse through the wires to the receiving set. There the pulse actuates an electromagnet, which works a sounding device, such as a buzzer.

telemetry Measuring from a distance. A system in which remote instruments record and transmit data, particularly that in a spacecraft.

telephone An instrument that transmits the spoken word by means of electrical impulses through wires. Alexander Graham Bell invented the telephone in the United States in 1876. When you speak into the transmitter (mouthpiece) of the telephone, the sound waves vibrate a diaphragm. This presses against grains of carbon, whose electrical resistance varies according to the pressure upon it. An electric current passes through the carbon grains, and varies in strength according to the pressure exerted upon them by the vibrating diaphragm. The variable current signals thus set up travel to the receiver (earpiece) of the telephone of the person you are calling. There they pass through an electromagnet over which is located another diaphragm. The diaphragm vibrates in sympathy with the signals and reproduces the sounds that originally went into the transmitter.

Telephone handset

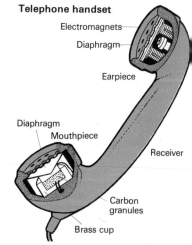

Electromagnets

Diaphragm

Earpiece

Diaphragm

Mouthpiece

Receiver

Carbon granules

Brass cup

Anglo-Australian Telescope

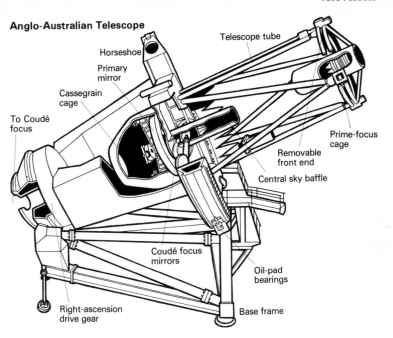

Horseshoe

Telescope tube

Primary mirror

Cassegrain cage

To Coudé focus

Prime-focus cage

Removable front end

Central sky baffle

Coudé focus mirrors

Oil-pad bearings

Right-ascension drive gear

Base frame

telephoto lens A camera lens that produces a magnified image. It has a long focal length and a narrow angle of view.

teleprinter A kind of typewriter that converts typed words into electrical impulses that can be transmitted by wire. A similar machine receives the impulses and converts them back into words, which are then typed out. Telex is an international telegraph service that uses teleprinters to pass messages. Subscribers are connected by dialling.

telescope An instrument that produces a magnified image of distant objects, such as the heavenly bodies. Hans Lippershey, a Dutch optician, built the first simple telescope in 1608. A year later Galileo built a much better one and used it to view the heavens. These were both simple lens telescopes, or refractors. In the 1670s Isaac Newton built the first practical mirror telescope, or reflector. All big telescopes are now reflectors. In 1931 Karl Jansky discovered that the heavenly bodies emit radio signals, and this led to the development of the radio telescope. See **reflector; refractor.**

teletext An electronic information system that displays data on an ordinary television receiver. The system was pioneered in Britain by the BBC, whose teletext system is called Ceefax; by the IBA, whose system is called Oracle; and by the Post Office, whose system is called Prestel. In Prestel, the viewer calls up information on his screen via the telephone, which connects him with an extensive data bank stored in a computer.

television The transmission of moving pictures by radio waves (in broadcasting) or by cable (in closed-circuit television). John Logie Baird in Britain developed the first practical television system in 1925 using a mechanical scanning process. But this process was soon replaced by the modern method of electronic scanning, pioneered by Vladimir Zworykin in the United States.

In a simple black and white television camera, a lens focuses an image on a light-sensitive plate in the camera tube. Electrons are emitted from the various parts of the plate according to the

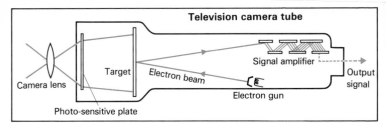

Television camera tube

Camera lens · Target · Electron beam · Photo-sensitive plate · Signal amplifier · Electron gun · Output signal

amount of light reaching them. They travel to a target plate and make it electrically charged. The pattern of charges on the target is an electrical representation of the optical image formed on the light-sensitive plate by the lens. The target is then scanned rhythmically from left to right and line by line, by a beam of electrons from an electron gun. The electron beam interacts with the charges on the target to produce an electric current, which varies according to the amount of charge there —that is, on the brightness of each part of the picture. This variable current comprises the video signals, which are then transmitted, together with a synchronizing signal, on a radio carrier wave.

A television receiver picks up the transmitted picture signals, via an aerial, and first separates the video signals from the carrier wave. The signals are then fed to a cathode-ray tube, where they are made to vary the strength of an electron beam. The beam strikes the fluorescent viewing screen, causing it to glow more or less according to the strength of the beam. At the same time the beam is made to scan the screen side to side and line by line in the same way as the beam did in the camera tube.

(625 lines are common in Britain.) In this way a pattern of light and dark is built up on the screen, which reproduces the original scene viewed by the television camera.

In colour television, the coloured light entering the camera is split up by filters into different combinations of the three primary colours red, blue and green. Each colour is then fed to a separate camera tube, which produces its own signals. These signals are then transmitted and picked up by the receiver. There they are fed to three separate electron guns in the picture tube. Beams of electrons representing red, blue and green light are then directed onto phosphors on the screen that glow respectively red, blue and green. The phosphors are so close together that their colours merge to reproduce the original colour.

Telex See **teleprinter.**

tellurium (**Te**) A semi-metallic element, or metalloid (rd 6·2, mp 450°C), chemically related to sulphur. It is used in some alloys, including stainless steels.

Telstar The first active communications satellite, launched in 1962 and used to transmit television programmes across the Atlantic.

Television receiver tube

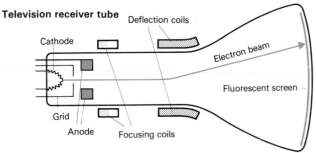

Cathode · Deflection coils · Electron beam · Fluorescent screen · Grid · Anode · Focusing coils

temperature A measure of the hotness or coldness of a body; of the kinetic energy of its molecules. It is measured by a thermometer on a scale calibrated in degrees. Commonly temperatures are expressed in degrees Celsius (or centigrade) or in degrees Fahrenheit. On the Celsius (C) scale, water boils at 100° and freezes at 0°. On the Fahrenheit (F) scale, water boils at 212° and freezes at 32°. Scientists now often use an absolute scale of temperature which uses Celsius degree units, but begins at the absolute zero of temperature, $-273 \cdot 16°C$. The unit is called a kelvin (K). Thus the freezing point of water is $273 \cdot 16K$ and the boiling point $373 \cdot 16K$. The average temperature of the human body is $36 \cdot 9°C$ ($98 \cdot 4°F$).

tempering A kind of heat treatment, used especially for steel, which can make the metal much harder and tougher. The steel is first heated to a certain temperature and then quenched, or plunged into oil or water.

template A pattern used to guide a cutting tool.

tensile strength The stress per unit area of cross-section needed to pull a body apart. It is expressed in such units as newtons per sq metre or tons per sq inch.

terbium (Tb) One of the rare-earth elements (rd $8 \cdot 3$, mp $1,360°C$), not widely used.

terminal velocity The maximum velocity reached when a body moves through a medium under a constant force. When a free-fall parachutist falls through the air under the constant pull of gravity, he reaches a terminal velocity of about $320 \, km/h$ ($200 \, mph$).

terminator The line on the surface of a planet or moon (particularly the Moon) that separates the dark and light hemispheres. Certain features, such as craters, can be seen best when they are on or near the terminator.

terpenes A group of hydrocarbons that occur widely in plants, for example, in essential oils. They are generally fragrant-smelling liquids. They have the empirical formula of isoprene, C_5H_8. They include pinene, $C_{10}H_{16}$, the main ingredient in a mixture of terpenes we know as the solvent turpentine.

terrestrial Relating to, or resembling the Earth.

Tertiary Period The geological time span that lasted from about 65,000,000 years ago to 2,500,000 years ago, which saw the rise of the mammals in general and latterly the emergence of Man.

Terylene Trade name for a common polyester fibre, made by polymerizing ethylene glycol and terephthalic acid, developed in 1941 by the British chemists J. R. Whinfield and J. T. Dickson.

tesla The SI unit of magnetic induction, equal to 1 weber per sq metre and equivalent to 10^4 gauss.

testosterone The hormone produced in the male testes, which is responsible for the development of male characteristics. It is a steroid, formula $C_{19}H_{28}O_2$.

tetraethyl lead See **lead tetraethyl**.

tetrahedron A four-sided figure whose sides are triangles; a triangular pyramid.

textiles Fabrics made from natural and synthetic fibres, generally by spinning the fibres into yarn and then weaving the yarn on looms. The textile industry spearheaded the Industrial Revolution in the 1700s.

Th The chemical symbol for thorium.

thallium (Tl) A metallic element (rd $11 \cdot 85$, mp $304°C$) related to tin and lead, but much rarer and much less useful.

theodolite An instrument used in land surveying for measuring angles horizontally and vertically. It consists essentially of a sighting telescope that can be moved over horizontal and vertical scales. See also **triangulation**.

Theodolite

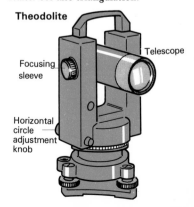

Telescope

Focusing sleeve

Horizontal circle adjustment knob

theorem A statement or proposition to be proved or demonstrated.

therm A unit of heat, used for example to measure the heating value of gas. It equals 100,000 Btus or 25,200,000 calories.

thermal In gliding, a warm, upward-rising air current.

thermal reactor A nuclear reactor in which low-energy, or thermal neutrons bring about fission. It uses a moderator to slow down the neutrons.

thermionic valves Electron tubes in which electrons are emitted from a heated electrode (cathode). The British physicist John Ambrose Fleming developed the first valve in 1904 for use in radio receivers. It was a two-electrode valve, or diode, containing a cathode and an anode in an evacuated glass bulb. Two years later the American Lee de Forest produced a triode, which had a third electrode (grid) between the cathode and the anode. In the triode, a small change of voltage on the grid has a great effect on the electron flow, or current passing between cathode and anode. In this way signals can be amplified. Thermionic valves with further grids are available for different uses. For many applications these days, however, thermionic valves have been replaced by semiconductor devices, such as transistors.

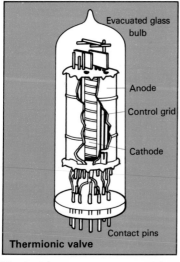

Evacuated glass bulb

Anode

Control grid

Cathode

Contact pins

Thermionic valve

thermistor A semiconductor device whose resistance varies as the temperature changes.

thermite A mixture of aluminium powder and iron oxide that reacts vigorously when ignited, to produce aluminium oxide and iron metal. The temperature rises to some 2,500°C; so the iron is molten. Thermite is used to weld iron and steel and in incendiary bombs.

thermochemistry A branch of physical chemistry concerned with the heat changes that occur during chemical reactions. Sometimes called energetics.

thermocouple A device used to measure temperatures, consisting of two different metals (eg antimony and bismuth, copper and iron) joined together. It uses the principle of the Seebeck effect. Generally, a number of thermocouples are joined together, forming a thermopile.

thermodynamics The branch of physics that deals with the inter-relationship between heat, work and other forms of energy and the conservation of energy. Two basic laws of thermodynamics are that energy can be neither created nor destroyed, and that heat cannot flow from a colder to a hotter body.

thermoelectricity A branch of physics dealing with the conversion of heat directly into electricity and vice versa. It involves the application of the Seebeck and Thomson effects.

thermometer A device for measuring temperature. Galileo made a crude thermometer in 1593, called a thermoscope. G. D. Fahrenheit made the first mercury thermometer—the type most used today—in 1714. The mercury thermometer consists of a column of mercury in a glass tube, whose length increases or decreases as the temperature rises or falls. The temperature is read on a scale at the side. The thermometer is calibrated by noting the height of the column of mercury at two reproducible temperatures— commonly the freezing point and boiling point of water, 0° and 100° respectively on the Celsius (centigrade, C) scale; and 32° and 212° on the Fahrenheit (F) scale. Another common liquid-in-the-glass thermometer contains alcohol. Other types of thermometer include the

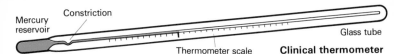

Mercury reservoir — Constriction — Glass tube — Thermometer scale — **Clinical thermometer**

resistance thermometer. It uses as a sensor platinum wire, whose resistance varies with temperature. See also **maximum and minimum thermometer; pyrometer; thermocouple.**

thermonuclear reaction A nuclear fusion reaction that takes place at very high temperatures ($20,000,000°C$) and is then self-sustaining.

thermopile See **thermocouple.**

thermoplastic A plastic that softens when reheated; such as nylon and polyethylene. Their molecules consist of long straight chains.

thermosetting plastics Or thermosets; plastics that set rigid when heat moulded and do not soften when reheated. Their molecules cross-link during the moulding process. Bakelite and other formaldehyde resins are thermosets.

Thermos flask See vacuum flask.

thermostat An automatic temperature-regulating device, which works on the principle of feedback. The simplest type consists of a bimetallic strip whose movement as the temperature changes is made to make or break contact in the power-supply circuit. When the temperature rises above the desired level, the strip moves and breaks the contact, shutting off the heating supply. When the temperature falls below the desired level, the strip moves back to make contact again and restore the heating.

thiamine Vitamin B_1, chemical formula $C_{12}H_{17}ON_4SCl$. Deficiency of thiamine causes beri-beri.

thiols Or mercaptans; organic compounds of the general formula RSH, where R is a hydrocarbon group. They are equivalent to the alcohols and phenols, but containing a sulphur atom instead of an oxygen atom. Thiols are generally foul-smelling substances, found in onions and the scent of skunks.

thiosulphates Salts or esters of the unstable thiosulphuric acid, $H_2S_2O_3$. The best known is sodium thiosulphate, $Na_2S_2O_3.5H_2O$, better known to photographers as hypo.

thixotropy A phenomenon displayed by certain gels. When undisturbed they remain relatively firm, but when pressure is applied to them they flow freely. This is made use of in 'jelly' paints, which are firm in the can and 'non-drip'; but flow readily under the pressure of brushing.

Thomson effect Also called Kelvin effect; when electric current passes between two points in a conductor that are held at different temperatures, then heat is either liberated or absorbed. It is named after its discoverer (1854), William Thomson, later Lord Kelvin.

thorium (Th) A naturally occurring radioactive element (rd $11·5$, mp $1,700°C$), extracted from the mineral monazite. It is useful in the nuclear power industry because its main isotope Th-23 is readily converted by irradiation into the fissionable isotope U-233.

thrust The propulsive force developed by a jet or rocket engine, usually expressed in newtons or pounds.

thulium (Tm) One of the rarest of the rare-earth elements (rd $9·3$, mp $1,600°C$).

thunder The sound that accompanies a lightning discharge. It is the shock wave produced when the discharge heats the surrounding air, causing it to expand violently. Thunder is heard after the lightning flash because it travels at the speed of sound, while the flash travels at the speed of light. For every kilometre you are distant from a lightning flash, you hear the thunder three seconds later (or for every mile, five seconds).

Ti The chemical symbol for titanium.

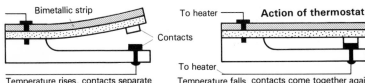

Action of thermostat

Bimetallic strip — To heater — Contacts — To heater

Temperature rises, contacts separate — Temperature falls, contacts come together again

tidal power

tidal power Power produced by harnessing the energy in the ebb and flow of the tides; a form of hydro-electricity. The most successful tidal power scheme (completed in 1966) is located in the Rance estuary in Brittany, France. Water surges through the 24 reversible turbines at high and low tide, producing a power output of some 240 megawatts.

tidal wave See **tsunami.**

tides The regular movement of the seas back (ebb) and forth (flow), caused by the gravitational attractions of the Moon, and to a lesser degree of the Sun. High tides occur about every 12 hours 25 minutes. The highest (spring) tides occur when the Sun and the Moon are in line, as at full and new Moon. The lowest (neap) tides occur at the first and third quarters when the directions of the Sun and Moon lie at right-angles. See also **phases of the Moon.**

time measurement The basic unit of time for scientific purposes is the second, defined in terms of the frequency of radiation given out by caesium atoms (see **second**). Time so measured is called atomic time. The time kept by clocks and watches is solar time. It is measured in terms of the rotation of the Earth relative to the Sun. The solar day of 24 hours is the length of time it takes for the Earth to return to the same position relative to the Sun. Because its orbit is elliptical, the Earth travels at different speeds throughout the year, which makes the actual solar day vary. For convenience clocks and watches show an average value—mean solar time. Astronomers use sidereal time. The sidereal day is the time it takes the Earth to return to the same position relative to the stars. It is about 4 minutes shorter than the mean solar day.

Places at different longitudes on the Earth have different local times. When it is noon (Sun highest in the sky) in one place, it is before noon in places to the west and after noon in places to the east. In other words local time changes continuously with longitude. To prevent confusion a time standard has been adopted (since 1884), which splits the world into 24 time zones, each spanning $15°$ of longitude $(24 \times 15 = 360°)$. In each zone the time is one hour ahead of the time zone to the west and one hour

behind the time zone to the east. The world standard time is taken as the time on the meridian at Greenwich $(0°$ longitude), this being called Greenwich Mean Time (GMT) or Universal Time (UT).

tin (Sn) One of the most useful metals (rd $7 \cdot 3$, mp $231 \cdot 9°C$), exploited in the form of its copper alloy bronze since the dawn of civilization. Its main ore is the oxide cassiterite, SnO_2. It is a soft, ductile and malleable metal with good corrosion resistance because of the presence of an oxide film. It has two allotropes; the normal white tin tends to form powdery grey tin below about $13°C$. In compounds tin exhibits valencies of two (stannous compounds) and four (stannic). Tin is used plated on steel as tinplate and in a variety of alloys including bronze, pewter, solder, type metal and bearing alloys.

tinfoil Originally a very thin sheet of tin, but nowadays usually of aluminium.

tinplate The usual material for making cans, being sheet mild steel coated, by electrolysis or hot-dipping, with a thin layer of tin.

Titan The largest moon of Saturn and indeed of the whole solar system (diameter 5,150 km, 3,200 miles). It is orange in colour and notable because it has a thick atmosphere, which consists mainly of nitrogen. It orbits Saturn at a distance of about 1,222,000 km (759,000 miles) in a little under 16 days.

titanium (Ti) A strong, light and corrosion-resistant transition metal (rd $4 \cdot 5$, mp $1,680°C$). It is obtained mainly from its dioxide ore, rutile, TiO_2. When purified, rutile makes an excellent white pigment for paints. The metal is widely used in aerospace construction by itself and in alloys.

titration An operation performed routinely in volumetric chemical analysis. In titration a measured volume of a solution of one substance is added to a measured volume of a solution of another until a reaction between the two substances is complete. Usually a solution of one substance of known strength (standard solution) is added gradually by burette to a solution of the other whose strength is to be determined. An indicator is added to show by a colour change the completion, or end-point of the reaction involved.

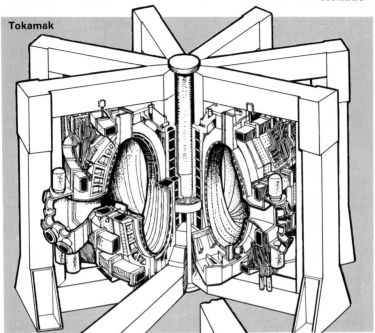

Tokamak

Tl The chemical symbol for thallium.

Tm The chemical symbol for thulium.

TNT The powerful high-explosive tri-nitrotoluene, $C_7H_5N_3O_6$, widely used in conventional bombs. Made by treating toluene with nitric and sulphuric acids, it is relatively stable at temperatures below about 250°C and can be melted (mp 82°C) and cast readily.

tokamak A machine used in nuclear-fusion research to confine high-temperature plasma. It consists of a toroidal (doughnut-shaped) chamber in which powerful magnetic fields are produced by field coils and transformers. The Joint European Torus being built at Culham, in Berkshire, is one of the biggest tokamaks.

tolerance In engineering, the allowable range in the dimensions of, say, a machined part. Where a product is assembled from separate parts, all such parts must be made to a certain tolerance or else they will not fit together accurately.

toluene Also called toluol and methyl-benzene, $C_6H_5CH_3$; an aromatic liquid obtained originally from coal tar but now mainly from petroleum. It is used to make a variety of useful products, including saccharin, dyes, drugs and TNT.

ton(ne) A common unit of weight. 1 ton (Imperial) = 20 hundredweight (cwt) or 2,240 lb; 1 short (American) ton = 2,000 lb; 1 tonne (metric ton) = 1,000 kg or 2,204·6 lb.

topaz A transparent gemstone that may be tinted yellow, blue, brown and pink. It is an aluminium silicate containing fluorine.

topography In geography, all the natural and physical features of a region.

topology A branch of mathematics that deals with the shapes and surfaces of geometrical figures rather than their dimensions.

tornado A very violent wind storm that may form during a severe thunderstorm. It descends from a thundercloud as a dark funnel, in which winds spiral at 500 km/h (300 mph) or more. Tornadoes last for a relatively short time but they are immensely destructive. At sea they are called waterspouts.

Tractor

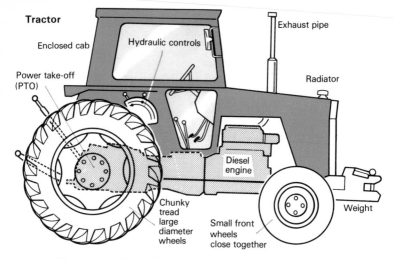

Enclosed cab

Hydraulic controls

Exhaust pipe

Power take-off (PTO)

Radiator

Diesel engine

Chunky tread large diameter wheels

Small front wheels close together

Weight

torque The turning effect of a force which tends to rotate a body. Also called the moment of a force, it is equal in magnitude to the product of the force applied and the perpendicular distance from the line of action of the force to the axis of rotation.

torque converter A hydraulic coupling in a car's automatic transmission system that transmits power between the engine flywheel and the automatic gearbox. It consists of a vaned impeller, attached to the flywheel, and a turbine attached to the gearbox shaft, with fluid in between. When the impeller spins, the fluid whirls round and spins the turbine. By means of an additional rotor in between, the unit can increase the torque transmitted.

torsion The strain set up in a body when it is twisted.

torsion balance A delicate device used to measure small forces, for example, gravitational attraction. It consists essentially of a fine wire or fibre at the end of which is suspended a short horizontal rod. The forces to be measured are made to act at the ends of the rod to form a couple that twists the wire or fibre. The extent of twisting is a measure of the forces acting.

torus The shape of a ring doughnut.

tourmaline A complex silicate mineral containing boron that forms crystals often used as gemstones. These crystals are piezoelectric and exhibit dichroism.

tractor The most important farm machine, which developed from the steam traction engines used on farms in the late 1800s. The prototype modern tractor was the British Ivel, built by D. Albone in 1902. Today's tractors are powered by paraffin, petrol or diesel engines and are fitted with power-take-off and hydraulic gears for raising and lowering implements.

trajectory The path through the air or space of a missile or spacecraft.

tranquillizer A drug that relieves nervous tension without causing undue drowsiness. Such tranquillizers as Valium and Librium, made from benzodiazepines, are widely used but may be harmful when used in excessive doses or for long periods.

transducer A device that converts one kind of energy or signal into another. A microphone is a transducer, as is a piezoelectric crystal.

transformer An electrical device that can alter the voltage of alternating current. It consists essentially of two coils of wire wrapped around an iron core. One coil (the primary) has fewer turns than the other coil (the secondary). When alternating current is passed through the primary coil, it induces a current in the secondary coil. Because the secondary coil has more turns, the output voltage in it is proportionately higher than the input voltage.

transistor A semiconductor device (usually silicon) that can generate, amplify or control elecrical signals. It performs similar functions to the thermionic valves it superseded. The invention of the transistor by Bell Telephone engineers John Bardeen, Walter H. Brattain and William B. Shockley in 1948 sparked off the electronics revolution. The great advantage of transistors over valves is that they are much smaller, much tougher and require much less power. See **semiconductor**.

transit In astronomy, the passage of a star across an observer's meridian; or the passage of Mercury or Venus across the face of the Sun.

transition elements A large group of elements occupying the middle of the periodic table, includes some of our most important metals, such as iron, copper, silver and gold. One of their most notable characteristics is the similarity between one element and the elements before and after it in the table. This similarity can be explained in terms of their electronic structure. In general they have a similar number of electrons in their outer electron shell, which makes them chemically similar. They differ in the number of electrons in the previous shell, which has less effect on their properties. Other features of the transition elements are that they tend to display variable valencies; many of their compounds are coloured; and they are good catalysts.

translucent Allowing the passage of light, though not clear vision. Contrast **transparent**.

transmission system Of a motor vehicle; the system that transmits power from the engine to the driving wheels. It commonly includes a clutch, gearbox, propeller shaft and final drive, incorporating a differential.

Simple transformer

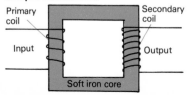

Transistor

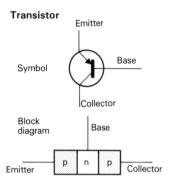

Symbol

Block diagram

transmitter In broadcasting, the equipment for generating radio waves, modulating them and feeding them to a transmitting aerial (see **carrier wave; modulation**). The telephone transmitter is a carbon microphone.

transmutation In nuclear physics, the changing of one element into another as a result of nuclear bombardment or natural radioactive decay.

transparent Allowing the passage of light and clear vision. Contrast **translucent**.

transponder An electronic device that is triggered into transmitting a signal by the receipt of another signal.

transuranium element An element beyond uranium in the periodic table, artificially produced by nuclear bombardment.

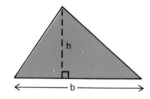

triangle A figure with three sides. The interior angles of a triangle add up to 180°. An equilateral triangle has equal sides and equal interior angles of 60°. An isosceles triangle has two equal sides and the angles opposite these sides are equal. The area of a triangle is equal to: ½ (base × height) or ½ bh (see diagram). See also **Pythagoras's theorem**.

triangulation A method of calculating distances used in surveying. It relies on the principle that when one side and two angles of a triangle are known, the lengths of the other two sides can be calculated. Surveyors establish a baseline of known length and take theodolite readings (angles) of a distant landmark from either end. They can then calculate the distance to the landmark.

Triassic Period A span of geological time lasting from about 225–190 million years ago, during which small dinosaurs became common and pine-like trees flourished.

tribology A branch of science concerned with friction, wear and lubrication.

trichloroethanol See **chloral.**

trichloromethane See **chloroform.**

trigonometry A branch of geometry concerned with the properties of the triangle, particularly the right-angled triangle. It involves the use of trigonometric ratios—various ratios of the sides of the triangle. The trig ratios are sine (abbreviated to sin), cosine (cos) and tangent (tan); and their respective reciprocals are cosecant (cosec), secant (sec) and cotangent (cotan). In the right-angled triangle ABC (see diagram) the sine of angle B ($\angle ABC$), or sin $B = b/c$; cos $B = a/c$; tan $B = b/a$.

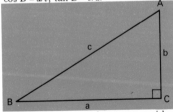

trilobite Extinct marine creature with a distinctive three-lobed body, whose fossils are abundant, particularly in Cambrian rocks.

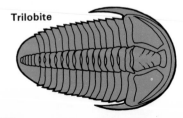

Trilobite

trinitrotoluene See **TNT.**

triode See **thermionic valve.**

triple point The point at which all three physical phases of a substance—gas, liquid and solid—are in equilibrium.

tritium One of the heavy isotopes of hydrogen, containing two neutrons as well as a single proton in the nucleus. It is radioactive and used with the other heavy hydrogen isotope deuterium as 'fuel' in nuclear fusion reactions.

tropics The region of the Earth's surface lying between latitudes $23\frac{1}{2}°$ North and South and centred on the equator. The $23\frac{1}{2}°N$ latitude line is called the Tropic of Cancer; and the $23\frac{1}{2}°S$ line, the Tropic of Capricorn. They represent the limits of latitude at which the Sun can be directly overhead at local noon. The Sun is directly overhead there at the times of the solstices. In the tropics temperatures remain high. In equatorial regions the rainfall is heavy all year. Elsewhere there is usually a wet season, when heavy rain falls, and a dry season.

troposphere The bottom layer of the Earth's atmosphere, where the air is densest and in which most of our weather occurs. It is bounded by the stratosphere, and the boundary between the two is called the tropopause. The tropopause is located between about 8–16 km (5–10 miles) high, depending on latitude. The temperature steadily decreases with increasing height, reaching about −50°C at the tropopause.

tsunani Misleadingly called tidal wave; a wave often of enormous height, which is generated by submarine earthquakes, volcanic eruptions or even hurricanes. Tsunanis race across the open sea at speeds of up to 800 km/h (500 mph). When they reach the shore, they can tower up to 30 m (100 ft) in height and have enormous destructive power.

tungsten (W) Once called wolfram; a valuable transition metal (rd 19·3, mp 3,380°C), which has a higher melting point than any other element. It has many uses where its heat resistance is exploited—most commonly as the resistance wire that forms the filament in electric-light bulbs. It is incorporated in tool steels for high-speed machining. Its main ores are mixed oxides with iron (as wolframite, $FeWO_4$) and calcium (scheelite, $CaWO_4$).

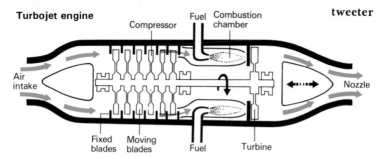

Turbojet engine

Compressor — Fuel — Combustion chamber

Air intake — Nozzle

Fixed blades — Moving blades — Fuel — Turbine

tuning In radio, altering the frequency of an electronic circuit so that it resonates with an incoming signal, a process called detection.

tuning fork A two-pronged steel fork that emits a pure note, without overtones, when struck.

tunnels Underground passages excavated for such purposes as mining, water supply and transport. The first major transport tunnels were built in the 1700s when canal systems were being expanded. The following century the pace of tunnel building increased with the coming of the railways. The first great Alpine tunnel (rail) was the 14-km (8·5-mile) long Mt Cenis, or Fréjus (completed 1857). One of the latest (1980) is the 16-km (10-mile) long St Gotthard road tunnel. Rock tunnelling is accomplished by repeated drilling, shot placing and blasting operations. Tunnelling through soft ground is done inside the protection of a tunnelling shield. The modern shield excavates and removes the soil automatically and lines the tunnel as it is bored with steel or concrete segments.

turbine A device, in principle rather like a windmill, which is set turning by the passage through it of a fluid. It consists essentially of a shaft that is free to rotate inside a casing. On the shaft are many vanes. When fluid flows through the vanes, the vanes and thus the rotor spin by reaction. Turbo-machinery is some of the most powerful in the world. See **gas turbine; steam turbine; water turbine.**

turbocharger Or supercharger; a type of compressor used to force more air into an internal combustion engine, such as the diesel and petrol engine. It is generally powered by a turbine in the engine's exhaust.

turbofan Also called fan jet and bypass turbojet; a type of turbojet engine that has a huge fan at the front. Some of the air from the fan by-passes the main part of the engine, leading to a more efficient propulsive jet.

turbogenerator A turbine coupled to an electricity generator.

turbojet A type of gas turbine, or jet engine, that derives its thrust wholly from a rearward jet of hot gases. In a simple turbojet (see diagram), air is taken in through the front of the engine and compressed by a compressor. The compressed air then feeds through nozzles into a combustion chamber, fuel (kerosene) is sprayed in, and the mixture is ignited. The hot gases produced expand through a turbine (which drives the compressor) before escaping at high speed through a rear nozzle. Reaction to the rearward jet produces a forward propulsive thrust. Practical turbojets have two or more compressor and turbine stages.

turboprop A type of gas turbine, or jet engine that derives part of its thrust from a jet and part from a propeller driven by a turbine in the jet exhaust (see **gas turbine**).

turbulent flow A type of fluid flow characterized by uneven velocity and eddying. Contrast **streamline flow;** see also **Reynolds number.**

turpentine See **terpenes.**

turquoise An opaque blue or blue-green gemstone. It is a hydrated copper aluminium phosphate, $CuAl_6(PO_4)_4(OH)_8 . 4H_2O$.

tweeter See **loudspeaker.**

two-stroke cycle   An operating cycle of an internal combustion engine which is repeated every two piston strokes. Many motorcycles and lawnmowers have a two-stroke petrol engine; some large marine diesel engines are two-stroke. In a typical petrol engine two-stroke cycle (see diagram), fuel is drawn into the crankcase via the inlet port as the piston moves up to compress mixture already in the cylinder. The mixture is then ignited and the piston is forced down on its power stroke. Near the bottom of the stroke, the exhaust port and transfer port are uncovered. Fresh fuel mixture enters the cylinder and forces out the burnt gases through the exhaust port.

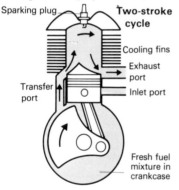

Sparking plug

Two-stroke cycle

Cooling fins

Exhaust port

Transfer port

Inlet port

Fresh fuel mixture in crankcase

type metal   The alloy used for casting printing type, made from lead, antimony and tin. The antimony confers hardness and causes the metal to expand slightly on solidifying, ensuring a sharp impression.

typewriter   A writing machine operated mechanically, electrically or electronically by means of a keyboard. The first practical typewriter was developed in 1867 by the Americans C. L. Sholes, S. W. Soule and C. Glidden. In many models, striking a key causes a pivoted arm bearing the type character to strike an inked ribbon against paper. Some electric machines have a 'golf-ball' that carries the type characters. In the latest electronic models a 'daisy-wheel' carries the type.

typhoon   The name given in Pacific regions to the violent tropical storms that are called hurricanes in the Caribbean.

UV

U   The chemical symbol for uranium.

UFO   The abbreviation for 'unidentified flying object'. UFOs are objects or phenomena seen in the atmosphere that cannot be immediately explained. Many sightings of UFOs can eventually be traced to such things as weather balloons, lightning, fireballs and other astronomical phenomena. But a few defy explanation, and popular opinion suggests that they originate out of this world.

UHF   The abbreviation for 'ultra-high frequency' radio waves, which are used, for example, for television transmissions.

ultrasonics   The study of sound waves of pitch, or frequency beyond that which the human ear can detect. Our ears cannot detect sounds of frequencies greater than about 20,000 hertz (cycles per second). Ultrasonic waves are used, for example, in sonar equipment. They are usually generated by means of piezoelectric crystals.

ultraviolet radiation   Electromagnetic radiation of shorter wavelength than visible light, beyond the violet end of the visible spectrum.

uncertainty principle   See **Heisenberg's uncertainty principle.**

underground railway   Also called subway; a railway that runs mainly underground. The first underground railway built just below street level opened in London in 1863, and was operated by steam locomotives. The first deep-level tube, operating in bored tunnels by electric locomotives, opened in London in 1890. The Paris Métro, opened in 1900, differs from most other underground systems in having pneumatic tyres. Underground systems are becoming increasingly popular in large cities that suffer from traffic congestion in the streets. Underground trains have no separate locomotives, being powered by electric motors installed in the passenger cars. They pick up electricity either from a third rail by the track (eg London Underground) or an overhead wire (eg Métro).

unit operations In chemical engineering, typical physical processes that are carried out on substances during processing. They include evaporation, distillation, crushing, filtration and crystallization.

unit processes In chemical engineering, typical chemical reactions between substances that are carried out during processing. They include oxidation, reduction, esterification and polymerization.

universal joint Also called Hooke joint; a joint that is free to bend in several directions while at the same time transmitting power. Universal joints are fitted, for example, to each end of a car's propeller shaft so that gearbox and rear axle can move independently.

universe All that exists—space and all the bodies therein. Study of the origins and development of the universe, such as the big-bang theory and steady-state theory, form part of the discipline of cosmology. The former theory is now favoured for it appears that the universe is expanding, as if from a 'big bang'.

unsaturated compound An organic compound that contains one or more double or triple bonds between some of the carbon atoms. Ethylene is an unsaturated compound with a double bond ($H_2C=CH_2$); acetylene, one with a triple bond ($HC\equiv CH$). Unsaturated compounds are generally more reactive than saturated compounds.

uranium (U) A heavy radioactive metal (rd 19·05, mp 1,133°C) obtained from the mineral pitchblende. It is notable because its atoms can be made to split (fission) with the release of enormous energy. It is the most important 'fuel' in nuclear reactors. Of the two main natural isotopes of uranium, U-238 (over 99·2%) and U-235 (over 0·7%), only the latter undergoes fission.

Uranus The seventh planet in the solar system going out from the Sun, which it orbits at a distance of about 2,870 million km (1,783 million miles) every 84 years. It is the fourth largest planet (diameter 49,000 km, 30,500 miles) with at least five moons, and is made up primarily of gaseous and probably liquid hydrogen. It is unusual in having its equatorial plane nearly at right-angles (actually 98°) to the plane of its orbit. It is surrounded by a faint ring system.

Universal joint

Spider

yokes

urea Or carbamide, $CO(NH_2)_2$; an organic compound found in animal urine. It is important historically because it was the first organic compound prepared artificially from inorganic materials. The German chemist Friedrich Wöhler prepared it from ammonium cyanate in 1828, disproving the 'vital-force' theory which held that organic compounds could originate only in living things. Urea is synthesized today from carbon dioxide and ammonia, and is used for example in fertilizers and to make synthetic resins.

V The chemical symbol for vanadium.

V-1, V-2 Hitler's revenge ('Vergeltungswaffe' in German) weapons used against the Allies in World War 2. The V-1, also called buzz bomb and flying bomb, was a kind of drone propelled by pulse jet. The V-2, or rocket bomb, was a ballistic missile propelled by rocket. The rocket, which used propellants of alcohol and liquid oxygen, was first designated A-4 and tested successfully at the Baltic Island of Peenemünde in 1942. Two years later it was used to bomb London.

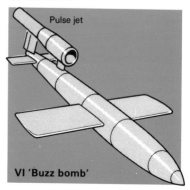

Pulse jet

VI 'Buzz bomb'

vaccination Inoculating a person with dead or reduced activity viruses or microorganisms so as to stimulate the body to make antibodies which give subsequent immunity to infection. Pioneered by Edward Jenner in the 1790s, it is so named because serum obtained from cows (Latin, vacca) infected with cowpox was used to immunize against smallpox.

vacuum Strictly, a space that contains no matter. In normal usage, however, the term refers to a space at very low pressure, which should properly be termed a partial vacuum. Even the space between the stars is not quite devoid of matter. On Earth, scientists can achieve partial vacuums down to pressures millions of millions of times less than atmospheric pressure by using special vacuum pumps.

Vacuum flask

Cork stopper

Evacuated double-walled glass container

Inner walls silvered

Outer casing

vacuum flask Also called Dewar flask and Thermos flask; a vessel used for keeping liquids hot (or cold) for long periods. It is so called because it consists of a glass container with a double wall, in which there is a partial vacuum. The vacuum prevents heat being transferred by conduction or convection across the gap from the inner wall (in contact with the liquid) to the outer. The inside walls of the cavity are silvered to reflect any heat radiated across.

vacuum pump A pump for reducing the pressure in a vessel so as to create a (partial) vacuum. A low vacuum can be achieved by means of a vane-type rotary pump. For higher vacuums different techniques are required. In a vapour pump, for example, a vapour is used to remove the gas. The mercury diffusion pump is of this type. See also **getter.**

valency Or valence; the combining power of an element. It can be expressed numerically as the number of atoms of hydrogen which an atom of the element can combine with or displace in chemical combination. For types of combination, see **covalency; electrovalency.** See also **oxidation number.**

valleys Long depressions in the Earth's surface, formed as a result of erosion by a river or glacier, or the collapse of land between faults in the Earth's crust (rift valley). Very deep, narrow and steepsided valleys are called canyons, and similar but smaller valleys, gorges. Hanging valleys are valleys branching off a main (trunk) valley, but at a higher altitude.

valve In engineering, a mechanical device for controlling the flow of a fluid.

valve, electronic See **thermionic valve.**

vanadium (V) A transition metal (rd 6·1, mp 1,920°C), widely used in steel alloys to confer hardness and durability. Vanadium pentoxide, V_2O_5, is an excellent catalyst, used for example in the contact process for making sulphuric acid.

Van Allen belts Doughnut-shaped regions of intense radiation around the Earth, discovered by the first American satellite Explorer 1 (1958) and named after James A. Van Allen. The inner belt dips to within 400 km (250 miles) of the Earth's surface; the outer belt extends to a height of about 50,000 km (31,000 miles).

van de Graaff generator An electrostatic generator for accelerating charged particles, devised by the American R. J. van de Graaff in 1931. It consists of a tall metal dome, at the bottom of which is a device that 'sprays' electric charge onto a rubber belt. The belt carries the charge to the top of the dome and transfers it to the dome's surface. Charges of several million volts can be built up in this way.

van der Waals forces The attractive forces that exist between molecules, named after the Dutch physicist J. D. van der Waals who developed an equation of state that allowed for this attraction.

vapour A gas below its critical temperature, in which state it can be liquefied by the application of pressure alone.

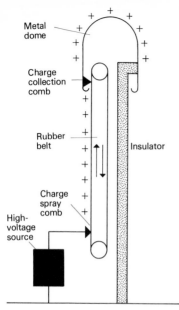

Van de Graaff generator

variable star One that varies in brightness regularly or irregularly. True, or intrinsic variables fluctuate in brightness because of processes going on in their interior. They include such stars as Mira Ceti and the Cepheids. Another type of variable is the eclipsing binary (eg Algol), in which the periodic variation in brightness is caused by the two stars in a binary system regularly eclipsing one another (see **binary**). See also **nova; pulsar; supernova.**

varnish A type of protective coating, which in its simplest form consists of a natural or synthetic resin dissolved in a solvent such as alcohol or white spirit. See also **lacquer.**

vector In mathematics, a quantity that represents not only magnitude but also direction. Velocity is a vector, being speed in a certain direction.

Vega (α Lyrae) The brightest (mag 0·04) star in the constellation Lyra (the Lyre) and fifth brightest in the heavens. Blue-white in colour, it lies about 26 light-years away.

velocity Speed in a certain direction; it is a vector quantity.

veneer A thin sheet of wood. Expensive veneers (mahogany, walnut) are used to face cheaper woods, while cheap veneers (softwoods) are glued together to make plywood.

venturi tube A tube with a constriction in the middle, used in devices for measuring fluid flow and in carburettors. When fluid flows through the constriction, its speed increases but its pressure decreases, according to Bernoulli's principle.

Venus The second planet in the solar system going out from the Sun. It is a near twin of Earth in size (diameter 12,140 km, 7,545 miles) but is covered with a very thick atmosphere of mainly carbon dioxide. The thick atmosphere traps the heat from the Sun and raises the surface temperature to 450°C or more. It is the planet that comes closest (within about 42 million km, 26 million miles) to the Earth and is the brightest in the night sky (mag −4·4). The planet has a slow clockwise (retrograde) rotation, turning on its axis in 244 days, while orbiting the Sun in about 225 days. Like the Moon it shows phases, first observed by Galileo.

verdigris A greenish-blue deposit that forms on copper when it corrodes; it is basic copper carbonate or sulphate.

vernier A device used to permit more accurate reading of a scale. It consists of a short scale with 10 divisions that slides along the main scale. The 10 vernier divisions are equal to 9 main scale divisions. On the main scale shown below, a vernier helps give a more accurate reading of 30·6.

VHF An abbreviation for 'very high frequency' radio waves, which are used for interference-free radio broadcasting.

Vernier scale

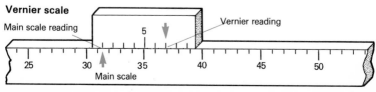

Vernier reading

Main scale reading

5

Main scale

25 30 35 40 45 50

Viking spacecraft

Viking spacecraft Two American spacecraft sent to Mars, where they arrived in 1976. They consisted of two parts, an orbiter which mapped the planet from orbit, and a lander which soft-landed on the surface. The landers set down in regions known as Chryse (Viking 1) and Utopia and returned spectacular pictures of a barren orange, rock-strewn landscape as well as other valuable data.

vinegar A dilute solution (about 5–10%) of acetic acid (CH_3COOH), made by oxidizing wine, beer or other fermented products and used for flavouring and preserving foods.

vinyl compounds Organic compounds containing the vinyl group $CH_2{=}CH{-}$. They include vinyl acetate, $CH_2CHOOCCH_3$; and vinyl chloride, or chlorethene, CH_2CHCl; which can both be polymerized into useful plastics (PVA and PVC).

Virgo The Virgin; a zodiacal constellation lying between Leo and Libra. Its most prominent star is the 1st magnitude Spica.

virtual image An image produced by a lens or mirror that can be seen by the eye but not received on a screen, as opposed to a real image that can be seen and received.

viscose See **rayon**.

viscosity The property of a fluid that makes it resist flowing. Treacle has a high viscosity—it is highly viscous. Water has a comparatively low viscosity. A common unit of viscosity is the poise.

vitamins Complex organic compounds needed in relatively small amounts by the body in order to remain healthy. Vitamin deficiency leads to disease. Of major importance among the 20 or so vitamins required by living things are vitamins A, B-complex, C and D. Vitamin A ($C_{20}H_{29}OH$) is found in milk and green vegetables. Two important B vitamins are B_1 (see **thiamine**) and B_2 (see **riboflavin**). Vitamin C is found abundantly in citrus and other fruits (see **ascorbic acid**). D vitamins may be found in fish-liver oils and also in skin exposed to sunlight.

volatile A term describing a substance that vaporizes readily, due to it having a low boiling point.

volcanic glass See **obsidian**.

Volcano — Parasitic cone, Crater, Throat, Ash & lava, Conduit, Rock layers, Hot magma reservoir

volcano An opening in the Earth's crust through which molten rock, or magma escapes, often with explosive violence. The emerging stream is called lava. Repeated volcanic eruptions at a site give rise to a typical conical structure as layers of ash and solidified lava build up. Most volcanoes are located at the boundaries of the plates that make up the Earth's crust (see **plate tectonics**).

volt The main unit of electric potential difference and emf, being the potential difference between two points in a conductor carrying a current of one ampere when the power dissipated between those points is one watt. It is named after the Italian electrical pioneer Alessandro Volta. Voltage is the potential difference, or emf measured in volts.

voltmeter An instrument for measuring electric potential difference. It is constructed in much the same way as an ammeter.

volume A measure of the space occupied by an object. It is measured in such units as litres, cubic centimetres and cubic feet.

volumetric analysis A major method in quantitative chemical analysis which involves measuring the volumes of reacting substances (see **titration**).

vortex A whirling mass of fluid, as in a whirlpool, tornado and smoke ring.

Vostok spacecraft The type of spacecraft that took pioneering cosmonaut Yuri Gagarin into space on April 12 1961. It consisted of an equipment module and a spherical re-entry module some 2·3 m (7·5 ft) in diameter.

Voyager spacecraft Two very successful American spacecraft which visited Jupiter and Saturn in the late 1970s and early '80s. They sent back some of the finest pictures ever taken of space, spying erupting volcanoes on the Jovian moon Io, and resolving Saturn's three ring into thousands of individual ringlets.

VTOL Abbreviation for 'vertical take-off and landing'. VTOL aircraft include helicopters and fighters like the Harrier, which manoeuvres by means of swivelling jets.

vulcanism Volcanic activity, from the name of the Roman god of fire, Vulcan.

vulcanization Improving the properties of rubber by treatment with sulphur or other chemicals. The process was accidentally discovered by the American inventor Charles Goodyear in 1839.

Voyager spacecraft

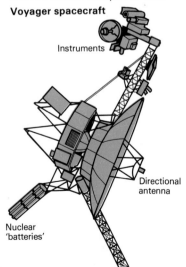

Instruments

Directional antenna

Nuclear 'batteries'

W

W The chemical symbol for tungsten; from its original name, wolfram.

Wankel engine A novel kind of rotary petrol engine invented by German engineer Felix Wankel in the 1950s. It uses a triangular rotor moving eccentrically in a figure-of-eight chamber. The four stages of the standard four-stroke petrol-engine cycle take place in different segments of the chamber. The Wankel engine has been used successfully in a few cars, including the Mazda RX7.

warfarin An organic chemical ($C_{19}H_{16}O_4$) widely used as a rat poison. It is an anticoagulant and causes death by internal bleeding.

watch A portable clock which may be regulated mechanically, electrically or electronically. The ordinary mechanical watch is powered by a spring and regulated by an oscillating balance wheel. Electric watches are powered by batteries and are regulated by the vibrations of a tiny tuning fork. Electronic watches are regulated by vibrations of a thin quartz crystal (see **quartz watch**).

water Hydrogen oxide, H_2O; perhaps after oxygen the most important compound on Earth, without which life as we know it would be impossible. Over 70% of the Earth is covered with water and more is locked in the polar ice caps and glaciers. There is estimated to be some 1·5 million million million tonnes of water in the Earth's crust. There is a constant exchange of water between the ground and the atmosphere, this being known as the water cycle (see page 242). The presence of more or less water vapour in the atmosphere has a profound effect on the weather. Chemically, water is an excellent solvent. It has a pH when pure of 7, being neither acidic nor basic. It freezes at 0°C and boils at 100°C under normal pressure. It is chemically similar to hydrogen sulphide, H_2S, and should theoretically also be a gas at room temperature. But it is less volatile due to hydrogen bonding.

water cycle

Water cycle

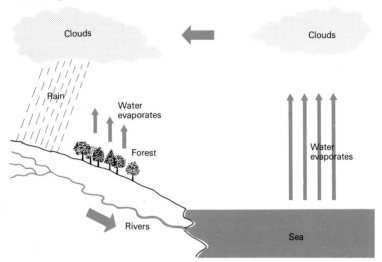

water cycle The never-ending exchange of water between the ground and the atmosphere; also called the hydrological cycle. Water evaporates into the atmosphere from surface waters and growing plants. It returns to the ground mainly as rain.

water gas Or blue gas; a fuel gas consisting of a mixture of carbon monoxide and hydrogen. It is made by passing steam over red-hot coke, the reaction being $H_2O + C \rightarrow CO + H_2$.

water-glass A syrupy solution of sodium silicate, Na_2SiO_3, used to preserve eggs as a bonding agent, and to make silicon 'gardens'.

watermark A translucent pattern seen in paper when it is held up to the light. It is made by wires of the so-called dandy roll pressing against the wet pulp sheet on the paper-making machine.

water of crystallization Molecules of water combined chemically with a substance in the crystalline state. Sodium carbonate (washing soda) crystals contain 10 molecules of water of crystallization, $Na_2CO_3.10H_2O$; copper sulphate crystals contain five molecules of water of crystallization, $CuSO_4.5H_2O$.

water power See **hydroelectric power; tidal power; water turbine.**

water softening Removing the hardness from water (see **hard water**). Temporary hardness may be removed simply by boiling the water. Permanent hardness requires treatment by agents such as zeolites or ion-exchange resins. These substances remove the ions that cause the hardness (such as magnesium and calcium) and replace them with soluble sodium ions.

water turbine Turbines whose rotors are spun by flowing water. There are two basic types—impulse and reaction. The Pelton wheel is an example of an impulse turbine, being turned when a stream of water strikes the buckets around its edge. The widely used Kaplan turbine, which looks rather like a ship's propeller, is by contrast spun by reaction as the water flows through the vanes on its rotor.

waterwheel The original means of harnessing the power of flowing water and ancestor of the water turbine.

watt The main unit of power, being the power dissipated between two points in a conductor carrying a current of one ampere when the potential difference between them is one volt. Named after Scottish engineer James Watt, it is equal to a work output of 1 joule per second. 1 horsepower = 746 watts.

wave A rhythmic disturbance in a medium, or variation in certain properties in space. In a medium such as air or water, a wave is set up and propagated when air or water molecules vibrate back and forth about an equilibrium position. The wave moves outwards from the source of the original disturbance (like the ripples on a pond), but the medium does not—it merely vibrates. Electromagnetic waves require no medium for propagation—they are electrical and magnetic disturbances in space. Characteristic features of wave motions are wavelength—the distance between two successive crests (see diagram); the frequency—the number of complete waves passing a certain point each second (expressed as cycles per second, or hertz); and amplitude—the extent of the disturbance about the equilibrium level. There are two basic wave forms—longitudinal (or compressional) and transverse. Sound waves are longitudinal. The particles causing the wave vibrate in the same direction as the wave is being propagated. Electromagnetic waves are transverse. The vibrations travel at right-angles to the direction in which the wave is moving. See also **diffraction; interference.**

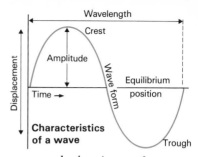

Characteristics of a wave

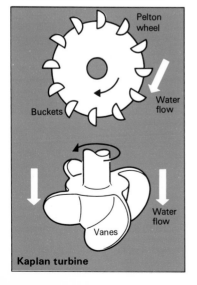

Kaplan turbine

wave mechanics A part of quantum mechanics concerned with the wave-like properties of travelling atomic particles.

wax A pliable substance obtained from animal, vegetable or mineral sources. It is widely used in polishes and for making candles. Common waxes include beeswax, palm wax and paraffin wax, obtained from petroleum distillation.

weather The prevailing atmospheric conditions—temperature, pressure, humidity, wind velocity and so on. The typical weather experienced by a region is the climate. The study of weather is meteorology. Meteorologists take regular readings of atmospheric conditions and, guided by experience, try to forecast what the weather will be like in the future. They are now aided by data and pictures returned by weather satellites.

weathering In geology, erosion of the landscape by the elements—rain, wind, frost, sunlight.

weather map See **synoptic chart.**

weather satellite A space satellite that records weather data and takes pictures of cloud cover and transmits them back to Earth. Weather satellites regularly scan large areas of the Earth, enabling meteorologists to get a much better overall picture of the way the weather is developing. Successful weather satellites have included the American Nimbus and Tiros series, the Russian Molniya, and the European Meteosat.

weaving The second of the two main textile-making processes, following spinning. Spinning collects short fibres into a long, firm yarn. Weaving interlaces yarns at right-angles to one another to make fabric, the process taking place on a loom. The lengthwise threads during weaving are called the warp; and the crosswise threads, the weft (see **loom**).

weber

Wheatstone bridge

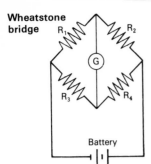

weber The SI unit of magnetic flux. It is the flux that produces in a circuit of one turn (loop of wire) an emf of one volt as it drops to zero at a uniform rate in one second.

wedge One of the simplest machines, being a solid with a triangular cross-section. Nails and axes may be considered types of wedge.

weight A measure of the gravitational attraction on an object. Weight must not be confused with mass. Mass is the amount of matter in an object, which always remains the same. Weight varies according to the force of gravity acting upon it. The SI unit of weight (given by mass × acceleration due to gravity) is the newton.

weightlessness A condition experienced by a body in orbit. In orbit a body is in free fall and the force of gravity is effectively neutralized. Astronauts readily adapt to weightless conditions, but they must take regular exercise to prevent their muscles wasting away.

welding Joining metals together by fusing them. In a typical welding operation, the edges of the metals to be joined are heated by an oxyacetylene torch or electric-arc until they soften. Then hot molten metal from a filler rod is run into the joint, fusing with both pieces. When the metal cools, a very strong joint is formed.

Weston cell A standard primary cell that produces a steady emf of $1 \cdot 018$ volts at $20\,°C$ for long periods. The anode is a cadmium amalgam coated with crystals of cadmium sulphate. The cathode is mercury coated with crystals of mercurous sulphate. The electrolyte is cadmium sulphate solution.

wetting agent See **surface-active agent.**

Wheatstone bridge A device that measures electrical resistance by balancing the currents flowing through circuits containing known and the unknown resistances. In the diagram (left) when the galvanometer shows no reading, the ratios of the resistances in the two arms are equal: $R_1/R_2 = R_3/R_4$. If three of them are known, then the fourth can be determined.

white dwarf A very dense stellar body that represents a late stage in the evolution of a star, when it has run out of nuclear fuel and collapsed. A white dwarf is comparable in size with the Earth, but it is a million times more dense than the densest metals on Earth. The first white dwarf identified was the companion of Sirius.

white lead A white pigment in paints, usually basic lead carbonate, $2PbCO_3.Pb(OH)_2$.

white light Ordinary sunlight, which actually consists of a mixture of light of different colours, or wavelengths. These can be separated by means of a prism (see **spectrum**).

white spirit A common paint solvent derived from petroleum, consisting of a mixture of paraffin hydrocarbons (alkanes) of boiling point between about $150° – 200°C$.

wide-angle lens A camera lens that has a wide field of view and short focal length. A common wide-angle lens for a 35-mm camera is the 28 mm, which has a viewing angle of 74°. A very wide angle lens like the 8 mm is called a 'fish-eye' lens. It can view 180° or more but with considerable distortion.

Wilson cloud chamber See **cloud chamber.**

Wimshurst machine A device for generating static electricity, invented by the British engineer James Wimshurst. It consists of two glass discs rotating in opposite directions. Electric charge is transferred from the discs via metal plates on the sides.

winch A mechanism used for hoisting or hauling loads, found for example in cranes. It is a form of the simple machine called the wheel and axle and consists of a drum with a cable wound around it. A simple crank-operated hand winch is called a windlass. See also **capstan.**

To prepare themselves for the weightlessness they will experience in orbit, astronauts train in weighted spacesuits in a huge water tank.

wind Air in motion. Winds are caused basically by temperature differences over the Earth's surface. A major circulation pattern is set up as air over the equator becomes hot and rises. This creates a low-pressure region, known to early sailors as the doldrums, where the air is still and waters calm for days on end. As the hot air rises, cool air from polar regions moves in to take its place. The hot air in turn cools and sinks down near the poles. If the earth were not rotating, the winds would blow north-south. But the Earth is rotating, and the winds are deflected by the so-called Coriolis effect. This makes the winds blow towards the equator from the north-east in the northern hemisphere and from the south-east in the southern. Such winds, called the trade winds, blow towards the equator from about the 30th parallels north and south. Between about the 30–60th parallels, by contrast, the winds blow predominantly from the west—from the south-west in the northern hemisphere and from the north-west in the southern. These winds are called the prevailing westerlies. See also **Beaufort scale.**

windmill A mechanical device for harnessing the power 'blowing in the wind'. It had its origins in Persia (now Iran) in the AD 600s and was an

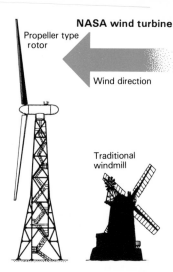

NASA wind turbine

Propeller type rotor

Wind direction

Traditional windmill

important power source for grinding corn until the beginning of this century. It also had wide application, particularly in the Netherlands, for pumping water in land drainage schemes. In its modern guise, the wind turbine, it is making a comeback, and experiments are proceeding in the use of large machines for community power generation.

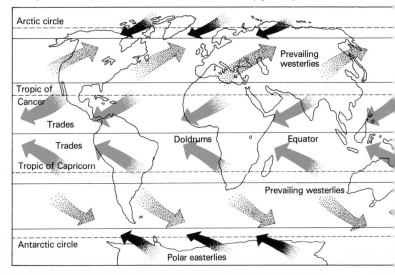

Arctic circle

Prevailing westerlies

Tropic of Cancer

Trades

Trades

Doldrums

Equator

Tropic of Capricorn

Prevailing westerlies

Antarctic circle

Polar easterlies

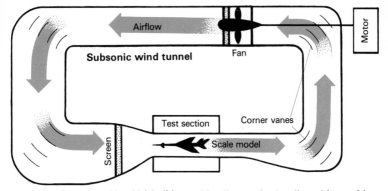

wind tunnel A tunnel in which builders of aircraft, vehicles and structures such as bridges, test scale models of their designs. In a wind tunnel, air is blown or drawn over the model and instruments attached to the model record how it behaves under different air-flow conditions. From this information designers can predict how their full-scale design will behave in practice.

wine An alcoholic drink made by fermenting the juice of the grape, particularly varieties of *Vitis vinifera*. The fermentation of sugars in the grape juice is brought about by natural yeasts on the skin of the grapes.

winter The coldest of the four seasons, which lasts in the northern hemisphere from the winter solstice on about December 21 to the vernal equinox on about March 21; and in the southern hemisphere from about June 21 to September 23.

wire Metal that has been drawn out into a thread by drawing a bar through successively narrower dies.

wireless The original name for what we now term radio. It was introduced to make the distinction between telegraphy by wire and telegraphy via electromagnetic waves, which requires no wires between sender and receiver.

wolfram See **tungsten.**

wood alcohol See **methanol.**

woodpulp The raw material from which paper, rayon and cellulose plastics are made. It is obtained either by mechanically shredding logs (ground woodpulp) or by treating wood chips with chemicals, such as sodium sulphite or caustic soda.

Wood's metal An alloy with a melting point (71°C) below that of boiling water. It contains bismuth (50%), lead (25%), tin (12½%) and cadmium (12½%).

woofer See **loudspeaker.**

wool The original fibre used by man to make textiles, obtained from the fleece of the sheep. It consists of the protein keratin, whose coiled structure gives wool its springiness. The finest wool is obtained from the Australian Merino breed, which may grow a fleece as heavy as 12 kg (25 lb). Other animal fleeces are used to make textiles, including those of the Angora and Cashmere goats. From these we get mohair and cashmere fabrics.

word processor An electronic typewriter with a coupled video display unit and memory facility. Words are typed first onto the screen, where they can be corrected and moved about. The final version is then automatically typed out, perfectly.

work The energy expended when a force moves through a certain distance. The amount of work W done when a force F moves through a distance d is given by: $W = Fd$. The SI unit of force is the joule; in the Imperial system, it is the foot-pound.

work hardening What happens when a metal undergoes such processes as rolling or hammering and becomes harder and stronger.

wow See **flutter.**

wrought iron A relatively pure form of iron made by the puddling process. It contains threads of slag and is strong, tough and easy to machine. Before the advent of cheap steel, wrought iron was the major structural metal.

XYZ

xanthate An organic salt produced when an alcohol reacts with carbon disulphide in the presence of an alkali. They have the general formula $ROCS_2M$, where R is an organic group and M is a metal. The most important xanthate is sodium cellulose xanthate, an intermediate in rayon production.

Xe The chemical symbol for xenon.

xenon One of the noble gases (bp $-108 \cdot 1°C$), used in high-intensity flash lamps. The hexafluoroplatinate of xenon was the first noble gas compound produced (1962).

xerography A photocopying process devised by the American C. F. Carlson in 1938. In the process an electrostatic image of the document to be copied is formed on a selenium-coated drum. This is then dusted with ink powder ('toner'), which adheres only to the areas where the charged image is. The ink image is then transferred to copy paper and fused to it by heat.

X-rays Also called Roentgen rays after their discoverer (1895) W. K. Roentgen; invisible radiation of the electromagnetic spectrum of very short wavelengths, produced in commercial X-ray machines by bombarding a tungsten metal target with a beam of electrons from a heated cathode. X-rays are very penetrating and are used in medicine to 'see inside' the human body. The rays are passed through the body onto a photographic plate, on which bones and flesh can be distinguished. The latest computerized X-ray scanners can also reveal details of soft body tissues. In crystallography, X-rays are used to investigate atomic structure. The ordered crystal lattice diffracts X-rays in a characteristic way, from which the crystal structure can be inferred.

xylene Also called xylol, the dimethyl derivative of benzene, $C_6H_4(CH_3)_2$, which exists in three isomeric forms: ortho-, meta- and para-xylene (see **ortho-, meta- and para-**). Xylenes are useful solvents and raw materials for making dyes and synthetic fibres.

xylol See **xylene**.

Y The chemical symbol for yttrium.

yard A measure of length in the Imperial system of units, being equal to 3 ft or 36 in. 1 yard (yd) = $0 \cdot 914$ m; 1 m = $1 \cdot 094$ yd.

Yb The chemical symbol for ytterbium.

year The length of time it takes the Earth to travel once in its orbit around the Sun. The sidereal year—the time it takes for completion of the orbit with respect to the stars—is $365 \cdot 2564$ mean solar days. The calendar, or civil year has $365 \cdot 2425$ mean solar days. The solar, or tropical year of $365 \cdot 2422$ mean solar days is the time it takes the Earth to complete one orbit measured in relation to the equinoxes. See also **leap year**.

Young's modulus A measure of the elasticity of a material, being the ratio of the stress applied to the strain produced in a wire or rod under tension or compression.

ytterbium (Yb) One of the scarcest of the rare-earth elements (rd $7 \cdot 0$, mp $824°C$).

yttrium (Y) One of the transition elements (rd $4 \cdot 6$, mp $1,500°C$), whose most common application is in the red phosphors used on colour television screens.

Zeeman effect The splitting of the lines in the spectrum of a light source when that source is located in a magnetic field. Named after its discoverer, Pieter Zeeman, it is used in astronomy to determine the strength of a star's magnetic field.

zenith In astronomy, the point on the celestial sphere directly above an observer. Compare **nadir**.

zeolites A group of complex aluminium silicate minerals containing alkali metals and alkaline-earths, such as mordenite, $(Na_2,K_2,Ca)Al_2Si_{10}O_{24} \cdot 7H_2O$. They are light and have open structures that allow certain molecules to pass through. They have widespread application as ion-exchange media.

zinc (Zn) A relatively soft metal (rd $7 \cdot 1$, mp $419 \cdot 5°C$) that has a host of uses. It is alloyed with copper to form brass and with other metals to form die-casting alloys. It is coated on iron and steel to protect them against corrosion, a process known as galvanizing. It is used as the negative electrode in dry batteries. Its main ore is the sulphide, blende, ZnS.

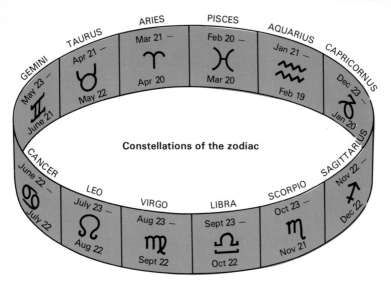

Constellations of the zodiac

zinc blende See **blende**.

zinc oxide A widely used oxide of zinc, ZnO, found in nature as the mineral zincite. It is used as a white pigment in paints, ceramics and cosmetics; and in medicine as an ingredient in soothing ointments. In crystalline form it can function as a semiconductor and has thermoluminescent, photoelectric and photochemical properties.

zinc sulphide Found in nature as the mineral blende, ZnS. It has commercial applications as the white pigment (with barium sulphate) lithopone. By heating it to high temperatures and adding traces of copper and silver, it becomes luminous and is used in that form on clock, watch and instrument dials.

zircon Zirconium silicate, $ZrSiO_4$, the principal ore of zirconium, which when transparent is valued as a gemstone. Though not as hard as diamond, it has a similar brilliance.

zirconium (Zr) A transition metal (rd 6·5, mp 1,850°C) that is corrosion resistant and strong at high temperatures. It has widespread uses in the nuclear-power industry because it is a poor absorber of neutrons. It is used for example in alloys for fuel-rod assemblies. Zirconium dioxide, or zirconia, is a useful pigment and refractory.

Zn The chemical symbol for zinc.

zodiac A belt through the heavens in which the Sun, the Moon and the planets (except Pluto) appear to move during the year. It extends 9° on either side of the ecliptic. The zodiac goes through 12 constellations—the constellations of the zodiac, which are illustrated in the diagram. The dates show the times of the year when the Sun passed through the constellations when the system was devised over 2,000 years ago. Because of precession, the constellations are now out: between mid-March and mid-April, for example, the Sun is now in Pisces rather than Aries. The signs of the zodiac play an important part in astrology.

zodiacal light A faint glowing band sometimes seen in the night sky along the ecliptic, probably caused by sunlight being reflected by dust particles. It is best seen in the tropics.

zoom lens A lens whose focal length can be progressively altered, thus changing the magnification and field of view, without losing focus. Long used in cinematography, zoom lenses are being increasingly used in photography.

Zr The chemical symbol for zirconium.

zwitterion A molecule that carries both positive and negative electric charges in different parts, often at each end. Many protein molecules are zwitterions.

PERIODIC TABLE OF THE ELEMENTS

A modern version of the periodic table, which arranges the elements in horizontal periods and vertical groups so as to bring out similarities in chemical properties. This version gives the symbols of the elements, their atomic number, and their relative atomic mass (atomic weight). Most relative atomic masses are not whole numbers. Elements are usually made up of two or more isotopes, and the relative atomic mass given reflects the relative abundance of the various isotopes in nature. This table shows hydrogen (H) in two places, because it can at various times behave like a metal and at others like a non-metal.

PERIODS

1A	2A	3B	4B	5B	6B	7B	8B		8B
1 1.01 **H** 1									
2 6.94 **Li** 3	9.01 **Be** 4								
3 22.99 **Na** 11	24.31 **Mg** 12								
4 39.09 **K** 19	40.08 **Ca** 20	44.96 **Sc** 21	47.9 **Ti** 22	50.94 **V** 23	52 **Cr** 24	54.94 **Mn** 25	55.84 **Fe** 26	58.93 **Co** 27	
5 85.47 **Rb** 37	87.62 **Sr** 38	88.91 **Y** 39	91.22 **Zr** 40	92.91 **Nb** 41	95.9 **Mo** 42	97 **Tc** 43	101 **Ru** 44	102.91 **Rh** 45	
6 132.91 **Cs** 55	137.34 **Ba** 56	Rare Earths Lanthanides 57-71	178.4 **Hf** 72	180.95 **Ta** 73	183.8 **W** 74	186.2 **Re** 75	190.2 **Os** 76	192.2 **Ir** 77	
7 223 **Fr** 87	226.03 **Ra** 88	Actinides 89-103	**Rf** 104	**Ha** 105	106				

GROUPS 1A 2A 3B 4B 5B 6B 7B 8B

Lanthanide Series—The Rare Earths

138.91 **La** 57	140.12 **Ce** 58	140.91 **Pr** 59	144.2 **Nd** 60	(147) **Pm** 61	150.4 **Sm** 62	151.96 **Eu** 63	157.2 **Gd** 64	158.93 **Tb** 65	162.5 **Dy** 66	164.93 **Ho** 67

167.2 **Er** 68	168.93 **Tm** 69	173 **Yb** 70	174.97 **Lu** 71

KEY

Light Metals   Heavy Metals

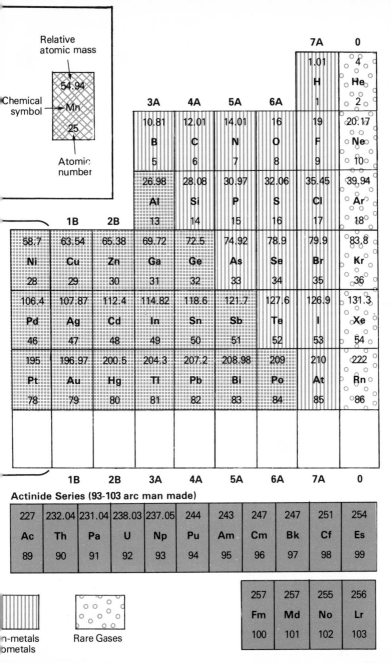

Relative atomic mass

		54.94
Chemical symbol	→	**Mn**
		25

Atomic number

							7A	0
							1.01 **H** 1	4 **He** 2
		3A	**4A**	**5A**	**6A**			
		10.81 **B** 5	12.01 **C** 6	14.01 **N** 7	16 **O** 8	19 **F** 9		20.17 **Ne** 10
1B	**2B**	26.98 **Al** 13	28.08 **Si** 14	30.97 **P** 15	32.06 **S** 16	35.45 **Cl** 17		39.94 **Ar** 18
58.7 **Ni** 28	63.54 **Cu** 29	65.38 **Zn** 30	69.72 **Ga** 31	72.5 **Ge** 32	74.92 **As** 33	78.9 **Se** 34	79.9 **Br** 35	83.8 **Kr** 36
106.4 **Pd** 46	107.87 **Ag** 47	112.4 **Cd** 48	114.82 **In** 49	118.6 **Sn** 50	121.7 **Sb** 51	127.6 **Te** 52	126.9 **I** 53	131.3 **Xe** 54
195 **Pt** 78	196.97 **Au** 79	200.5 **Hg** 80	204.3 **Tl** 81	207.2 **Pb** 82	208.98 **Bi** 83	209 **Po** 84	210 **At** 85	222 **Rn** 86
1B	**2B**	**3A**	**4A**	**5A**	**6A**	**7A**		**0**

Actinide Series (93-103 arc man made)

227 **Ac** 89	232.04 **Th** 90	231.04 **Pa** 91	238.03 **U** 92	237.05 **Np** 93	244 **Pu** 94	243 **Am** 95	247 **Cm** 96	247 **Bk** 97	251 **Cf** 98	254 **Es** 99

257 **Fm** 100	257 **Md** 101	255 **No** 102	256 **Lr** 103

Non-metals / Metalloids

Rare Gases

DATA FOR SOME INORGANIC COMPOUNDS

Substance	Formula	MP($^\circ$C)	BP($^\circ$C)	Density (g/cm^3)
Aluminium oxide	Al_2O_3	2024	d	3·99
Ammonia	NH_3	$-77\cdot8$	$-33\cdot4$	0·77g/dm^3
Ammonium chloride	NH_4Cl	sub 340		1·52
Ammonium nitrate	NH_4NO_3	169·6	d 210	1·73
Ammonium sulphate	$(NH_4)_2SO_4$	513		1·77
Barium sulphate	$BaSO_4$	1350		4·48
Calcium carbide	CaC_2	2300		2·22
Calcium fluoride	CaF	1418	2500	3·18
Calcium oxide	CaO	2600	3000	3·35
Cupric oxide	CuO	d		6·3
Cuprous oxide	Cu_2O	1229		6
Hydrogen chloride	HCl	$-114\cdot2$	$-85\cdot1$	1·64g/dm^3
Hydrogen cyanide	HCN	$-13\cdot2$	25·7	0·90g/dm^3
Hydrogen peroxide	H_2O_2	-2	158	1·46/0°
Hydrogen sulphide	H_2S	$-85\cdot5$	$-60\cdot3$	1·54g/dm^3
Iron oxide (ic)	Fe_2O_3	1457		5·24
Iron sulphide (ous)	FeS	1195	d	4·84
Lead acetate	$Pb(C_2H_3O_2)_2$	280		3·25
Lead chloride	$PbCl_2$	498	954	5·9
Lead sulphate	$PbSO_4$	1087		6·2
Lead sulphide	PbS	1114	1280	7·6
Magnesium oxide	MgO		sub 2770	3·58
Magnesium sulphate	$MgSO_4$	1127		2·65
Manganese chloride	$MnCl_2$	650	1190	2·98
Mercuric chloride	$HgCl_2$	277	304	5·44
Molybdenum disulphide	MoS_2	1185		4·7
Nickel carbonyl	$Ni(CO)_4$	-25	42·4	1·31
Nitric oxide	NO	$-163\cdot6$	$-151\cdot8$	1·34g/dm^3
Nitrogen dioxide	NO_2	$-11\cdot2$	21·2	1·49
Potassium bromide	KBr	735	1383	2·75
Potassium carbonate	K_2CO_3	898		2·43
Potassium chloride	KCl	772	1407	1·98
Potassium cyanide	KCN	610		1·52
Potassium dichromate	$K_2Cr_2O_7$	398		2·69
Potassium hydroxide	KOH	400	1327	2·04
Potassium iodide	KI	685	1330	3·12
Potassium nitrate	KNO_3	337		2·11
Potassium sulphate	K_2SO_4	1069		2·67
Silicon chloride	$SiCl_4$	-68	57	1·49
Silicon dioxide	SiO_2	1610		2·65
Silver bromide	$AgBr$	430	1533	6·47
Silver chloride	$AgCl$	455	1564	5·56
Silver iodide	AgI	558	1504	5·67
Silver nitrate	$AgNO_3$	210		4·35
Sodium carbonate	$Na_2CO_3\cdot10H_2O$	32		1·45
Sodium chloride	$NaCl$	801	1465	2·17
Sodium hydroxide	$NaOH$	320	1390	2·13
Sodium nitrate	$NaNO_3$	310	d 380	2·26
Sodium sulphate	Na_2SO_4	890		2·69
Sulphur dioxide	SO_2	$-75\cdot5$	$-10\cdot0$	2·93g/dm^3
Sulphuric acid	H_2SO_4	10·4		1·83
Titanium dioxide	TiO_2	1920		4·26
Tungsten carbide	WC	2870		15·65
Uranium dioxide	UO_2	2500		10·97
Zinc oxide	ZnO	1975	d	5·47

MP = melting point   BP = boiling point   d = decomposes

DATA FOR SOME ORGANIC COMPOUNDS

Substance	Formula	MP(°C)	BP(°C)	Density (g/cm³)
Acetaldehyde	CH_3CHO	− 121	21	0·79/16°
Acetamide	CH_3CONH_2	82	222	1·16
Acetic acid	CH_3COOH	16·7	118	1·05
Acetone	CH_3COCH_3	− 95	56·5	0·79/25°
Acetylene	C_2H_2	− 81·5	sub − 84	0·61/ − 80°
Aniline	$C_6H_5NH_2$	− 6	184	1·03/15°
Benzene	C_6H_6	5·5	80·1	0·88
Benzoic acid	C_6H_5COOH	122	249	1·26/15°
n-Butane	C_4H_{10}	− 138·4	− 0·5	0·58
n-Butyl alcohol	C_4H_9OH	− 79·9	117·4	0·81
Camphor	$C_{10}H_{16}O$	176	204	0·99/10°
Carbon disulphide	CS_2	− 111	46·3	1·29/0°
Carbon tetrachloride	CCl_4	− 23	76·7	1·63/0°
Chloroform	$CHCl_3$	− 63·5	61	1·50/15°
Cholesterol	$C_{27}H_{45}OH$	148·5	360 d	1·07
Citric acid	$C_6H_8O_7$	153	d	1·54/18°
Cyclohexane	C_6H_{12}	6·6	80·7	0·78
Ethane	C_2H_6	− 183·3	− 88·6	
Ethanol	C_2H_5OH	− 117	78·5	0·79
Ether (diethyl)	$(C_2H_5)_2O$	− 116	34·5	0·71/20°
Ethyl acetate	$CH_3COOC_2H_5$	− 83·6	77·1	0·92
Ethyl chloride	C_2H_5Cl	− 142·5	12·5	0·92/0°
Ethylene	C_2H_4	− 169·2	− 103·7	
Ethylene glycol	$(CH_2OH)_2$	− 15·6	197·3	1·12
Formaldehyde	$HCHO$	− 92	− 21	0·82/ − 20°
Formic acid	$HCOOH$	8·4	100·5	1·22
Fumaric acid	$(CHCOOH)_2$	300	sub	1·64
Glucose	$C_6H_{12}O_6$	142		1·54/25°
Glycerol	$C_3H_8O_3$	20	290	1·26/17°
n-Hexane	C_6H_{14}	− 95·3	68·7	0·66
Isobutane	C_4H_{10}	− 159·6	− 11·7	0·56
Isooctane	C_8H_{18}	− 107·4	99·2	0·69
Lactic acid	$C_3H_6O_3$	25	d	1·25
Maleic acid	$(CHCOOH)_2$	130	d	1·59
Menthol	$C_{10}H_{20}O$	31–44	216	0·90/15°
Methane	CH_4	− 182·5	− 161·5	
Methanol	CH_3OH	− 93·9	64·1	0·79
Methyl chloride	CH_3Cl	− 93	− 24·1	0·99/ − 25°
Naphthalene	$C_{10}H_8$	80·2	218	1·14
Nitrobenzene	$C_6H_5NO_2$	5·9	210·9	1·17/25°
n-Octane	C_8H_{18}	− 56·8	125·7	0·70
Phenol	C_6H_5OH	41·0	181·7	1·07
Phosgene	$COCl_2$	− 104	8·3	
Picric acid	$C_6H_3O_7N_3$	122·5	exp >300	1·76
Propane	C_3H_8	− 187·7	− 42·1	
Propylene	C_3H_6	− 185·3	− 47·7	
Pyridine	C_5H_5N	− 42	115·3	0·98
Resorcinol	$C_6H_4(OH)_2$	111	276·5	1·28/15°
Salicylic acid	$C_7H_6O_3$	159		1·44
Styrene	C_8H_8	− 30·6	145·2 d	0·91
Sucrose	$C_{12}H_{22}O_{11}$	184 d		1·59/15°
Tartaric acid	$C_4H_6O_6$	170		1·76
Toluene	$C_6H_5CH_3$	− 95·0	110·6	0·87
Trinitrotoluene	$C_7H_5(NO_2)_3$	80·8		1·65
Urea	$CO(NH_2)_2$	132		1·34

exp = explodes   sub = sublimes